AF411222

Early Detection of Faults in Induction Motors

Editors

Daniel Morinigo-Sotelo
Rene Romero-Troncoso
Joan Pons-Llinares

Basel • Beijing • Wuhan • Barcelona • Belgrade • Novi Sad • Cluj • Manchester

Editors

Daniel Morinigo-Sotelo
Institute of Advanced
Manufacturing Technologies
(ITAP), Research Group
ADIRE-HspDigital
Universidad de Valladolid
Valladolid, Spain

Rene Romero-Troncoso
HSPdigital–CA Mecatronica
Universidad Autónoma
de Querétaro
San Juan del Río, Mexico

Joan Pons-Llinares
Departamento de
Ingeniería Eléctrica
Universidad Politécnica
de Valencia
Valencia, Spain

Editorial Office
MDPI
St. Alban-Anlage 66
4052 Basel, Switzerland

This is a reprint of articles from the Special Issue published online in the open access journal *Energies* (ISSN 1996-1073) (available at: https://www.mdpi.com/journal/energies/special_issues/induction_motors).

For citation purposes, cite each article independently as indicated on the article page online and as indicated below:

Lastname, A.A.; Lastname, B.B. Article Title. *Journal Name* **Year**, *Volume Number*, Page Range.

ISBN 978-3-0365-9335-7 (Hbk)
ISBN 978-3-0365-9334-0 (PDF)
doi.org/10.3390/books978-3-0365-9334-0

Contents

About the Editors

Daniel Morinigo-Sotelo

Daniel Morinigo-Sotelo received the B.S.and Ph.D. degrees in electrical engineering from the University of Valladolid (UVA), Valladolid, Spain, in 1999 and 2006, respectively. From 2000 to 2015, he was a Research Collaborator on Electromagnetic Processing of Materials with the Light Alloys Division of CIDAUT Foundation. He is currently with the Research Group on Analysis and Diagnostics of Electrical Grids and Installations, which belongs to the ITAP Institute, UVA, and HSPdigital Research Group, México. His research interests include fault detection and the diagnostics of induction machines, power quality, and smart grids.

Rene Romero-Troncoso

Rene de Jesus Romero-Troncoso (Senior Member, IEEE) received his Ph.D. degree in mechatronics from the Autonomous University of Queretaro, Queretaro, Mexico, in 2004. He is currently a National Researcher level three with the Mexican Council of Science and Technology, CONACYT. He is a Head Professor with the Autonomous University of Queretaro, Queretaro, Mexico. He has been an advisor for more than 200 theses, the author of two books on digital systems (in Spanish), and coauthor of more than 130 technical papers published in international journals and conferences. His research interests include hardware signal processing and mechatronics. Dr. Romero-Troncoso was the recipient of the 2004 Asociación Mexicana de Directivos de la Investigación Aplicada y el Desarrollo Tecnológico Nacional Award on Innovation for his work in applied mechatronics and 2005 the IEEE ReConFig Award for his work in digital systems. He is part of the Editorial Board of the Hindawi's *International Journal of Manufacturing Engineering*.

Joan Pons-Llinares

Joan Pons-Llinares received hisM.Sc. degree in industrial engineering and hisPh.D. degree in electrical engineering from the Universitat Politécnica de Valéncia (UPV), Valéncia, Spain, in 2007 and 2013, respectively. He is currently an Associate Professor with the Electric Engineering Department, UPV. His research interests include time–frequency transforms, condition monitoring, and efficiency estimation of electrical machines.

Preface

The idea of Industry 4.0 was first introduced at the Hanover Fair in 2011 and has since become a reality thanks to advancements in numerous fields, including digital technology, robotics, artificial intelligence, nanotechnology, biotechnology, the Internet of Things, 3D printing, and autonomous vehicles. The primary goal of Industry 4.0 is to improve industry asset management by combining all of these technologies. The design principles of Industry 4.0 were established in 2016 by Herman, Penntek, and Otto and include interoperability, information transparency, technical assistance, and decentralized decision-making.

This Special Issue of *Energies* focuses on the importance of detecting faults in induction motors, which are crucial industrial components and are also present in many different areas, such as transportation, service, and utilities. Although these machines are typically robust, they can fail and cause significant damage if not detected early. This can result in unplanned production stops, destroyed facilities, and service interruptions, which can be costly for companies. Therefore, there is a growing interest in developing early detection systems to prevent these failures from becoming catastrophic. Early fault detection techniques are essential for implementing predictive maintenance systems, which are a critical element of Industry 4.0.

Information transparency and decentralized decision-making are two principles of Industry 4.0 that will shape the future of maintenance systems for these machines. This means that maintenance systems for electric motors must include more valuable context information beyond the raw data provided by sensors. In other words, maintenance systems must have instrumentation and monitoring systems and provide automatic device diagnosis and prognosis.

Traditionally, electrical machine condition monitoring has relied on analyzing stator current or motor vibrations. However, this Issue presents papers that propose using other signals, such as stray flux, sound, speed, and thermographs. The authors also present different techniques to process the signals of these physical variables (apFFT Time-shift phase difference spectrum, Prony methods, Min-Norm techniques, etc.) and obtain valuable information for automatic diagnosis based on models or machine learning techniques.

Daniel Morinigo-Sotelo, Rene Romero-Troncoso, and Joan Pons-Llinares
Editors

Review

Early Detection of Faults in Induction Motors—A Review

Tomas Garcia-Calva [1], Daniel Morinigo-Sotelo [2], Vanessa Fernandez-Cavero [3] and Rene Romero-Troncoso [4,*]

[1] HSPdigital-Electronics Department, University of Guanajuato, Salamanca 36700, Mexico
[2] HSPdigital-ITAP-ADIRE, University of Valladolid, 47002 Valladolid, Spain
[3] Department of Electrical Engineering, University of Valladolid, 47002 Valladolid, Spain
[4] HSPdigital-Mechatronics Department, Autonomous University of Querétaro, San Juan del Río 76806, Mexico
*   Correspondence: troncoso@hspdigital.org

Abstract: There is an increasing interest in improving energy efficiency and reducing operational costs of induction motors in the industry. These costs can be significantly reduced, and the efficiency of the motor can be improved if the condition of the machine is monitored regularly and if monitoring techniques are able to detect failures at an incipient stage. An early fault detection makes the elimination of costly standstills, unscheduled downtime, unplanned breakdowns, and industrial injuries possible. Furthermore, maintaining a proper motor operation by reducing incipient failures can reduce motor losses and extend its operating life. There are many review papers in which analyses of fault detection techniques in induction motors can be found. However, all these reviewed techniques can detect failures only at developed or advanced stages. To our knowledge, no review exists that assesses works able to detect failures at incipient stages. This paper presents a review of techniques and methodologies that can detect faults at early stages. The review presents an analysis of the existing techniques focusing on the following principal motor components: stator, rotor, and rolling bearings. For steady-state and transient operating modes of the motor, the methodologies are discussed and recommendations for future research in this area are also presented.

Keywords: artificial intelligence; condition monitoring; early detection; fault diagnosis; fault severity; frequency analysis; incipient fault; induction motor; machine learning; signal processing

Citation: Garcia-Calva, T.; Morinigo-Sotelo, D.; Fernandez-Cavero, V.; Romero-Troncoso, R. Early Detection of Faults in Induction Motors—A Review. *Energies* **2022**, *15*, 7855. https://doi.org/10.3390/en15217855

Academic Editor: Sheldon Williamson

Received: 16 September 2022
Accepted: 20 October 2022
Published: 23 October 2022

Publisher's Note: MDPI stays neutral with regard to jurisdictional claims in published maps and institutional affiliations.

1. Introduction

Electric motor failures in industrial systems often result in unplanned downtime, loss of production, higher operating costs, and loss of profits [1]. The most common family of electric motors used in homes, businesses, and industry is the induction motor (IM) [2], for which there is a variety to choose from, depending on the power source, load requirements, mechanical interface, operating cost, energy efficiency, and reliability. IMs are often preferred over other kinds of motors since they are significantly less expensive, more robust, and capable of reliable operation in harsh ambient conditions, even in an explosive atmosphere. Induction motors, particularly those of the squirrel cage type, have been for almost a century the principal workhorse in industry [3]. Although IMs are more reliable than other types of motors, these machines are not exempt from developing faults in their structure or components, and these failures could lead to motor malfunction. Therefore, reliable condition monitoring for induction motors is of great value to avoid catastrophic unscheduled downtime [4]. An unexpected failure might lead to the loss of valuable human life or a costly standstill in industry, which needs to be prevented by precisely detecting the fault. The induction motor consists of many mechanical and electrical parts, such as a motor frame, stator windings, rotor cage, rolling bearings, fan, rotor shaft, among others. Despite induction motors are designed to be robust machines, they are exposed to external situations such as unstable supply voltage, unstable supply current sources, overloads, unbalanced loads, and electrical stresses. Due to the above-mentioned situations, damages in operation and natural deterioration of the material parts (or manufacturing defects), the motor will eventually develop a faulty condition.

Different condition states of an IM component can be characterized as healthy or faulty condition. In the healthy condition, the internal components have no degradation, and the induction motor operates with maximum energy efficiency. There are three states that characterize a faulty condition in IM. The first one consists in an incipient fault, known as the early stage, where the degradation begins to develop in one or several of the internal components. Although the motor component has damage (i.e., partially broken rotor bars), the induction motor can continue to operate with no apparent symptoms. In the second faulty state, known as the developed fault, the damage on the motor component is advanced (i.e., one or more broken rotor bars). In this condition, the IM still operates; however, the damaged component severely affects the motor performance. Finally, the third faulty condition, known as catastrophic fault, occurs when the developed failure has propagated to other components, and the IM is no longer operating. Three stages of fault growth can be considered for an IM: the incipient fault with a steady propagation, the advanced fault with an accelerated propagation, and the catastrophic stage with a very accelerated propagation. The first stage is from the healthy state of the component until the very incipient fault state. This stage covers most of the portion of the IM component's useful life. After the incipient fault is present, the degradation propagates slowly until the developed fault state is reached. Once the IM component presents developed fault symptoms, the last stage is the accelerated propagation, where the fault grows rapidly until a breakdown. The early detection of IM faults is carried out before an internal component exhibits a developed fault stage. According to the literature, the tendency in industry and academia is to consider incipient fault of the rotor or bearing when there exists only a partial fracture of an internal component. Whereas for the stator, it considers incipient fault when there exists a short circuit among less than 3% of the total turns winding.

Some surveys [5–7] revealed that the occurrence rate of bearing faults can reach around 40% of the total failures in induction motors, while other studies indicated that this rate can be even higher for small motors. On the other hand, as per the study by the Institute of Electrical and Electronics Engineers (IEEE) and the Electric Power Research Institute (EPRI), faults that occur in the stator of an IM are 36% and 28%, respectively. Finally, it is reported in the literature that rotor faults are responsible for 8–10% of failures in induction motors [8]. Bearing defects usually lead to an increment in the sound and vibration levels as well as high temperatures. A severe damage can even provoke catastrophic failures (i.e., rotor-stator rubbing, insulation damage). Moreover, stator failures during motor operation lead to reduced efficiency of the machines. Once begun, the stator fault provokes progressive degradation of insulating materials, ultimately leading to electrical breakdown. Despite rotor failures have a lower rate of occurrence compared to the above-mentioned ones, these faults are just as important because they can lead to shaft vibrations and winding damages, and thus bearing or stator failures. The existence of damages or anomalies at incipient stage implies additional losses in that part of the machine due to their improper operation. IM faults usually progress from incipient to a very advanced stage in a lapse of time, depending on the type of failure. Unless detected early enough, a motor failure may lead to costly standstill in industry or fatal consequences such as fire, explosion, and even loss of human lives. The constant need for reducing industrial injuries, unscheduled shutdowns, and operational costs has led to developing new techniques for early fault detection before they become prominent to cause machine failures. Some of the advantages of early fault detection in induction motors are as follows:

- Cost savings which are realized by estimating potential failures before they occur [9].
- Facilitate pre-planned preventive machine schedules.
- Better maintenance activities instead of replacing components [10].
- Prevent unexpected stop in the production lines.
- Prevent an extended period of down-time caused by extensive machine failure.
- Improve the induction motor efficiency [11].

In the recent literature, there are some excellent reviews on fault detection techniques and their implementation in induction motors [12–16]; however, these reviews are limited to detecting developed or advanced faults. This paper presents a review of diagnosis techniques and methods for the detection of faults at incipient severity stages, due to its important role in the condition monitoring field, and decision-making in industrial maintenance activities. The existing literature on the subject is categorized into two approaches based on the operational mode of the IM: steady-state and transient regime. In the steady-state regime, the reported techniques apply analysis in the time-domain or frequency-domain to extract fault indicators and evaluate the motor condition. In the transient-state regime, the reported methods obtain fault indicators from time-frequency maps that allow evaluating the present state of fault indicators as well as their evolution over time. The organization of this paper is as follows. Section 2 presents the type of faults that induction motors develop in their main components and briefly describes the causes and consequences for each failure with special focus on early detection. A brief description of different signals used in the monitoring condition field for fault detection is presented in Section 3 including practical considerations for proper signal analysis and remarking advantages and limitations of each monitoring signal. In Section 4, a review of various techniques that are used for early fault detection in induction motors is presented; including those that are based on signal processing and knowledge-based. Section 5 is dedicated to the analysis of the different methodologies, whose strengths and weaknesses are described and discussed. According to the reviewed techniques, the conclusions are shown in Section 6. Lastly, in Section 7, directions and a future perspective are presented.

2. Faults in Induction Motors

Induction motor faults are commonly categorized as mechanical faults and electrical faults. Despite the existence of many fault classifications, this work categorizes motor failures according to the component that develops the fault for the sake of simplicity. The most common failures occur in three principal components of the rotatory machine. Figure 1 shows these main parts: the stator, the rotor, and rolling bearings.

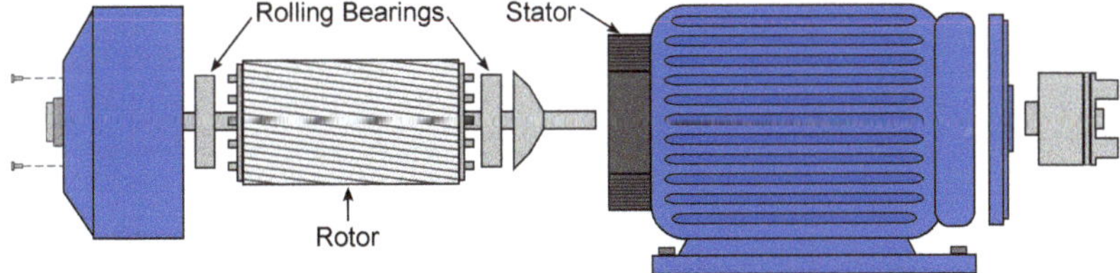

Figure 1. Three fundamental components of an induction motor: stator, rotor, and rolling bearings.

2.1. Stator Faults

The stator consists of a laminated core, an outer frame, and insulated electrical windings. Its components are subjected to electrical and environmental stresses, which severely affect the stator condition leading to faults [17]. Stator faults (SF) can be categorized based on their localization as failure in the stator frame, fault in the stator winding, and failure in the laminations of the stator core. Among these, stator winding failures are the most severe faults and are often caused by failure of insulation of winding, which leads to local heating. If unnoticed, this local heating further damages the insulation of the stator winding until a catastrophic failure may occur. This fault is also known as the short circuit inter-turn fault. The appearance of stator faults depends on the size of the electrical machine [18]. According to [19], low-voltage IM stator faults account for only 9% of total failures. In medium-voltage IM, the percentage increases to 35–40%, whereas for high-voltage IM, it is more than 65%.

2.2. Rotor Faults

The rotor is the main driving shaft in induction motors through which the mechanical energy is transferred to the load. This component is placed inside the stator, and for a squirrel-cage type, it consists of the shaft, the aluminum or copper bars, and the end rings. Approximately 8 to 10% of all failures in IM are at the rotor. These faults can be classified as: broken rotor bars (BRB), cracked end-ring, and rotor eccentricities. Broken bars are caused by a combination of different stresses (mechanical, electrical, and thermal), manufacturing problems, dynamic stress from shaft torque, and fatigued mechanical parts [12]. This type of fault may not show any incipient symptoms, propagating to the next bars and leading to a sudden collapse of the rotor, producing damage in the stator and an abrupt interruption of the motor operation [20]. On the other hand, air-gap irregularities are produced by rotor eccentricities when the rotor axis of rotation does not coincide with stator geometrical axis. Manufacturing and constructive errors that generate a non-uniform air-gap or an incorrect positioning of the stator and rotor at the commissioning stage produce static eccentricity. When the center of the rotor is not at the center of rotation, then dynamic eccentricity is produced. The common causes of dynamic eccentricity are rotor shaft bending and bearing faults.

2.3. Rolling Bearing Faults

Rolling element bearings are the support of the shaft rotor in the induction motor in order to facilitate its rotation by reducing friction. In a rough manner, a bearing has four components: an inner raceway, an outer raceway, balls, and a cage that provides an equidistant arrangement between the balls. Bearing faults (BF) are classified as localized failures and distributed faults (roughness or non-cyclic) [7]. Distributed defects affect a whole region and their mathematical modeling is very difficult. In contrast, localized failures are single-point defects and can be classified according to the affected element: inner raceway defect, outer raceway defect, ball defect, and cage failure. Bearing wear can be caused by a wide variety of reasons, such as excessive or deficient lubrication (due to inadequate viscosity, excess or lack of grease, lubricant contamination, etc.), circulation of bearing currents (in power converter-fed motors), brinelling (due to punctual overloads, severe impacts), etc.

2.4. Early Detection of Faults

Induction motors are symmetric machines and the occurrence of any type of fault is linked to the harmonic content of its monitored signals. The existence of a fault in IM results in the appearance of specific frequency signatures. In general, fault detection techniques are based on the magnitude evaluation of these signatures. When there is an anomaly in the mechanical structure of the rolling bearing, characteristic frequencies emerge in the vibration spectrum as a consequence of the asymmetry. There are two bearing defects that are analyzed in the literature from the severity level point of view, the outer race defect and the inner race defect. The related frequencies are shown below [21].

$$\text{Inner raceway}: \; f_i = \frac{N_b}{2} f_r \left(1 + \frac{D_b}{D_c} \cos \beta \right), \tag{1}$$

$$\text{Outer raceway}: \; f_o = \frac{N_b}{2} f_r \left(1 - \frac{D_b}{D_c} \cos \beta \right), \tag{2}$$

where f_r is the rotational frequency, N_b is the number of balls in the bearing, D_c is the pitch or cage diameter, D_b the diameter of the balls, and β is the contact angle between a ball and the raceway. When a stator fault occurs, the current through the shorted winding affects the magnetic field and is reflected in the axial flux as follows [22].

$$\text{Stator fault}: \; f_{ss} = f_s \left[\left(k \pm n \frac{1-s}{p} \right) \right] \tag{3}$$

where k is the order of the time harmonic, n the order of the shorted coil space harmonic, s the machine slip, p the number of pole pairs, and f_s the supply frequency. The damage of broken bars in the rotor produces additional frequencies in the current spectrum. These signatures are characterized by [23]:

$$\text{Broken rotor bars}: \; f_{brb} = f_s(1 \pm 2s) \tag{4}$$

Additionally, any fluctuation in the load torque will produce oscillations in the stator currents at frequencies of

$$\text{Load oscillation}: \; f_{load} = f_s \pm mf_r = f_s\left[1 \pm m\left(\frac{1-s}{p}\right)\right] \tag{5}$$

where m is the order of the harmonic. Since the same fault harmonic is given by BRB, a low-frequency load-oscillation results in stator currents that can overlap those produced by the BRB fault [24].

The magnitude of each of these fault frequencies is directly related to the severity level of the fault. The higher the severity level, the higher the magnitude of the spectral component. Each failure type has a different evolution over time since the degradation of the component depends on its construction and the material from which it is made. Figure 2 illustrates the degradation of the motor components and its severity levels: bearing failure (Figure 2a), stator failure (Figure 2b), and rotor fault (Figure 2c). The tendency in industry and academia is to be able to make an early detection of the motor component degradation. For bearing failure, the early detection is considered when the diameter of the crack (λ) is less than $1/8$ the diameter of the bearing ball (Θ). For broken rotor bars, early detection is considered when one bar is partially broken and the depth of the breakage (ρ) is less than the total length of the bar (l). Whereas for stator faults, it is considered that an early fault occurs when there is a short circuit between less than $1/30$ times the total turns of the winding (Φ). Figure 2d–f illustrates the relationships of the fault with the construction of the motor components. As component degradation is very low at an early stage, the magnitude of fault signatures is also very low. Therefore, a condition diagnosis of the motor component when the degradation is incipient presents a challenge in the detection, identification, and evaluation of failure indicators.

The most important characteristic of any condition monitoring scheme is its quickness of detection. Different types of faults usually progress from incipient to a very advanced stage in a different manner, as is shown in Figure 2. This work only considers fault detection at an early stage. For the detection of early rotor failure, only partially broken rotor bars are taken into account, whereas fully broken rotor bars are considered a developed fault. This is because once a bar is completely fractured, the failure spreads rapidly to adjacent bars and subsequently damages the stator winding causing irreparable damage. In the case of bearing faults, the early fault condition is considered when the diameter of the fracture in the inner or the outer race is less than 12.5% of the diameter of the bearing ball, since from then on, the bearing stroke undergoes great alterations and alters the rotor symmetry, making the failure to be considered as developed or advanced. Finally, the stator failure is considered in an early condition before exceeding 3.33% of the turns in the stator phase. Once the short circuit begins, it propagates rapidly in a very short time. Unless detected early enough, it might lead to catastrophic consequences. Faults detected at advanced stages are far more likely to cause unplanned breakdowns in the line production than those detected while the failure is still at an early stage. Techniques that can detect faults at an early stage are very desirable for the possibility of correcting the faulty condition entirely with low impact to the production line. For early detection to be an effective and practical approach, techniques must satisfy three basic requirements. First, the detection analysis should be able to distinguish faulty IM from healthy IM cases with a high degree of accuracy, showing both low rates of false-positives and false negatives. Second, the detection should be possible before the fault progresses to a developed stage, when the propagation is accelerated, and preventive actions are less effective. Lastly, the diagnosis methodology should allow the assessment of the IM condition when the motor is fed by inverters. It must be noticed that inverters induce several spectral components to the voltage, current, and vibration signals, which overlap with the fault-related spectral signatures; moreover, the magnitude of the fault-related components are very close to the noise floor, making the evaluation of the fault severity difficult.

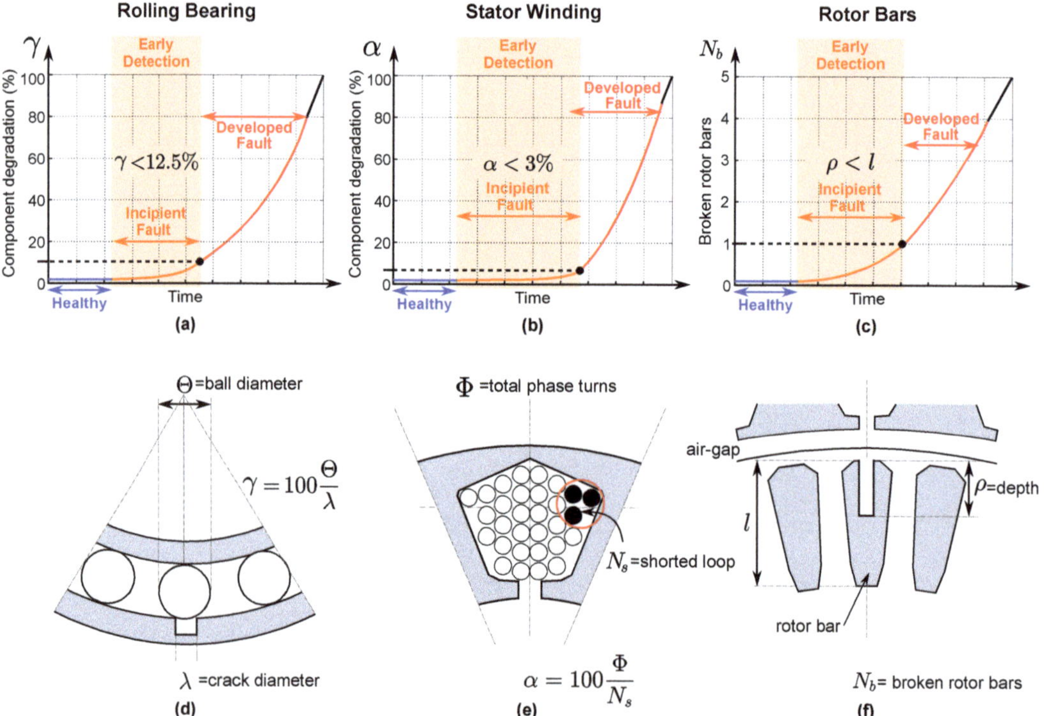

Figure 2. Degradation stages of induction motor components: (**a**) rolling bearing degradation, (**b**) stator winding degradation, (**c**) rotor bar degradation, (**d**) rolling bearing schematic, (**e**) stator winding schematic, and (**f**) rotor bars schematic.

3. Monitoring Signals for Fault Detection

When IMs are running, each type of fault generates characteristic features in the machine's behavior. Fault detection techniques can be based on the analysis of vibrations, stator current, shaft speed, acoustic emissions, voltage, internal air-gap flux, external stray flux, electric power, and temperature (usually trough infrared thermography).

3.1. Vibration Monitoring

Among the many condition monitoring techniques, vibration-based methods are the most widely preferred owing to their reliability, non-intrusiveness, and easy measurability. Vibration monitoring has been used for decades and utilized to detect mechanical faults in IM. In working mode, radial magnetic forces are produced between the rotor and stator surfaces and are proportional to the square of the flux density. These forces result in stator core, winding, and motor frame vibrations. As faults associated with the rotor, stator, and rolling bearings alter the machine symmetry, vibration signals which are a function of the symmetrical air-gap and symmetrical components will also change [25]. Most vibration measurements usually use sensors of vibration-acceleration that work based on the piezoelectric effect, whose output voltage is proportional to the force applied to the sensor [26]. The vibration signals need to be processed in order to extract the fundamental components and to filter out nonlinear effects due to the cover frame and noise from the environment.

3.2. Stator Current Monitoring

Even though vibration signals have been widely utilized, most of the work in the last decade has been directed toward stator current monitoring also known as motor current signature analysis (MCSA) [27]. The stator current monitoring can provide unique fault patterns for the effective detection of mainly electrical faults, i.e., the stator winding fault, broken rotor bars fault, phase unbalance, and single phasing [28]. Motor current monitoring provides a non-intrusive way to continuously monitor the health condition with the aim of using non-invasive sensors and possibly already existing in the drive for motor control.

3.3. Magnetic Flux Monitoring

An alternate approach based on magnetic flux monitoring has received the attention of many researchers and motor manufacturers during the last years. The great improvements and reduction in the costs and dimensions of the required transducers, the development of advanced signal-processing tools that are suitable for magnetic flux analysis, along with other inherent advantages provided by this technology, are important aspects that have permitted the proliferation of flux-based methods [29]. The magnetic flux in IM is broadly classified as internal magnetic flux and external stray magnetic flux. However, using magnetic flux as monitoring signal for fault detection is not new and there are works carried out decades earlier [30].

3.4. Others

Apart from locating specific features in the above-mentioned signals, other physical magnitudes, such as rotor position, rotor speed, torque, acoustic emissions, electric power, and temperature have been investigated in the recent years by researchers in the field. Sometimes, the combination of several monitoring signals is used by fault detection techniques to improve the detection rate. There are also other detection methods being developed on modeling and control techniques, such as state observers [31], state estimators [32], signal injection, and parameter estimators.

Despite the relevant advances obtained with the vast diversity of proposed approaches in the fault detection field, the effectiveness of the techniques usually depends on the induction motor size, loading, state operation (steady or transient), supply type, control mode, type of the fault, and to a great extent, on the type of the monitoring signal. Table 1 summarizes some of the main advantages and drawbacks of the most common monitoring signals.

Table 1. Main advantages and drawbacks of common monitoring physical magnitudes.

Signal	Advantages	Drawbacks	Sensor Cost
Vibration	high sensitivity to mechanical faults	environment noise, reverb effects	low
Stator Current (MCSA)	remote monitoring, clear fault patterns	high amplitude of the supply frequency	very low
Magnetic Flux	easy sensor installation	interference among patterns	very low
Shaft Speed	low noise level	high dependence on load inertia	high
Sound	very easy sensor installation	environment noise, echo effects	low

4. Review of Techniques and Methodologies

In general, early fault detection techniques based on signal analysis relies on: (1) acquiring one or various physical magnitudes of the induction motor, (2) processing the measured magnitudes (signals) with suitable tools to extract fault patterns, and (3) analyzing the patterns to determine the fault severity. According to the reviewed technical literature, two main approaches for early fault detection can be determined based on the IM operation mode: the steady-state analysis and transient-state analysis. The techniques employed in these approaches depend on the stationarity or non-stationarity of the monitored signals.

4.1. Early Fault Detection via Steady-State Analysis

When an induction motor operates at a constant speed, measured magnitudes from the machine can be described by periodic signals. Thus, in the literature, the best-known techniques for processing these signals belong to the frequency-domain analysis.

Although most IMs in the industry are fed directly from the grid, in the recent years, the use of inverters has become more widespread due to their capabilities in speed control and energy efficiency advantages. Nevertheless, their use also affects fault detection in IMs due to the higher noise and rich harmonic content they cause in the measured signals for condition monitoring. For these reasons, it is important to separate the analysis of early fault detection methods according to the motor power supply.

4.1.1. Direct Line-Fed Induction Motors

Condition monitoring methods for the early detection of induction motor faults based on signal spectral analysis are found in the literature. In [33], the authors presented a methodology for incipient broken rotor bar detection based on the spectral analysis of vibration signals with the FFT. The proposed approach is based on sparse signal representations [34] and the use of dictionaries trained with sets of signals with the fault to be detected. The incipient fault is simulated by drilling only 5% of the length of a rotor bar, and high detection accuracy is achieved even in unloaded motor operation but only for a line-fed induction motor.

As the authors of [35] pointed out, online detection of low-severity stator interturn faults is one of the most challenging electrical machine defects to detect. If this fault develops undetected, it can lead to a ground fault and complete motor burning. The authors used a modified induction motor that allows them to simulate very low severity interturn faults: 0.25%, 0.5%, and 0.75%. They proved that conventional monitoring methods, such as MCSA, Park's vector approach and negative sequence current analysis, are unreliable and insensitive to these very low-severity interturn faults. However, they were able to detect these faults by using three different coils placed in different motor locations to collect the motor stray flux. The Park vector of these three signals is calculated, and its modulus is analyzed with the FFT.

Recent condition monitoring approaches have benefited from advances in computational technologies that allow the combination of signal processing methods (time and frequency domain) with knowledge-based techniques (KB) such as machine learning (ML), genetic algorithms (GA), artificial intelligence (AI), surface learning and deep learning (DL). For instance, reference [36] dealt with the detection of partially broken rotor bars. It proposed a hybrid approach combining electrical-synchronous averaging (ETSA), Discrete Wavelet transform (DWT), and fuzzy logic algorithms. The first two techniques are used to identify a frequency band with fault-related components in the stator current, and the third is a classifier to assess the severity of the fault. It is worth noting that this methodology can detect a hole of only 2 mm in a 36 mm bar, even under no-load conditions.

In the paper [37], hybrid DL architectures were explored to solve the problem of classifying interturn faults at an incipient stage. This detection method uses a hybrid architecture based on a one-dimensional Convolutional Neural Network (1DCNN), long short-term memory, and gated recurrent unit. This approach can distinguish this fault from other conditions, such as voltage imbalance and load variations.

Reference [38] also showed very good results in detecting interturn faults at an early stage. The authors compared two different ANN-based approaches: (1) the first one is based on a multi-agent system with a classification behavior; (2) the second one uses a neural estimator. The input data of both methods are the analysis of the stator current in the time domain, which differentiates them from other approaches based on frequency-based signal processing. Their approach has been tested by simulating short circuits covering 1 to 10 percent of the stator winding and with a wide range of motor power ratings and voltage supplies. Reference [39] made another contribution to the same topic as the previous one. Their proposal is based on analyzing the negative sequence of the stator current to detect motor asymmetries produced by this type of fault. The results are based on simulation and corroborated by experimental measurements under varying motor load conditions. Most of these knowledge-based (KB) works achieve good detection accuracy of over 95%. However, it is important to note that these KB approaches require a large amount of data (from both the healthy and faulty induction motors), so they have limitations for generalization and are techniques with low scalability.

The detection of bearing failures has also attracted many researchers, and numerous papers have been published. However, few of them deal with their early detection or the diagnosis of these faults at an early stage before they develop into catastrophic breakdowns. The vast majority of these works is based on signal processing, knowledge-based techniques or a combination of both.

In [40], the processed signal is the stator current, which allows the detection of incipient cage and outer race-bearing faults in a 2.2 kW line-fed induction motor. The authors proposed a spectral frequency subtraction using several wavelet transforms (DWT, stationarity wavelet transform (SWT), and wavelet packet decomposition (WDP)) to suppress the dominant components in the stator current spectrum. This suppression makes it possible to observe fault-related frequency components whose amplitude is very small, especially at an early fault stage. Due to the impossibility of installing vibration sensors in many industrial applications, the authors in [41] also proposed using two stator current signals for early bearing damage detection. Their interesting contribution pre-processes these two signals with fractional B-spline wavelet transforms for denoising them. Next, the overlapping group shrinkage (OGS) algorithm reconstructs two signals that will be used as the input to a convolutional neural network (CNN) and a long short-term memory (LSTM) algorithm for feature extraction. Next, information fusion and fuzzy c-means algorithms perform the fault diagnosis. Remarkably, this methodology is based on unlabeled learning, and it is tested in an actual industrial application with an induction motor driving a centrifugal pump.

4.1.2. Inverter-Fed Induction Motors

Even though much research has been done on incipient fault diagnosis in induction motors, all previous works still deal with line-fed motors under stationary conditions. Recently, in industry, it is more common to see IM fed by power electronic converters (also known as voltage source inverters) [42]. Furthermore, the stationary operation is quite unusual in the industrial environment due to voltage variations, speed oscillations, and load changes [43]. In this context, some techniques have been proposed and showed very good results for early fault detection in inverter-fed IM and some of them under non-stationary conditions. In [44], a combination of two-level hybrid hierarchical CNN with SVM was proposed for incipient interturn fault diagnosis. The authors showed that their proposal is fast and has significant performance improvement in accuracy in comparison with other proposals. The authors of [45] concentrated on the incipient inter-turn short circuit detection and estimation of its severity. Moreover, they studied the effect of load oscillations on the recognition of fault patterns in the stator current. Another combination of SP and KB techniques can be found in [46], the authors described a methodology combining the DWT for multi-resolution analysis, statistical features, and ML to detect incipient short circuits using axial leakage flux signals.

Several authors targeted the detection of one or more BRB in IMs controlled by electronic converters. Besides the fact that many of those works established a successful detection, other works improved the techniques to extract more incipient BRB fault features from the measurements. Interestingly, in [47], the authors presented a novel supervised classification approach for BRB fault diagnosis based on adaptive boosting algorithm with an optimized sampling method that deals with an imbalanced experimental dataset. Experimental results of the proposal provide accurate diagnosis of different intermediate severities before a full BRB. Another work that investigates various severity levels was presented in [48], where a combined approach with the fast Fourier transform (FFT) and the multiple signal classification (MUSIC) algorithm is proposed. This study exposes that incipient rotor faults in a squirrel-cage rotor, prior to the complete breaking of a rotor bar, are better identified when IMs are feed by some inverters than others. In addition, to detect incipient BRB, the authors of [49] proposed to use robust statistical techniques that are commonly applied in quality control applications. In this study, the proposal can detect partial bar breaks of 6.4 mm, 11.7 mm, and 17 mm (a full-broken bar).

Otherwise, different types of diagnosis methodologies have also developed for bearing failures in inverter-fed induction motors. The investigation in [50] presented a technique to attenuate multiple dominant harmonics with the aim to reduce spectral leakage, expose minute fault components, and improve the amplitude estimation of fault-related harmonics. Experimental outcomes of the methodology prove that the algorithm used for the spectral estimation of the vibration signal is adequate for an early determination of inverter-fed IM faults (inner raceway, outer raceway, cage train, rolling element, and in a single bar of the rotor). Another interesting technique for bearing failure recognition was presented in [51]. Its main contribution lies in the proposal of a condition monitoring strategy that is focused on the analysis and identification of five different fault severities of the outer race bearing (drill bits with diameters of 1 to 5 mm). The proposed approach is supported by fusing information from different sensors and the application of ML and AI. Apart from KB methods, reference [52] proposed a new approach based on a dynamic model in $d - q$ coordinate systems and analyzed BRB and turn-to-turn short fault. This approach uses a residual technique between model and measured signals. Experimental analyses show that the designed detection and isolation scheme provides high sensitivity and accurate isolation to incipient winding faults. To suppress the impact of the motor fault, an interesting control method was proposed in [53]. This control method includes a monitoring technique that can detect faults occurring due to stator winding short circuit at an incipient stage by means of a harmonic analysis of the magnetic air-gap flux. Methods based on steady-state analysis such as the above-mentioned are noteworthy in motor fault detection. However, these methods have severe limitations and may lead to false positives or false negatives. Table 2 shows a performance comparison of the above-mentioned applied methodologies.

Table 2. Comparison of early fault detection methodologies applied on IM under steady regime.

Ref.	Monitoring Signal	Power Source	Detected Fault	Applied Algorithm	Fault Severity
[33]	vibrations	line	broken bar	FFT + OMP + SVM	10%
[35]	magnetic flux	line	interturn	FFT + PV + EPV	0.25%
[36]	1-ϕ current	line	broken bar	WT + ETSA + FL	15.5%
[37]	3-ϕ currents + 3-ϕ voltages	line	interturn	1DCNN + LSTM + GRU	0.358%
[38]	3-ϕ currents	line	interturn	ANN	1%
[39]	3-ϕ currents + 3 voltages	line	shorted turn	Phasor Compensation	1.7%
[41]	3-ϕ currents	inverter	bearing	Fuzzy + C-means	10%
[44]	3-ϕ currents	inverter	interturn	HCNN + SVM	0.3%
[45]	1-ϕ current	inverter	interturn	DWT	1%

Table 2. *Cont.*

Ref.	Monitoring Signal	Power Source	Detected Fault	Applied Algorithm	Fault Severity
[46]	axial flux	inverter	interturn	DWT + SF + ML	1.41%
[47]	1-ϕ current	inverter	broken bar	AB + OS	25%
[48]	1-ϕ current	inverter	broken bar	MUSIC + FFT	50%
[49]	1-ϕ current	inverter	broken bar	QCC	33.33%
[50]	1-ϕ current	inverter	bearing	RQS	10%
[51]	1-ϕ current + 2 vibration	inverter	bearing	SF + ANN	11.1%
[52]	3-ϕ voltages + speed	inverter	interturn	Robust Observer	3.83%
[53]	1-ϕ current + 1-ϕ voltage + 1 flux	inverter	interturn	DNN	2.8%

4.2. Early Fault Detection via Transient-State Analysis

On the other hand, the analysis of IM data with non-stationary behavior mostly relies on time-frequency transforms computed from signals that are measured when motors are running under transient regimes. The most common time-frequency decompositions are the short-time Fourier transform (STFT) and the wavelet transforms (WT). The use of a combination of one variant of wavelet transform, the recursive wavelet transform, and a widely used tool in quality control, the statistical process control was presented in [54] to detect incipient stages of rotor fault. The methodology offers high accuracy in broken bar detection, starting from a deep hole of 3 mm in one bar to 2 fully broken bars. The detection was implemented during a steady-state operation mode and for short transient events of the motor. Another methodology using the wavelet transform has been reported in [55]. The fault diagnosis system is based on an empirical neuro-predictor and the application of wavelet analysis to residual signals between the model and the measured physical magnitudes. The method reports effective accuracy in detecting the most widely encountered electrical and mechanical faults. The motor anomalies consist of variations in the balance of the electric power supply and the driven mechanical load level when the IM experiences short transients as the load is varied from 0% up to 120% of the rated load. The investigation carried out in [56] examined the impact of incipient rotor faults on the shaft speed of an inverter-fed induction motor. This work used a tachogenerator to measure the rotor speed and applying a high-resolution spectral analysis (MUSIC algorithm) which detects and quantifies the fault severity in the time-frequency domain. This methodology can identify 5 health condition levels (healthy, $\frac{1}{4}$ BRB, $\frac{1}{2}$ BRB, $\frac{3}{4}$ BRB, and 1 BRB deep hole of 13 mm) during startup transients of 10 s. In [57], the stator current of a DOL-fed motor starting was used to extract statistical features, and using homogeneity as a classification index. This methodology can identify and classify differences between distinct fault severity conditions of the rotor bars (e.g., healthy, $\frac{1}{2}$ BRB, 1 BRB, and 2 BRB). The low computational complexity of the homogeneity index makes the method suitable for hardware implementation. The authors of [58] provided interesting time-frequency results for detecting various motor conditions such as stator winding interturn shorts, and phase to ground faults. In this work, the Stockwell transform (STW) was used to analyze the starting current of a DOL startup transient, the resulting time-frequency matrix was used to extract fault features and fed two different support vector machine (SVM) models. An average classification accuracy of 96% was achieved for both types o faults. Other researchers proposed in [59] a fault diagnosis technique based on the acquisition of signals from multiple sensors in order to assess the occurrence of single, combined, and simultaneous fault conditions in an induction motor. The proposal performs principal component analysis (PCA) to each signal, then joined as input to a decision tree method. The considered early fault stage was a rough hole of 2 mm diameter, at a depth of 14 mm, into the rotor bars of a DOL-fed motor. As another example of the transient analysis, in [60], the tooth-FFT algorithm was introduced to track time-varying frequency components. Half and full broken bars were considered; experimental results obtained a percentage of detection of 97.35% for all motor conditions. In Table 3, a comparison of the applied methodologies is presented.

Table 3. Comparison of early fault detection methodologies applied on IM under transient regime.

Ref.	Monitoring Signal	Power Source	Detected Fault	Applied Algorithm	Fault Severity
[54]	1-ϕ current	line	broken bar	WPD + SPC	75%
[55]	3-ϕ currents 3-ϕ voltages + speed	line	interturn	WPD	0.92%
[56]	speed	inverter	broken bar	short-time Minimum-Norm	25%
[57]	1-ϕ current	line	broken bar	Homogeneity	25%
[58]	3-ϕ currents	line	interturn	Stockwell transform	1.6%
[59]	3-ϕ currents + 3-ϕ voltages + 3-axis vibrations	inverter	bearing	PCA + decision trees	12.5%
[60]	1-ϕ current	line	broken bar	Tooth-FFT	50%

5. Discussion

According to the literature, three monitored signals are principally used for early fault detection in IM: mechanical vibrations, stator current, and magnetic flux. The vibration signal is widely used because it is sensitive to internal asymmetries in the machine during its operation. Despite this advantage, it is difficult to determine specifically the internal fault that produces the asymmetry. Furthermore, used sensors are sensitive to external vibrations also; thus, it is common that vibration analysis suffers from external interference. Important to note that monitoring vibration signals requires a short sampling period at the acquisition stage due to fault patterns appearing in the high-frequency band. On the other hand, the stator current contains specific fault patterns for each type of fault and the measurement is not very sensitive to external interference. Furthermore, most of the spectral patterns related to a fault type are generated in the low-frequency band of the spectral content. Therefore, a short sampling period in the acquisition stage is not necessary. Despite the benefits of current monitoring, when the induction motor is powered by an electronic converter, additional frequency components are induced in the current spectrum, thereby adding many spectral components foreign to the MI behavior that can alter the magnitude of the fault indicators.

Some works use magnetic flux signals. This type of magnitude is spectrally dense, so it has many frequency components interacting with each other, which makes it difficult to evaluate the magnitude of one spectral component without interference from another. One of the advantages presented by the authors when monitoring this signal is that the magnetic flux is not sensitive to external mechanical vibrations and depending on the location of the sensor; the flux signal makes it possible to locate faults in specific places of the motor.

Most of the reviewed works focus on analyzing signals from IM operating at steady state. This analysis has many advantages compared to the transient-state analysis because in the steady-state analysis, several fault indicators can be extracted by classic signal processing techniques. On the contrary, the analysis of non-stationary signals requires more complex tools and advanced signal processing techniques. Despite the complexity of the analysis, the study of IM operating in transients permits locating and tracking the behavior of specific fault indicators for each internal motor component. This allows more accurate fault identification and severity diagnosis compared to the analysis of an IM operating at steady-state.

In the literature, it is possible to find three groups of techniques that are mainly used for early fault detection: classical, modern, and heuristic. In Table 4, a list of the applied methodologies is presented. First, in the classical techniques, statistical tools have been used mainly to extract fault features from the behavior of signals in the time-domain. Meanwhile, frequency-domain tools have also been used to extract fault components from the spectral content of the analyzed signal, the best known is the Fourier transform. Although these classical techniques have worked to detect incipient faults in IM operating at steady-state, this group of techniques is not appropriate for monitoring time-varying systems. Some authors used modern techniques, which allow the analysis of physical magnitudes in a simultaneous time-frequency domain. The tools used for a time-frequency analysis are the STFT, the wavelet transform, the Winner–Ville transform, the Hilbert–Huang transform, and high-resolution decompositions (short-time MUSIC), among others. These modern techniques have positioned as great alternatives in the field to reduce diagnostic errors that classic techniques have. Despite the advantages of this type of techniques, they require a greater computational burden and a higher level of interpretation of results than classical techniques. In most recent works, the application of heuristic methodologies such as GA, pattern recognition, ML and AI techniques can be found. These works report high levels of accuracy in the detection of some faults at incipient stage. Despite the good results reported, these methodologies require a very large and diverse database.

Table 4. List of the different applied techniques.

Technique	Year	Detected Fault	References	Advantages
Adaptive Boosting	2017	BRB	[47]	improves the predictive accuracy of classifiers
Artificial Neural Network	2017	SF	[38]	improves efficiency in process decision
C-means	2021	BF	[41]	improves the performance by suitable changes in regularization, cluster shape, and cost function
Convolutional Neural Network (1D)	2021	SF	[37]	extracts deep features maps
Decision trees	2020	BF	[59]	reduces specific parameters information for diagnosis
Deep neural network	2017	SF	[53]	high accuracy in classification and estimation
Discrete wavelet transform	2021, 2021	SF	[45,46]	multi-scale analysis
Down-sampling	2017	BRB	[47]	reduces data length
Electrical time synchronous averaging (ETSA)	2022	BRB	[36]	improves classification
Extended Park	2021	SF	[35]	improves the relation of harmonics with the severity of the fault
Fourier transform	2001, 2018, 2021	BRB, BRB, SF	[33,35,47]	reduces sensitivity to noise
Fuzzy logic	2022	BRB	[36]	improves classification features
Gated recurrent Unit	2021	SF	[37]	processes extra features
Hierarchical CNN	2021	SF	[44]	improves the severity simultaneously level features/higher classification accuracy with lesser testing time
Homogeneity	2017	BRB	[57]	improves classify differences on distinct operational conditions
Linear discriminant analysis	2021	BF	[51]	reduces the dimensionality of features
Long short-term memory	2021	SF	[37]	improves the classification of long term and nonlinear time data
Multiple signal classification	2017	BRB	[48]	high-resolution frequency analysis
Orthogonal matching pursuit	2018	BRB	[33]	improves the classifier criterion /majority decision classifier
Optimized sampling	2017	BRB	[47]	improves the original imbalanced dataset
Principal component analysis	2020	BF	[59]	reduces the dimension of attributes
Phasor compensation	2022	SF	[39]	improves the calibration of disturbances effect
Quality control charts	2017	BRB	[49]	reduces the problem of classification/improves the robustness
Rayleigh quotient spectrum	2021	BF	[50]	reduces complexity, accuracy in frequency estimation

Table 4. *Cont.*

Technique	Year		Detected Fault		References	Advantages
short-time minimum norm	2021		BRB		[56]	improves the frequency resolution analysis avoiding spurious components
Statistical features	2021		SF		[46]	improves the general accuracy of the prediction system
Statistical process control	2018		BRB		[54]	improves the learning process
SVM	2018, 2022	2020,	SF, SF	BRB,	[33,44,58]	better generalization of nonlinear classification
Tooth-FFT	2018		BRB		[60]	higher sensitive of non-stationary signals
Wavelet packet decomposition	2018		BRB		[54]	multi-resolution analysis

6. Conclusions

This paper has reviewed the most recent contributions related to the early fault detection in induction motors. These contributions are classified into two main groups according to the operational mode of the motor: steady-state and transient-state. In this paper, it is shown that most of the research work is focused on the steady-state analysis. Despite the high-level of accuracy reported in the fault severity techniques based on steady-state analysis, these proposals still suffer for diagnostic errors. This work also presents reported algorithms based on different type of monitoring signals used for fault detection, and some characteristics of each measured magnitude. According to the developed review, it can be concluded that the techniques most used for fault detection at incipient stages are heuristic methods (knowledge-based), and sometimes a combination of signal processing methods and KB. The major problems with heuristic methods are the required computational resources and the diverse and large amount of data. Despite the amount of work carried out in detection of incipient faults of IM, just few works analyze transients and just some of them analyze inverter-fed IM transients. Regarding the fault type, most of the research work is focus on the detection of partially broken rotor bars.

7. Future Perspective

Despite constant research activity in the fault detection field, it can be observed that incipient fault detection and severity fault evaluation in IM is still an open challenge. As most of the existing research has been focused on the detection of incipient faults in IM operating under the steady regime, the study of IM operating under non-stationary regime is a natural trend. More importantly, as the use of inverters is increasing in the industry, an approach for future research is the development of new techniques that can diagnose incipient faults in the inverter-fed IM under transient operation. In addition, although some methods to separate load-oscillations from BRB signatures have been proposed [61–66], the effectiveness of early fault detection methods against this kind of external oscillation still needs to be validated. Several techniques have been applied to current and vibration signals; therefore, the reliability of other monitoring signals it could be explored. In order to obtain robust and practical solutions, various corner cases and environmental conditions should also be analyzed, such as the effects of combined multiple faults, realistic degradation, and industrial measurement inference. In in this paper, it is shown how KB techniques have been an increasing activity over recent years, opening new opportunities in which the application of these new techniques and the combination with modern SP can bring interesting advantages. Finally, new indicators are required to know the severity stage of a fault and to extract its features for improving the classification and the remaining useful life of the internal element. This necessity will encourage the development of new techniques able to filter out external interferences and, at the same time, quantify the fault indicator features. In addition, new techniques to be proposed should include the benefits of the existing works in the developed fault detection field, such as reliability, feasibility for implementation (low computational burden), portability, online detection, detection of multiple and combined faults, and so on.

Author Contributions: T.G.-C. performed the investigation of several of the revised works in the state of the art, performed analysis, and wrote a part of the paper; D.M.-S. performed the investigation of several of the revised works in the state of the art and wrote a part of the paper; V.F.-C. performed the investigation of several of the revised works in the state of the art and wrote a part of the paper; R.R.-T. conceived and developed the idea for this research, performed analysis, and wrote some of the sections. All authors have read an agreed to the published version of the manuscript.

Funding: This research received no external funding.

Institutional Review Board Statement: Not applicable.

Informed Consent Statement: Not applicable.

Data Availability Statement: No new data were generated.

Conflicts of Interest: The authors declare no conflicts of interest.

Nomenclature

The following abbreviations are used in this manuscript:

AI	Artificial Intelligence
ANN	Artificial Neural networks
BRB	Broken Rotor Bar
BF	Bearing Fault
CNN	Convolutional Neural Network
DL	Deep Learning
DOL	Direct on Line
DWT	Discrete Wavelet Transform
EMD	Empirical Mode Decomposition
ETSA	Electrical-time-Synchronous Averaging
FFT	Fast Fourier Transform
FFNN	Feed-forward Neural Network
FPGA	Field Programmable Array
GA	Genetic Algorithms
IM	Induction Motors
KB	Knowledge-based
LDA	Linear Discrimination Analysis
LSTM	Long Short-Term Memory
MCSA	Motor Current Signature Analysis
MUSIC	Multiple Signal Classification
ML	Machine Learning
PCA	Principal Component Analysis
SF	Stator Faults
SP	Signal Processing
SPC	Statistical Process Control
STFT	Short-time Fourier Transform
SVM	Support Vector Machine
STW	Stationary Wavelet Transform

References

1. Toliyat, H.A.; Nandi, S.; Choi, S.; Meshgin-Kelk, H. *Electric Machines: Modeling, Condition Monitoring, and Fault Diagnosis*; CRC Press: Boca Raton, FL, USA, 2012.
2. Toliyat, H.A.; Kliman, G.B. *Handbook of Electric Motors*; CRC Press: Boca Raton, FL, USA, 2018; Volume 120.
3. Trzynadlowski, A. *The Field Orientation Principle in Control of Induction Motors*; Springer Science & Business Media: Berlin/Heidelberg, Germany, 1993; Volume 258.
4. Zhang, P.; Du, Y.; Habetler, T.G.; Lu, B. A survey of condition monitoring and protection methods for medium-voltage induction motors. *IEEE Trans. Ind. Appl.* **2010**, *47*, 34–46. [CrossRef]

5.   Nandi, S.; Toliyat, H.; Li, X. Condition Monitoring and Fault Diagnosis of Electrical Motors—A Review. *IEEE Trans. Energy Convers.* **2005**, *20*, 719–729. [CrossRef]

6.   Frosini, L. Novel diagnostic techniques for rotating electrical machines—A review. *Energies* **2020**, *13*, 5066. [CrossRef]

7.   Cerrada, M.; Sánchez, R.V.; Li, C.; Pacheco, F.; Cabrera, D.; de Oliveira, J.V.; Vásquez, R.E. A review on data-driven fault severity assessment in rolling bearings. *Mech. Syst. Signal Process.* **2018**, *99*, 169–196. [CrossRef]

8.   Gundewar, S.K.; Kane, P.V. Condition monitoring and fault diagnosis of induction motor. *J. Vib. Eng. Technol.* **2021**, *9*, 643–674. [CrossRef]

9.   Chow, M.; Yee, S. Methodology for on-line incipient fault detection in single-phase squirrel-cage induction motors using artificial neural networks. *IEEE Trans. Energy Convers.* **1991**, *6*, 536–545. [CrossRef]

10.   Zhou, W.; Lu, B.; Habetler, T.G.; Harley, R.G. Incipient Bearing Fault Detection via Motor Stator Current Noise Cancellation Using Wiener Filter. *IEEE Trans. Ind. Appl.* **2009**, *45*, 1309–1317. [CrossRef]

11.   Garcia, M.; Panagiotou, P.A.; Antonino-Daviu, J.A.; Gyftakis, K.N. Efficiency Assessment of Induction Motors Operating Under Different Faulty Conditions. *IEEE Trans. Ind. Electron.* **2019**, *66*, 8072–8081. [CrossRef]

12.   Hassan, O.E.; Amer, M.; Abdelsalam, A.K.; Williams, B.W. Induction motor broken rotor bar fault detection techniques based on fault signature analysis–a review. *IET Electr. Power Appl.* **2018**, *12*, 895–907. [CrossRef]

13.   Liang, X.; Ali, M.Z.; Zhang, H. Induction Motors Fault Diagnosis Using Finite Element Method: A Review. *IEEE Trans. Ind. Appl.* **2020**, *56*, 1205–1217. [CrossRef]

14.   Gangsar, P.; Tiwari, R. Signal based condition monitoring techniques for fault detection and diagnosis of induction motors: A state-of-the-art review. *Mech. Syst. Signal Process.* **2020**, *144*, 106908. [CrossRef]

15.   de Jesús Rangel-Magdaleno, J. Induction Machines Fault Detection: An Overview. *IEEE Instrum. Meas. Mag.* **2021**, *24*, 63–71. [CrossRef]

16.   Zhang, S.; Zhang, S.; Wang, B.; Habetler, T.G. Deep learning algorithms for bearing fault diagnostics—A comprehensive review. *IEEE Access* **2020**, *8*, 29857–29881. [CrossRef]

17.   Siddique, A.; Yadava, G.; Singh, B. A review of stator fault monitoring techniques of induction motors. *IEEE Trans. Energy Convers.* **2005**, *20*, 106–114. [CrossRef]

18.   Sumislawska, M.; Gyftakis, K.N.; Kavanagh, D.F.; McCulloch, M.D.; Burnham, K.J.; Howey, D.A. The Impact of Thermal Degradation on Properties of Electrical Machine Winding Insulation Material. *IEEE Trans. Ind. Appl.* **2016**, *52*, 2951–2960. [CrossRef]

19.   Tavner, P.J. Review of condition monitoring of rotating electrical machines. *IET Electr. Power Appl.* **2008**, *2*, 215–247. [CrossRef]

20.   Garcia-Calva, T.A.; Morinigo-Sotelo, D.; de Jesus Romero-Troncoso, R. Non-Uniform Time Resampling for Diagnosing Broken Rotor Bars in Inverter-Fed Induction Motors. *IEEE Trans. Ind. Electron.* **2017**, *64*, 2306–2315. [CrossRef]

21.   Blodt, M.; Granjon, P.; Raison, B.; Rostaing, G. Models for Bearing Damage Detection in Induction Motors Using Stator Current Monitoring. *IEEE Trans. Ind. Electron.* **2008**, *55*, 1813–1822. [CrossRef]

22.   Mirzaeva, G.; Saad, K.I. Advanced Diagnosis of Stator Turn-to-Turn Faults and Static Eccentricity in Induction Motors Based on Internal Flux Measurement. *IEEE Trans. Ind. Appl.* **2018**, *54*, 3961–3970. [CrossRef]

23.   Trujillo Guajardo, L.A.; Platas Garza, M.A.; Rodríguez Maldonado, J.; González Vázquez, M.A.; Rodríguez Alfaro, L.H.; Salinas Salinas, F. Prony Method Estimation for Motor Current Signal Analysis Diagnostics in Rotor Cage Induction Motors. *Energies* **2022**, *15*, 3513. [CrossRef]

24.   El Hachemi Benbouzid, M. A review of induction motors signature analysis as a medium for faults detection. *IEEE Trans. Ind. Electron.* **2000**, *47*, 984–993. [CrossRef]

25.   Cameron, J.; Thomson, W.; Dow, A. Vibration and current monitoring for detecting airgap eccentricity in large induction motors. *IEE Proc. (Electric Power Appl.)* **1986**, *133*, 155–163. [CrossRef]

26.   Iorgulescu, M.; Beloiu, R. Vibration and current monitoring for fault's diagnosis of induction motors. *Ann. Univ. Craiova, Electr. Eng. Ser.* **2008**, *32*, 102–103.

27.   Thomson, W.T.; Culbert, I. *Current Signature Analysis for Condition Monitoring of Cage Induction Motors: Industrial Application and Case Histories*; John Wiley & Sons: Hoboken, NJ, USA, 2017.

28.   Gangsar, P.; Tiwari, R. Comparative investigation of vibration and current monitoring for prediction of mechanical and electrical faults in induction motor based on multiclass-support vector machine algorithms. *Mech. Syst. Signal Process.* **2017**, *94*, 464–481. [CrossRef]

29.   Zamudio-Ramirez, I.; Osornio-Rios, R.A.; Antonino-Daviu, J.A.; Razik, H.; Romero-Troncoso, R.d.J. Magnetic Flux Analysis for the Condition Monitoring of Electric Machines: A Review. *IEEE Trans. Ind. Inform.* **2022**, *18*, 2895–2908. [CrossRef]

30.   Penman, J.; Hadwick, J.; Barbour, B. Detection of faults in electrical machines by examination of the axially directed fluxes. InProceedings ICEM; IEEE: Brussels, Belgium, 1978; Volume 78.

31.   Guezmil, A.; Berriri, H.; Pusca, R.; Sakly, A.; Romary, R.; Mimouni, M.F. Detecting inter-turn short-circuit fault in induction machine using high-order sliding mode observer: Simulation and experimental verification. *J. Control. Autom. Electr. Syst.* **2017**, *28*, 532–540. [CrossRef]

32.   Namdar, A.; Samet, H.; Allahbakhshi, M.; Tajdinian, M.; Ghanbari, T. A robust stator inter-turn fault detection in induction motor utilizing Kalman filter-based algorithm. *Measurement* **2022**, *187*, 110181. [CrossRef]

33. Morales-Perez, C.; Rangel-Magdaleno, J.; Peregrina-Barreto, H.; Amezquita-Sanchez, J.P.; Valtierra-Rodriguez, M. Incipient Broken Rotor Bar Detection in Induction Motors Using Vibration Signals and the Orthogonal Matching Pursuit Algorithm. *IEEE Trans. Instrum. Meas.* **2018**, *67*, 2058–2068. [CrossRef]

34. Gribonval, R.; Nielsen, M. Sparse representations in unions of bases. *IEEE Trans. Inf. Theory* **2003**, *49*, 3320–3325. [CrossRef]

35. Gyftakis, K.N.; Cardoso, A.J.M. Reliable Detection of Stator Interturn Faults of Very Low Severity Level in Induction Motors. *IEEE Trans. Ind. Electron.* **2021**, *68*, 3475–3484. [CrossRef]

36. Sabir, H.; Ouassaid, M.; Ngote, N. An experimental method for diagnostic of incipient broken rotor bar fault in induction machines. *Heliyon* **2022**, *8*, e09136. [CrossRef] [PubMed]

37. Husari, F.; Seshadrinath, J. Early Stator Fault Detection and Condition Identification in Induction Motor Using Novel Deep Network. *IEEE Trans. Artif. Intell.* **2021** , *3*, 809–818. [CrossRef]

38. Palacios, R.H.C.; da Silva, I.N.; Goedtel, A.; Godoy, W.F.; Lopes, T.D. Diagnosis of Stator Faults Severity in Induction Motors Using Two Intelligent Approaches. *IEEE Trans. Ind. Inform.* **2017**, *13*, 1681–1691. [CrossRef]

39. Bakhri, S.; Ertugrul, N. A Negative Sequence Current Phasor Compensation Technique for the Accurate Detection of Stator Shorted Turn Faults in Induction Motors. *Energies* **2022**, *15*, 3100. [CrossRef]

40. Deekshit Kompella, K.; Venu Gopala Rao, M.; Srinivasa Rao, R. Bearing fault detection in a 3 phase induction motor using stator current frequency spectral subtraction with various wavelet decomposition techniques. *Ain Shams Eng. J.* **2018**, *9*, 2427–2439. [CrossRef]

41. Barcelos, A.S.; Cardoso, A.J.M. Current-based bearing fault diagnosis using deep learning algorithms. *Energies* **2021**, *14*, 2509. [CrossRef]

42. Ferreira, F.J.; de Almeida, A.T. Reducing Energy Costs in Electric-Motor-Driven Systems: Savings Through Output Power Reduction and Energy Regeneration. *IEEE Ind. Appl. Mag.* **2018**, *24*, 84–97. [CrossRef]

43. Garcia-Calva, T.A.; Morinigo-Sotelo, D.; Garcia-Perez, A.; Camarena-Martinez, D.; de Jesus Romero-Troncoso, R. Demodulation Technique for Broken Rotor Bar Detection in Inverter-Fed Induction Motor Under Non-Stationary Conditions. *IEEE Trans. Energy Convers.* **2019**, *34*, 1496–1503. [CrossRef]

44. Husari, F.; Seshadrinath, J. Incipient Interturn Fault Detection and Severity Evaluation in Electric Drive System Using Hybrid HCNN-SVM Based Model. *IEEE Trans. Ind. Inform.* **2022**, *18*, 1823–1832. [CrossRef]

45. Almounajjed, A.; Sahoo, A.K.; Kumar, M.K. Diagnosis of stator fault severity in induction motor based on discrete wavelet analysis. *Measurement* **2021**, *182*, 109780. [CrossRef]

46. Guerreiro Carvalho Cunha, R.; da Silva, E.T.; Marques de Sá Medeiros, C. Machine learning and multiresolution decomposition for embedded applications to detect short-circuit in induction motors. *Comput. Ind.* **2021**, *129*, 103461. [CrossRef]

47. Martin-Diaz, I.; Morinigo-Sotelo, D.; Duque-Perez, O.; de Romero-Troncoso, R.J. Early Fault Detection in Induction Motors Using AdaBoost with Imbalanced Small Data and Optimized Sampling. *IEEE Trans. Ind. Appl.* **2017**, *53*, 3066–3075. [CrossRef]

48. Martin-Diaz, I.; Morinigo-Sotelo, D.; Duque-Perez, O.; Arredondo-Delgado, P.; Camarena-Martinez, D.; Romero-Troncoso, R. Analysis of various inverters feeding induction motors with incipient rotor fault using high-resolution spectral analysis. *Electr. Power Syst. Res.* **2017**, *152*, 18–26. [CrossRef]

49. García-Escudero, L.A.; Duque-Perez, O.; Fernandez-Temprano, M.; Morinigo-Sotelo, D. Robust Detection of Incipient Faults in VSI-Fed Induction Motors Using Quality Control Charts. *IEEE Trans. Ind. Appl.* **2017**, *53*, 3076–3085. [CrossRef]

50. Samanta, A.K.; Routray, A.; Khare, S.R.; Naha, A. Minimum Distance-Based Detection of Incipient Induction Motor Faults Using Rayleigh Quotient Spectrum of Conditioned Vibration Signal. *IEEE Trans. Instrum. Meas.* **2021**, *70*, 1–11. [CrossRef]

51. Saucedo-Dorantes, J.J.; Zamudio-Ramirez, I.; Cureno-Osornio, J.; Osornio-Rios, R.A.; Antonino-Daviu, J.A. Condition monitoring method for the detection of fault graduality in outer race bearing based on vibration-current fusion, statistical features and neural network. *Appl. Sci.* **2021**, *11*, 8033. [CrossRef]

52. Wu, Y.; Jiang, B.; Wang, Y. Incipient winding fault detection and diagnosis for squirrel-cage induction motors equipped on CRH trains. *ISA Trans.* **2020**, *99*, 488–495. [CrossRef]

53. Ghosh, E.; Mollaeian, A.; Kim, S.; Tjong, J.; Kar, N.C. DNN-Based Predictive Magnetic Flux Reference for Harmonic Compensation Control in Magnetically Unbalanced Induction Motor. *IEEE Trans. Magn.* **2017**, *53*, 1–7. [CrossRef]

54. Hmida, M.A.; Braham, A. An On-Line Condition Monitoring System for Incipient Fault Detection in Double-Cage Induction Motor. *IEEE Trans. Instrum. Meas.* **2018**, *67*, 1850–1858. [CrossRef]

55. Kim, K.; Parlos, A. Induction motor fault diagnosis based on neuropredictors and wavelet signal processing. *IEEE/ASME Trans. Mechatronics* **2002**, *7*, 201–219. [CrossRef]

56. Garcia-Calva, T.A.; Morinigo-Sotelo, D.; Fernandez-Cavero, V.; Garcia-Perez, A.; Romero-Troncoso, R.d.J. Early detection of broken rotor bars in inverter-fed induction motors using speed analysis of startup transients. *Energies* **2021**, *14*, 1469. [CrossRef]

57. Lizarraga-Morales, R.A.; Rodriguez-Donate, C.; Cabal-Yepez, E.; Lopez-Ramirez, M.; Ledesma-Carrillo, L.M.; Ferrucho-Alvarez, E.R. Novel FPGA-based Methodology for Early Broken Rotor Bar Detection and Classification Through Homogeneity Estimation. *IEEE Trans. Instrum. Meas.* **2017**, *66*, 1760–1769. [CrossRef]

58. Singh, M.; Shaik, A.G. Incipient Fault Detection in Stator Windings of an Induction Motor Using Stockwell Transform and SVM. *IEEE Trans. Instrum. Meas.* **2020**, *69*, 9496–9504. [CrossRef]

59. Juez-Gil, M.; Saucedo-Dorantes, J.J.; Arnaiz-González, Á.; López-Nozal, C.; García-Osorio, C.; Lowe, D. Early and extremely early multi-label fault diagnosis in induction motors. *ISA Trans.* **2020**, *106*, 367–381. [CrossRef] [PubMed]

60. Rivera-Guillen, J.R.; De Santiago-Perez, J.; Amezquita-Sanchez, J.P.; Valtierra-Rodriguez, M.; Romero-Troncoso, R.J. Enhanced FFT-based method for incipient broken rotor bar detection in induction motors during the startup transient. *Measurement* **2018**, *124*, 277–285. [CrossRef]
61. Marzebali, M.H.; Bazghandi, R.; Abolghasemi, V. Rotor Asymmetries Faults Detection in Induction Machines under the Impacts of Low-Frequency Load Torque Oscillation. *IEEE Trans. Instrum. Meas.* **2022**, 1–11. [CrossRef]
62. Garcia-Calva, T.A.; Morinigo-Sotelo, D.; de Jesus Romero Troncoso, R. Fundamental Frequency Normalization for Reliable Detection of Rotor and Load Defects in VSD-Fed Induction Motors. *IEEE Trans. Energy Convers.* **2022**, *37*, 1467–1474. [CrossRef]
63. Park, Y.; Choi, H.; Shin, J.; Park, J.; Lee, S.B.; Jo, H. Airgap Flux Based Detection and Classification of Induction Motor Rotor and Load Defects During the Starting Transient. *IEEE Trans. Ind. Electron.* **2020**, *67*, 10075–10084. [CrossRef]
64. Guellout, O.; Rezig, A.; Touati, S.; Djerdir, A. Elimination of broken rotor bars false indications in induction machines. *Math. Comput. Simul.* **2020**, *167*, 250–266. [CrossRef]
65. Bossio, G.R.; De Angelo, C.H.; Bossio, J.M.; Pezzani, C.M.; Garcia, G.O. Separating Broken Rotor Bars and Load Oscillations on IM Fault Diagnosis Through the Instantaneous Active and Reactive Currents. *IEEE Trans. Ind. Electron.* **2009**, *56*, 4571–4580. [CrossRef]
66. Puche-Panadero, R.; Martinez-Roman, J.; Sapena-Bano, A.; Burriel-Valencia, J. Diagnosis of Rotor Asymmetries Faults in Induction Machines Using the Rectified Stator Current. *IEEE Trans. Energy Convers.* **2020**, *35*, 213–221. [CrossRef]

Article

Stator ITSC Fault Diagnosis of EMU Asynchronous Traction Motor Based on apFFT Time-Shift Phase Difference Spectrum Correction and SVM

Jie Ma [1,2], Xiaodong Liu [1], Jisheng Hu [1], Jiyou Fei [1,*], Geng Zhao [1] and Zhonghuan Zhu [3]

[1] College of Locomotive and Rolling Stock Engineering, Dalian Jiaotong University, Dalian 116028, China; majie@lntdxy.com (J.M.); lxd@djtu.edu.cn (X.L.); zhaogeng1983@163.com (G.Z.)

[2] Liaoning Railway Vocational and Technical College, Jinzhou 121000, China

[3] Shenyang EMU Depot, China Railway Shenyang Group Co., Ltd., Shenyang 110179, China; zhuzhonghu@sydcd.ntesmail

* Correspondence: fjy@djtu.edu.cn

Abstract: EMU (electric multiple unit) traction motors are powered by converters whose output voltage increases the voltage stress borne by the insulation system, making the ITSC (inter-turn short-circuit) fault more prominent. An index based on short-circuit thermal power is proposed in the article to evaluate the non-metallic ITSC faults extent. The apFFT (all-phase FFT) time-shift phase difference correction with double Hanning windows is used to calculate fault features to train the SVM (support vector machine) fault diagnosis model whose hyper-parameters C and g are optimized using grid search methods. The experimental verification was carried out on the EMU electric traction simulation experimental platform. According to the fault extent index proposed in this article, the experimental samples were divided into three categories, normal, incipient and serious fault samples. The ITSC fault diagnosis accuracy was 100% on the training dataset and 93.33% on the test dataset. There was no misclassification between normal and serious ITSC fault samples.

Keywords: ITSC fault; traction motor; fault diagnosis; apFFT; SVM

Citation: Ma, J.; Liu, X.; Hu, J.; Fei, J.; Zhao, G.; Zhu, Z. Stator ITSC Fault Diagnosis of EMU Asynchronous Traction Motor Based on apFFT Time-Shift Phase Difference Spectrum Correction and SVM. *Energies* **2023**, *16*, 5612. https://doi.org/10.3390/en16155612

Academic Editors: Daniel Morinigo-Sotelo, Joan Pons-Llinares and Rene Romero-Troncoso

Received: 26 April 2023
Revised: 14 July 2023
Accepted: 17 July 2023
Published: 26 July 2023

1. Introduction

The AC–DC–AC transmission mode is used in modern EMU traction systems, and three-phase AC squirrel-cage asynchronous motors are used as traction motors [1]. Affected by mechanical stress, thermal stress, and electrical stress, EMU traction motors are prone to failure [2,3]. The fault types of three-phase AC asynchronous squirrel-cage motors in industrial applications mainly include stator insulation faults (37%), rotor broken bar faults (12%), bearing faults (41%), and other faults (10%) [4]. The high voltage stress generated by the inverter PWM (pulse width modulation) voltage accelerates the degradation of the traction motor insulation system [5–8]. The inter-turn insulation is the weakest part of the asynchronous traction motor insulation system [9]. When an ITSC fault occurs, an inter-turn current circulates between the short-circuit turns, quickly producing a large amount of heat [10], which weakens the insulation of the motor and results in inter-phase or ground short-circuit faults [11]. Timely maintenance can prevent the ITSC fault's further expansion and significantly reduce the maintenance cost. Since the ITSC fault of the asynchronous traction motor is more hidden than the main insulation system fault, it is more difficult to detect the incipient ITSC fault [12,13].

The ITSC fault diagnosis of three-phase asynchronous motors mainly includes model-based, signal process–based, and artificial intelligence–based diagnosis methods [14–16]. An accurate motor ITSC fault model is needed for model-based ITSC fault diagnosis. Model-based methods mainly include the parameter estimation method and residual estimation method. With the parameter estimation method, the model parameters related to the ITSC fault are estimated [17–19]. Based on the three-phase asynchronous motor ITSC fault model under the dq axis, the particle filter algorithm is used to estimate multiple parameters to detect the stator ITSC fault and assess the motor insulation residual life [20]. With this method, real-time and non-intrusive diagnosis can be easily achieved by just measuring motor phase voltage and phase current. A healthy motor model is necessary for the residual estimation method, which takes the detectable variables related to the ITSC fault as state variables. It uses the difference between the state variables of the healthy motor estimated by the model and the measured variables as the residual to detect the ITSC fault [21,22]. The three-phase current is taken as the state variable, and the high-order sliding mode observer is used to observe the three-phase current of the healthy motor [23]. The residual of the observed and measured values is taken as the index for the ITSC fault. The model-based ITSC fault diagnosis method can be easily affected by working conditions. For example, when the method of estimating stator resistance is used to diagnose ITSC faults, the diagnostic results are highly influenced by temperature and frequency (skin effect).

The diagnosis method for the ITSC fault based on signal processing is mainly based on the electrical, magnetic, thermal, vibration, and acoustic signals. The stator ITSC fault is diagnosed by analyzing and processing the above signals in the time, frequency, or time–frequency domains [24–26]. The voltage or current signals can be used to realize non-intrusive diagnosis [27], saving costs without the need for installing extra sensors. For the steady operation state, the FFT algorithm is generally used to calculate specific frequency components of the current or other signals to detect the ITSC fault [28]. With the continuous development of new signal processing methods, time–frequency analysis methods, such as wavelet transform, WVD (Wigner–Ville distribution), and HHT (Hilbert–Huang transform), are also applied to the motors' fault diagnosis [29,30]. The discrete wavelet is used to decompose the stator current, and the maximum norm of the detail coefficient is used to detect the incipient ITSC fault [31].

Shallow machine learning and deep learning methods are also applied to motor stator ITSC fault diagnosis [32–35]. The ITSC fault diagnosis method based on shallow machine learning is generally divided into three stages: data preparation, feature extraction, and model training. Particle swarm optimization and principal component analysis can be used for feature extraction. The BP neural network and SVM models can be used as diagnosis models. The BP neural network is trained based on the phase difference of the three-phase stator current, which can detect and locate the stator ITSC fault [36]. When a deep learning network is adopted, such as a convolution neural network, it is unnecessary to extract features artificially because the deep learning network automatically extracts them. The instantaneous value of the three-phase current is taken as the feature, the convolution neural network is taken as the diagnosis model, and the trained convolution neural network can accurately detect the ITSC fault of a three-phase asynchronous motor [37].

Research on the diagnosis method for ITSC fault of asynchronous motors stator has achieved many positive results, but the diagnosis of EMU traction motor stator ITSC fault has special needs. Firstly, in previous studies, the extent of an AC motor ITSC fault is generally evaluated based on the number of short-circuit turns when the inter-turn resistance is fixed. In the non-metallic short circuit, the resistance between short-circuit turns is directly related to the extent of the motor damage caused by the fault. Secondly, most of the previous studies are carried out in the condition of non-variable frequency speed regulation, and the steady speed of the motor is fixed. The traction motor operates stably at different speeds according to operating conditions. Finally, the traction motor of the EMU adopts vector control or direct torque control based on the current closed loop. The fundamental frequency of voltage and current signals cannot be directly obtained, and a spectral correction method is needed to achieve a more accurate fundamental frequency.

The article is mainly divided into five parts. After the introduction, it introduces the measurement method for the traction motor's ZSVC (zero-sequence voltage component) and the apFFT time-shift phase difference correction method. This method is used to calculate the traction motor's ZSVC fundamental frequency, the fundamental component amplitudes of the ZSVC and the three-phase current in a steady state. In the third part, an ITSC fault extent index related to the number of short-circuit turns and the inter-turn resistance is proposed. The index is based on the thermal power of the circulating current between short-circuit turns. In addition, the SVM and the hyper-parameter optimization method are introduced in this part. In the following section, SVM is used to diagnose the ITSC fault. The fourth part is the experimental part. The EMU electric traction simulation experimental platform is used to simulate the steady-state operation of the EMU. According to the fault extent index proposed in this article, the experimental samples are divided into normal, incipient, and serious fault samples, and the ITSC fault diagnosis model is trained and tested. The last part summarizes all the research contents and puts forward the follow-up work.

2. Calculation of Signals' Fundamental Components

2.1. ZSVC Measurement Method

The traction motor's current is measured for speed and torque control during the operation, and only the ZSVC needs to be measured additionally. The ZSVC of the three-phase asynchronous motor can effectively monitor the stator ITSC fault [38,39]. The measurement circuit is relatively simple, and installing sensors on the motor body is unnecessary. According to the ZSVC definition of a three-phase asynchronous motor [40,41], as shown in Formula (1), three voltage sensors are needed when measuring the three-phase voltage.

$$v_0 = \frac{1}{3}\left(v_{\text{an}} + v_{\text{bn}} + v_{\text{cn}}\right) \tag{1}$$

Directly measuring three-phase voltage and calculating ZSVC according to Formula (1) can be applied to a sinusoidal power supply. The EMU traction inverter generates the ZSVC inherent in the PWM voltage and related to the PWM modulation mode. Although the frequency produced by the inverter is far from the fundamental frequency, if reasonable compensation and filtering are not carried out, frequency aliasing occurs, and the measurement is affected. The ZSVC measurement of the traction motor in Figure 1 is adopted, wherein both v_{n} and v_{nR} contain the ZVSC produced by the inverter. v_0 is the difference between them and only contains the ZVSC caused by the stator ITSC fault and asymmetry, so the three balanced resistors can eliminate the influence of the inverter [42].

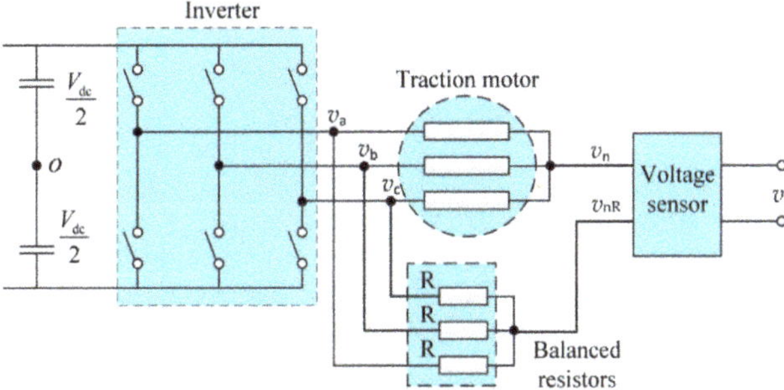

Figure 1. ZSVC measurement circuit.

2.2. ApFFT Time-Shift Phase Difference Correction Algorithm

The apFFT time-shift phase difference correction mainly includes two parts, i.e., apFFT and time-shift phase difference correction. The apFFT algorithm can effectively suppress spectrum leakage caused by data truncation [43–45]. As shown in Figure 2, the required data points for the N-order spectrum analysis are $x(-N+1)$, $x(-N+2)$, ..., $x(-1)$, $x(1)$, ..., $x(N-2)$, $x(N-1)$ with a total of $2N-1$ data points. W is a convolution window formed by a convolution operation with the front window W_1 and the flipped rear window W_2. When both W_1 and W_2 are rectangular windows, the spectrum analysis approach is referred to as windowless apFFT. When either W_1 or W_2 is a rectangular window, the spectrum analysis approach is referred to as single-window apFFT. When neither W_1 nor W_2 is a rectangular window, the spectrum analysis approach is referred to as double-window apFFT.

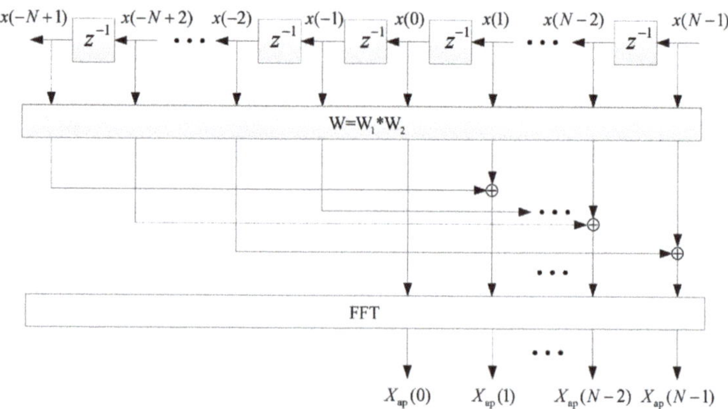

Figure 2. N-order apFFT calculation method.

The N-order apFFT spectrum analysis mainly includes the preprocessing of $2N - 1$ point data and the FFT calculation. If the data are processed by windowless apFFT, the following operations can be performed on the data. First, divide $2N - 1$ data points into N segments with length N according to Formula (2):

$$
\begin{aligned}
x_0 &= [x(0), x(1), x(2), \ldots, x(N-1)]^T, \\
x_1 &= [x(-1), x(0), x(1), \ldots, x(N-2)]^T, \\
x_2 &= [x(-2), x(-1), x(0), \ldots, x(N-3)]^T, \\
&\quad \cdots\cdots \\
x_{N-1} &= [x(-N+1), x(-N+2), \ldots, x(0)]^T.
\end{aligned}
\tag{2}
$$

Then, rotate the N segment data of Formula (2), taking the Nth data point, i.e., $x(0)$, as the first data point of the data segment. Formula (2) is transformed into Formula (3):

$$
\begin{aligned}
x_0 &= [x(0), x(1), x(2), \ldots, x(N-1)]^T, \\
x_1 &= [x(0), x(1), \ldots, x(N-2), x(-1)]^T, \\
x_2 &= [x(0), \ldots, x(N-3), x(-2), x(-1)]^T, \\
&\quad \cdots\cdots \\
x_{N-1} &= [x(0), x(-N+1), \ldots, x(-2), x(-1)]^T.
\end{aligned}
\tag{3}
$$

Finally, add the shifted N segments of the data separately and normalize them to obtain x_{ap} in Formula (4), which comprises the N data points obtained after the windowless apFFT preprocessing.

$$
x_{\text{ap}} = \frac{1}{N}[Nx(0), (N-1)x(1) + x(-N+1), \ldots, x(N-1) + (N-1)x(-1)]^T.
\tag{4}
$$

Perform N-point FFT on x_{ap}; that is, obtain the calculation result $X_{\text{ap}}(k)$ of windowless apFFT, and k is the spectrum index.

The second part of the apFFT time-shift phase difference correction algorithm is the time-shift phase difference correction [46]. The single-frequency complex exponential signal with frequency ω^*, initial phase θ_0, and amplitude A is $x(n)$, where n is the discrete time point.

$$
x(n) = Ae^{j(w^* n + \theta_0)},
\tag{5}
$$

The data points are divided into two segments of the same length, as shown in Figure 3. The data interval of the first segment is $[-N + 1, N - 1]$. Assuming the spectrum serial number is k^*, the phase value of apFFT main spectrum line is:

$$\varphi_X(k^*) = \theta_0, \tag{6}$$

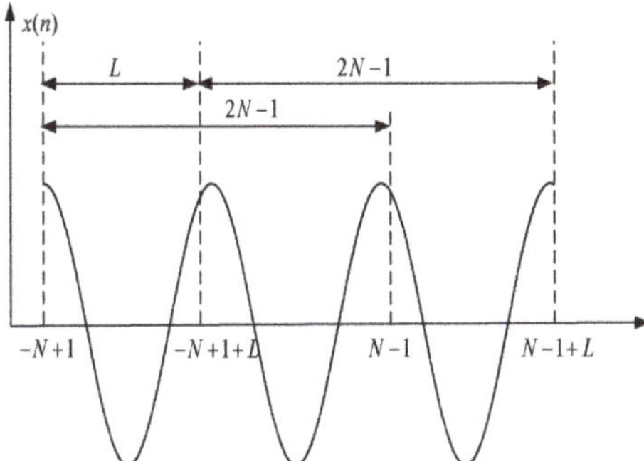

Figure 3. Data truncation for time-shift phase difference correction.

The second data segment starts after L data points of the first data segment. The data range is $[-N + 1 + L, N - 1 + L]$. The central data point of this data segment is $x(-L)$, as shown in Figure 3. If apFFT is performed on the second segment of data, the phase of apFFT main spectral line is $\varphi_{XL}(k^*)$, which is the approximate estimation of the phase of data point $x(-L)$; that is,

$$\varphi_{XL}(k^*) = \theta_0 - \omega^* L, \tag{7}$$

The estimation of signal frequency can be obtained from Formulas (6) and (7):

$$\hat{\omega}^* = [\varphi_X(k^*) - \varphi_{XL}(k^*)]/L = \Delta\varphi/L, \tag{8}$$

To eliminate the "phase ambiguity" phenomenon [47], the frequency estimation after phase compensation is [48]

$$\hat{\omega}^* = [\varphi_X(k^*) - \varphi_{XL}(k^*)]/L + 2k^*\pi/N, \tag{9}$$

For the double-window apFFT, the signal amplitude estimation can be obtained:

$$\hat{A} = \frac{|Y(k^*)|}{\left|F_g(k^*\Delta\omega - \hat{\omega}^*)\right|^2}. \tag{10}$$

In Equation (10), $Y(k^*)$ is the value of the double-window apFFT at point k^*; $F_g(k^*\Delta\omega - \hat{\omega}^*)$ is obtained from bringing $(k^*\Delta\omega - \hat{\omega}^*)$ into the Fourier transform of the window function. Generally, the window function is a cosine window, and its Fourier transform expression is determined.

3. Fault Diagnosis Method for Stator ITSC Fault of Traction Motor

3.1. Stator ITSC Fault Extent Index

In previous studies, only the metallic short circuit of windings is generally considered. The two windings are directly short-circuited without any resistance, and the motor's ITSC fault extent is evaluated based on the number of short-circuit turns. In most cases, the metallic ITSC fault is caused by the expansion of the non-metallic ITSC fault. The non-metallic ITSC fault means that there is some resistance left between short-circuit turns. In this case, the number of short-circuit turns alone is insufficient for evaluating the ITSC fault. The heat caused by the ITSC fault mainly results in damage to the traction motor.

If the heat generated by the inductance is ignored, the thermal power of the inter-turn resistance is

$$P_f = \frac{U_f{}^2}{R_f} = \frac{\left(N_f * \frac{U}{N_s}\right)^2}{R_f} = U^2 * \frac{N_f{}^2}{N_s{}^2 * R_f},\tag{11}$$

where P_f is the thermal power of the inter-turns resistance, U_f is the short-circuit turns voltage, R_f is the inter-turn resistance, U is the motor phase voltage, and N_s is the total number of turns of each phase winding. From Formula (11), it can be concluded that the heat generated by the short-circuit current after the ITSC fault is in direct proportion to $\frac{N_f{}^2}{N_s{}^2 * R_f}$.

Define the fault extent index of ITSC fault:

$$\lambda_f = \sqrt{\frac{N_f{}^2}{N_s{}^2 * R_f}} = \frac{N_f}{N_s} * \frac{1}{\sqrt{R_f}}.\tag{12}$$

According to Formula (12), the ITSC fault extent index λ_f is related to the number of short-circuit turns and the inter-turn resistance.

3.2. SVM Model for Diagnosis of ITSC Fault

The apFFT time-shift phase difference spectrum correction algorithm is used to calculate the fundamental frequency of the ZSVC, the fundamental component amplitudes of the traction motor's ZSVC, and the three-phase current. The SVM-based fault diagnosis model of ITSC fault is established with the five parameters as input. The traction motor ITSC condition is classified as a normal condition, incipient fault condition, and serious fault condition using the proposed index. SVM is a machine learning method based on statistical theory, mainly used to solve classification and regression problems [49–52]. Its core idea is to complete the model training based on the structural risk minimization principle. It has nonlinear solid approximation ability, good generalization performance, and good results in dealing with small samples and nonlinear problems. SVM uses nonlinear mapping $\phi(\mathbf{x})$ to map the original data to the high-dimensional space to deal with nonlinear regression problems of multidimensional data.

The C-SVC model is a relatively standard two-class SVM model. The train set is

$$\mathbf{T} = \{(\mathbf{x}_1, \mathbf{y}_1), \cdots, (\mathbf{x}_l, \mathbf{y}_l)\} \in (X \times Y)^l,\tag{13}$$

where $\mathbf{x}_i \in X = \mathbf{R}^n, \mathbf{y}_i \in Y = \{1, -1\}(i = 1, 2, \cdots, l)$, and $\mathbf{x}_i$ is the features vector.

Select kernel function $K(x, x')$ and appropriate parameter C. The standard kernel functions $K(x, x')$ mainly include linear, polynomial, and radial basis kernel functions. The Lagrange dual problem of the original problem is

$$\min_{\alpha} \frac{1}{2} \sum_{i=1}^{j} \sum_{j=1}^{l} y_i y_j \alpha_i \alpha_j K(x_i, x_j) - \sum_{j=1}^{l} \alpha_j,\tag{14}$$

$$s.t. \sum_{i=1}^{l} y_i \alpha_i = 0, \, 0 \leqslant \alpha_i \leqslant C, i = 1, \cdots, l$$

Obtain the optimal solution: $\boldsymbol{\alpha}^* = \left(\alpha_1^*, \cdots, \alpha_l^*\right)^{\mathrm{T}}$.

Select a positive component of $0 < \alpha_j^* < C$ from $\boldsymbol{\alpha}^*$, and calculate the threshold accordingly:

$$b^* = y_j - \sum_{i=1}^{l} y_i \alpha_i^* K(x_i - x_j),\tag{15}$$

The constructed decision function is

$$f(x) = \mathrm{sgn}\left(\sum_{i=1}^{l} \alpha_i^* y_i K(x, x_i) + b^* \right),\tag{16}$$

If the Gaussian radial basis function is used as the kernel function, g is the parameter of the Gaussian radial basis function:

$$K(\mathbf{x}_i, \mathbf{x}) = \exp\left(-\frac{\|\mathbf{x}_i - \mathbf{x}_j\|^2}{2\sigma} \right) = \exp\left(-g\|\mathbf{x}_i - \mathbf{x}_j\|^2 \right).\tag{17}$$

In the SVM classification model, the selection of the model penalty parameter C and Gaussian kernel function parameter g is directly related to the model performance. The K-CV (K-fold cross-validation) is a standard cross-validation algorithm that divides datasets into K sub-datasets evenly in model training. Each sub-dataset is used as the validation set in turn, and the remaining K − 1 sub-datasets are used as the training set to train K models. The average mean square error (MSE) of K models on the validation set is used as the performance index. The mean square error is

$$\delta_{\mathrm{MSE}} = \frac{1}{n} \sum_{i=1}^{n} (y_i - \overline{y}_i)^2.\tag{18}$$

In Formula (18), n is the number of samples, y_i is the predicted value, and $\overline{y}_i$ is the target value.

Grid search is used to select several discrete points on each dimension of the parameter space according to certain rules. The discrete points of different dimensions intersect in the parameter space to obtain the discrete solution. Calculate each discrete solution to obtain the optimal solution. Figure 4 is the flow chart of the hyper-parameter optimization using K-CV and the grid search method. Take the grid point as C = 2^a, g = 2^b, and initialize the range of a and b, which are positive and negative integers. Parameters a and b are selected from the minimum to the maximum using 1 step of their range. K-CV is used to calculate the average MSE of the K models on the K validation sets. After calculating all the combinations of C and g at all grid intersections, C and g at the minimum average MSE are the optimal solutions.

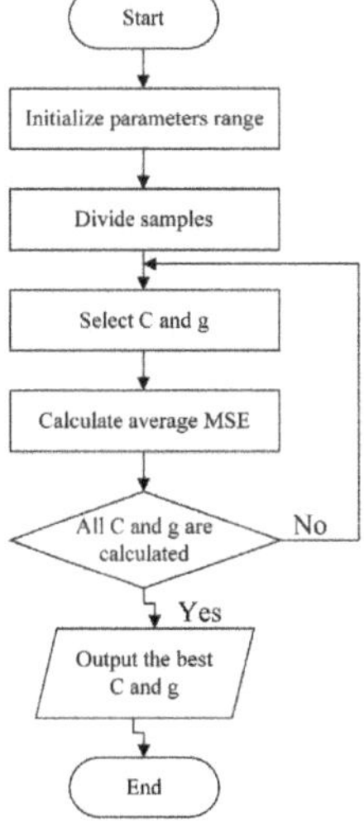

Figure 4. Flow chart of the hyper-parameters optimization.

3.3. ITSC Fault Diagnosis Procedure Based on SVM Model

As shown in Figure 5, the EMU traction motor ITSC fault diagnosis based on SVM includes two stages: model training and online diagnosis. In the model training stage, the ZSVC and three-phase current are first measured with the circuit proposed in the article. Second, the apFFT time-shift phase difference correction algorithm is used to calculate the ZSVC fundamental frequency, the fundamental amplitudes of ZSVC, and the three-phase current. Third, the ZSVC fundamental frequency, the amplitude of ZSVC, and the three-phase current are used as features. Based on the ITSC fault index λ_f, the samples are divided into three categories: normal, incipient, and serious faults. Fourth, the K-CV method divides all the samples into training and validation samples. K-CV and the grid search method are used to optimize the hyper-parameters. Last, the optimal ITSC fault diagnosis model is saved. In the online diagnosis stage, the ITSC fault features are acquired similarly to the training stage. The optimal ITSC fault diagnosis model is loaded, and the fault features are input into the SVM model to predict the ITSC category.

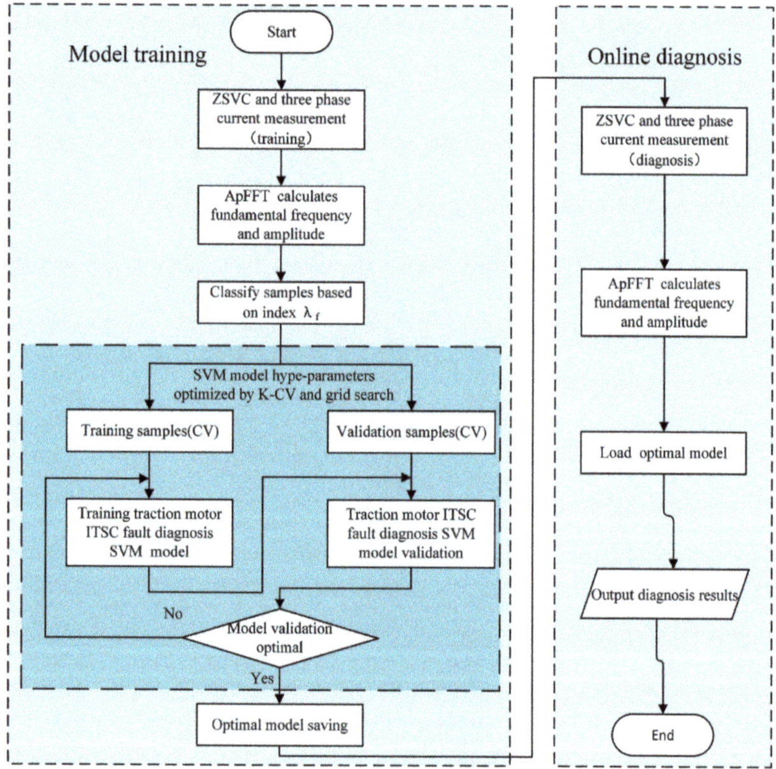

Figure 5. Procedure of the ITSC fault diagnosis based on SVM model.

4. EMU Electric Traction Simulation Experimental Platform

4.1. Overall Design of the Experimental Platform

The experimental data are acquired from the mutual-feed electric traction simulation experimental platform shown in Figure 6. The platform mainly includes the tested system and the load system. The tested system mainly includes an S120 variable frequency speed control system and the tested motor. The S120 controls the tested motor to operate according to the experimental conditions. The S120 system mainly includes the CU320-2PN control unit, ALM rectifier, and MM inverter modules. The load system mainly includes the braking motor and the H1000 converter. The PCI-6229 NI-DAQ gives the H1000 converter work instructions. The DC power supply of the tested system is obtained from the DC link of the load system. When the tested motor works in the motor state, the braking motor works in the generator state. The load system feeds electric energy back to the DC link to realize DC energy mutual feedback.

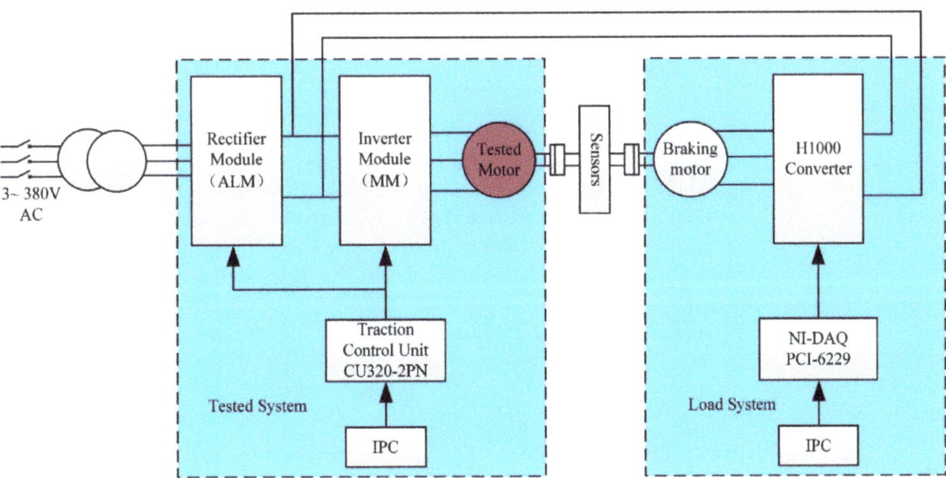

Figure 6. Energy mutual-feed electric traction simulation experimental platform.

Figure 7 shows the main parts of the experimental platform. The tested motor is a three-phase AC asynchronous squirrel-cage motor with three-phase winding taps pulled out, whose parameters are shown in Table 1. The braking motor is a normal motor with the same type and power.

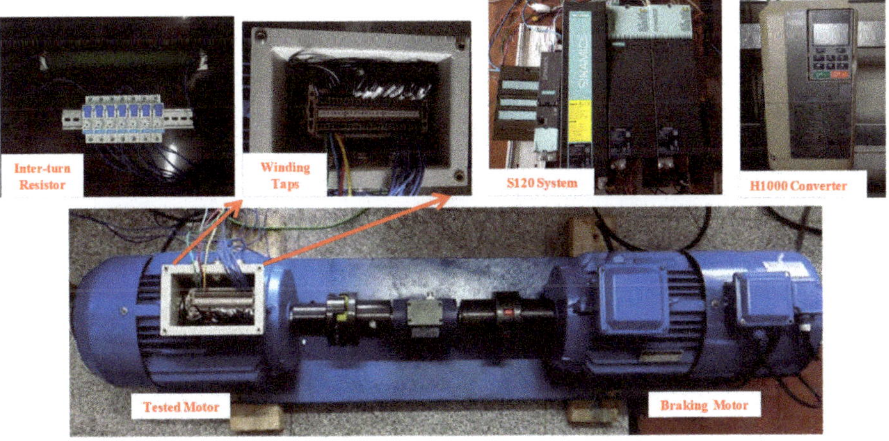

Figure 7. Main parts of the EMU electric traction simulation experimental platform.

Table 1. Rated parameters of the tested motor.

Parameter	Value	Parameter	Value
Power	5.5 kW	Frequency	50 Hz
Voltage	380 V	Speed	1445 rpm
Current	11.7 A	Turns per phase	162
Poles	4	Connection mode	Y
Magnetizing inductance	205.2 mH	Stator resistance	1.061 Ω
Rotor resistance	0.6269 Ω	Stator leakage inductance	3.217 mH
Rotor leakage inductance	7.349 mH	Inertia	0.1367 kg·m^2

4.2. Setting ITSC Faults on Tested Motor

Figure 8 shows that the winding taps are pulled out at different stator winding turns during manufacturing to simulate the ITSC fault. The taps can be connected externally to simulate the short-circuit fault between different turns. The power resistor simulates the inter-turn resistor between non-metallic short-circuit turns. The vacuum breaker conveniently controls the short circuit of different turns loop.

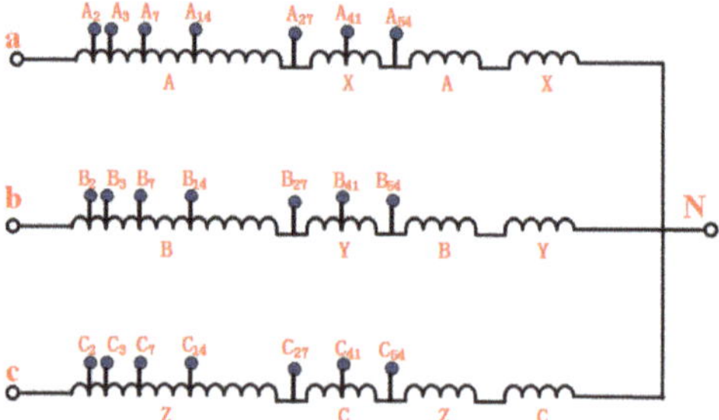

Figure 8. Stator winding taps pulled out of the tested motor.

4.3. Signal Measurement of the Experimental Platform

The signal measurement is shown in Figure 9. The DL850E ScopeCorder produced by Yokogawa corporation in Japan is used for signal measurement whose LPF (low pass filter) is set to 400 Hz, and the sampling frequency is 2000 Hz. The A621 passive current probe is used to collect the inter-turn current, which cannot be measured in the actual application. If the inter-turn current is too large, it generates heat quickly to burn the motor. E3N active current probe is used to measure the three-phase current of the tested motor. The DP-50 voltage probe is used to measure the ZSVC using the measurement circuit shown in Figure 2. The ZSVC measurement balanced resistors are three 15 kΩ (1 kW) power resistors.

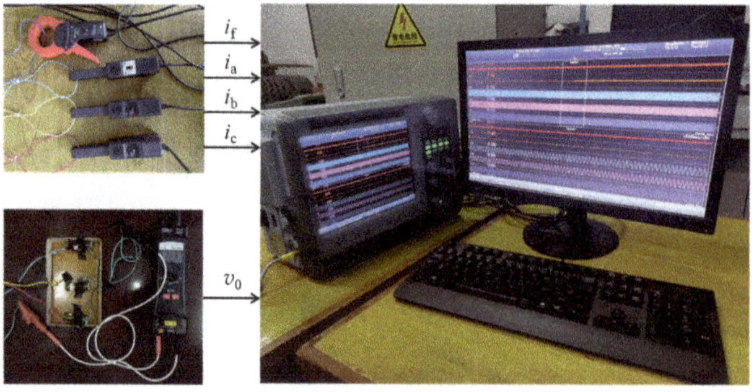

Figure 9. Signal measurements of the tested motor.

5. Analysis of ITSC Fault Diagnosis Model Based on Experimental Samples

During the experiment, the S120 converter system controlled the tested motor to operate in the torque control mode outputting a fixed electromagnetic torque. The H1000 converter controlled the braking motor according to the speed control mode running at a fixed speed. This experimental operation mode could simulate the steady operation conditions of EMU traction or electric braking at different speeds and torques.

5.1. Analysis of Motor Signals with ITSC Fault

The tested fault motor ran at 900 rpm rotating speed with 10 Nm electromagnetic torque, 12 short-circuit turns ITSC fault in the a-phase stator winding, and a 1 Ω inter-turn resistor, as shown in Figure 10.

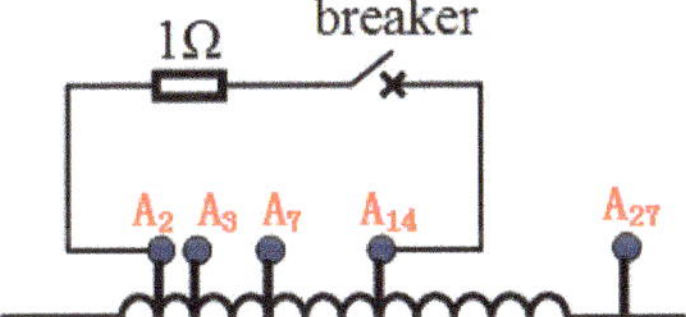

Figure 10. The stator ITSC fault set on the a-phase.

The insulation fault occurred at around 20 s. It can be seen from Figure 11a that when the stator winding ITSC fault occurs, a sinusoidal inter-turn current with the same fundamental frequency as the power supply is generated between the short-circuit turns. Figure 11b shows the three-phase current before and after the ITSC fault. Although the amplitude of the short-circuit current reached about 10 A, it had little impact on the three-phase current. Figure 11c shows the S120 system outputting three-phase voltage filtered by the 400 Hz LPF filter, and the outputting waveform conforms to the saddle waveform of SVPWM. Figure 11d shows the ZSVC before and after the ITSC fault, and the ZSVC will be studied and analyzed later.

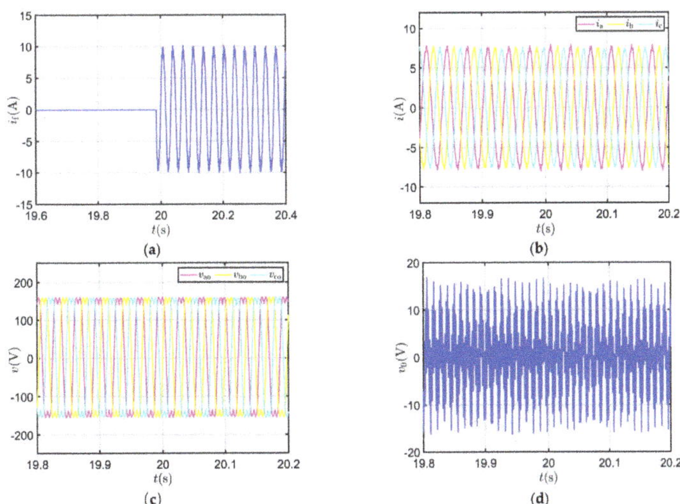

Figure 11. Signals of the tested system before and after ITSC fault: (**a**) inter-turn current of the tested motor; (**b**) three-phase current of the tested motor; (**c**) three-phase voltage of S120 inverter module; (**d**) ZSVC of the tested motor.

5.2. Analysis of ITSC Fault Features

The tested motor setting speed was 900 rpm, and the electromagnetic torque was set to 10 Nm. There was an ITSC fault in the a-phase winding. The frequency of the ZSVC fundamental component, the fundamental amplitudes of ZSVC, and the three-phase current were calculated using the apFFT time-shift spectrum correction algorithm with double Hanning windows. Based on Formula (12), 20 different indexes λ_f were calculated according to 5 different numbers of short-circuit turns and 4 different inter-turn resistances, as shown in Table 2.

Table 2. The ITSC fault set and the extent index λ_f.

Resistance (Ω)	Turns 5	7	12	20	25
1	0.03049	0.04268	0.07317	0.12195	0.15244
2	0.02156	0.03018	0.05174	0.08623	0.10779
4	0.01524	0.02134	0.03659	0.06098	0.07622
8	0.01078	0.01509	0.02587	0.04312	0.05390

It can be seen from Figure 12a that the ZSVC fundamental amplitude of the tested motor increased with the fault extent index λ_f. The fundamental amplitude of the ZSVC was about 0.2 V under normal conditions, mainly caused by the asymmetry of the three-phase winding. It can be seen from Figure 12b that the fundamental amplitude of the a-phase current increased with the ITSC fault extent. The b-phase and c-phase currents changed little. Similarly, because of the unbalance of the three-phase winding, the three-phase current was unbalanced under normal conditions.

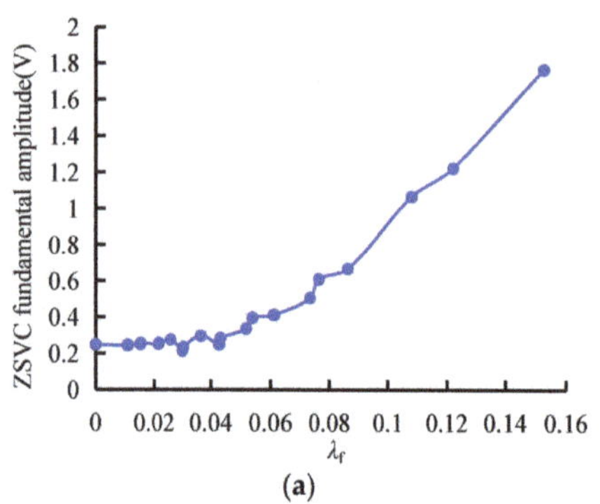

(a)

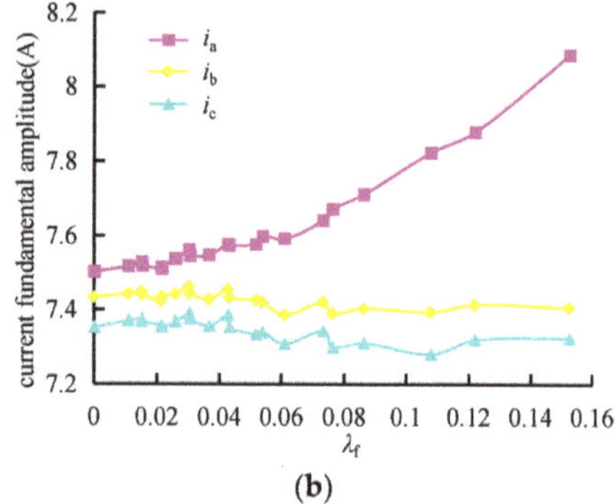

(b)

Figure 12. Influence of ITSC fault on the tested motor signal fundamental amplitude: (**a**) influence of ITSC fault on ZSVC fundamental amplitude; (**b**) influence of ITSC fault on three-phase current fundamental amplitude.

The electromagnetic torque was set to 10 Nm, and the ITSC fault extent index λ_f was 0.07317. It can be seen from Figure 13a that in the process speed regulation, the fundamental frequency changed with the experimental system setting speed. According to the control characteristics of variable frequency speed regulation, the three-phase voltage increased linearly with the increase in speed. The ZSVC also increased with the tested motor's fundamental frequency under the same λ_f. Figure 13b shows that the a-phase current did not change much, but the b-phase and c-phase currents decreased significantly with the increase of the fundamental frequency. The asymmetry of the three-phase current became more and more prominent.

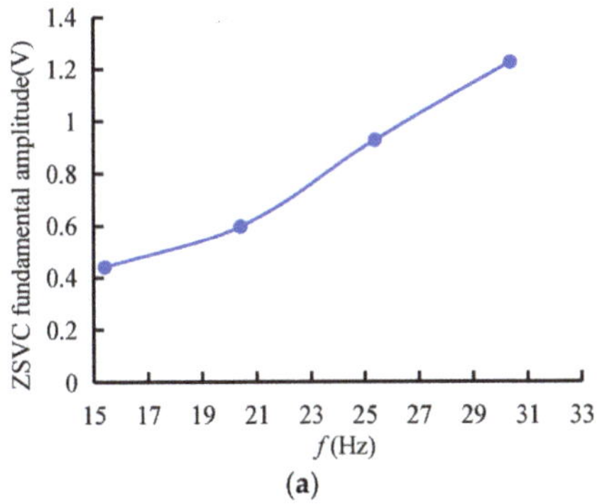 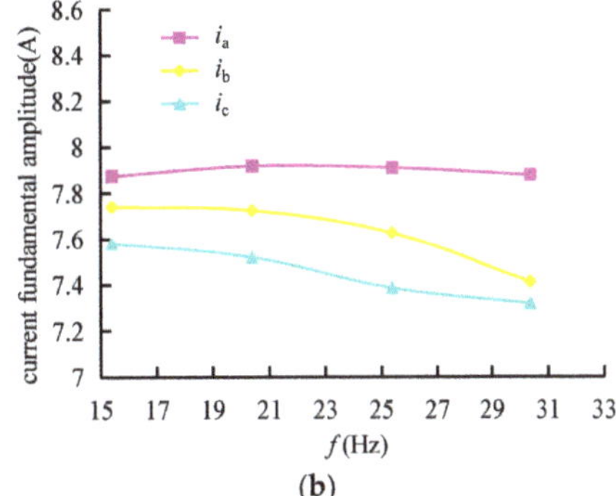

(a)   **(b)**

Figure 13. Influence of frequency on the tested motor signal fundamental amplitude: (**a**) influence of frequency on ZSVC fundamental amplitude; (**b**) influence of frequency on three-phase current fundamental amplitude.

According to the analysis above, the ZSVC fundamental component amplitude and the three-phase current asymmetry increases with the ITSC fault extent under fixed electromagnetic torque and speed. The three-phase current amplitude can reflect the electromagnetic torque value, and the speed is approximately linear with the fundamental frequency. Therefore, the ZSVC fundamental frequency, the fundamental amplitudes of ZSVC, and the three-phase current are selected as the features to establish the ITSC fault diagnosis model.

5.3. Analysis of SVM ITSC Fault Diagnosis Model Performance

The tested motor's data acquisition conditions are shown in Table 3. The tested motor with each fault extent index operated under four different speeds and electromagnetic torques. There were 16 working conditions in which the speeds or electromagnetic torques were different, and 7 normal samples needed to be acquired at each working condition to first obtain a total of 112 normal samples. There were 20 different ITSC fault extent λ_f samples under each speed and electromagnetic torque, as shown in Table 2. The classification of fault severity categories is determined by the application situation defined by users. In the experiment, the samples with $0.03018 \leq \lambda_f < 0.06098$ were defined as incipient ITSC fault samples because the thermal power caused by inter-turn short-circuit fault was small. Samples with $0.06098 \leq \lambda_f$ were considered serious ITSC fault samples because the thermal power was large. Thus the ITSC fault samples were divided into 112 incipient and 112 serious ITSC fault samples. The SVM-based ITSC fault diagnosis model was established by selecting 92 samples from each category as train samples and 20 samples from each category as the test samples. The grid search range was a = $[-5, 5]$, b = $[-5, 5]$. The parameter K was 3 in the K-CV method.

Table 3. The working condition and the ITSC fault setting.

Speed (rpm)	Torque (Nm)	Turns	Resistance (Ω)
450, 600, 750, 900	2, 10, 18, 26	5, 7, 12, 20, 25	1, 2, 4, 8

Figure 14 shows the prediction results of the SVM-based ITSC fault diagnosis model on the experimental samples. Figure 14a shows that there is no misclassification on the training dataset. Figure 14a shows that there is a sample misclassified among the normal and incipient samples and a sample misclassified among incipient and serious samples. There is no misclassification between the normal and the serious fault samples. The model's prediction accuracy is 100% on the training dataset and 93.33% on the test dataset, which indicates that the model can detect and evaluate the ITSC fault accurately.

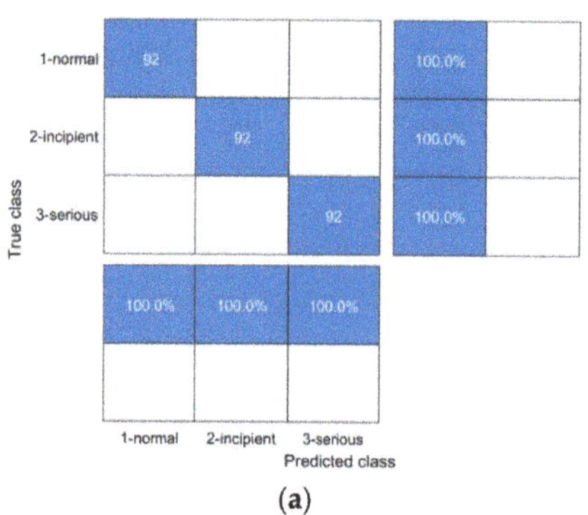

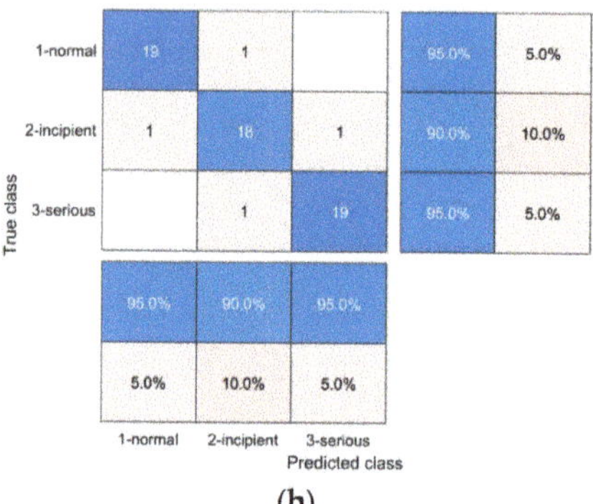

(a)   (b)

Figure 14. Confusion matrix of ITSC fault diagnosis SVM model: (**a**) confusion matrix of SVM model on the training dataset; (**b**) confusion matrix of SVM model on the test dataset.

6. Conclusions

The ITSC fault diagnosis of the asynchronous traction motor significantly ensures the EMU's safe operation and saves maintenance costs. The non-metallic ITSC fault extent evaluating index λ_f was proposed based on the short-circuit thermal power. The index λ_f is related to the number of short-circuit turns and inter-turn resistance. The apFFT time-shift phase difference spectrum correction with double Hanning windows was used to calculate five parameters used as fault features to train SVM model to diagnose the ITSC fault, and the SVM model hyper-parameters C and g were optimized using K-CV and the grid search method. The method proposed in the article can detect and evaluate the ITSC fault of traction motors in a speed control system under vector control or direct torque control conditions in which the fundamental frequency of supply voltage is unknown. EMU traction motors work at different speeds and torque points during operation. The prediction results of different steady-state operating points can be integrated to improve the accuracy of the fault diagnosis model. The method can be used when the traction motor is in a steady state, and it cannot be used if the traction motor accelerates or decelerates.

Author Contributions: Methodology, J.M.; software, G.Z.; validation, J.H.; formal analysis, J.M. and J.H.; investigation, Z.Z.; resources, J.F.; data curation, J.M.; writing—original draft preparation, J.M. and J.H.; writing—review and editing, X.L.; visualization, X.L.; supervision, J.F. All authors have read and agreed to the published version of the manuscript.

Funding: This research was funded by the 2021 Scientific Research Fund Project of Liaoning Provincial Department of Education (LJKZ1297).

Data Availability Statement: Not applicable.

Conflicts of Interest: The authors declare no conflict of interest.

References

1. Lee, S.-G. A Study on Traction Motor Characteristic in EMU Train. In Proceedings of the 13th International Conference on Control, Automation and Systems, Gyeonggi-do, Republic of Korea, 22–25 October 2014.
2. Zhang, K.; Jiang, B.; Chen, F. Multiple-Model-Based Diagnosis of Multiple Faults With High-Speed Train Applications Using Second-Level Adaptation. *IEEE Trans. Ind. Electron.* **2021**, *68*, 6257–6266. [CrossRef]
3. Chen, Z.P.; Wang, Z.; Jia, L.M.; Cai, G.Q. Analysis and Comparison of Locomotive Traction Motor Intelligent Fault Diagnosis Methods. *Appl. Mech. Mater.* **2011**, *97–98*, 994–1002. [CrossRef]
4. Al-Ameri, S.M.; Alawady, A.A.; Yousof, M.F.M.; Kamarudin, M.S.; Salem, A.A.; Abu-Siada, A.; Mosaad, M.I. Application of Frequency Response Analysis Method to Detect Short-Circuit Faults in Three-Phase Induction Motors. *Appl. Sci.* **2022**, *12*, 2046. [CrossRef]
5. Chao, C.; Wang, W.; Chen, H.; Zhang, B.; Shao, J.; Teng, W. Enhanced Fault Diagnosis Using Broad Learning for Traction Systems in High-Speed Trains. *IEEE Trans. Power Electron.* **2020**, *36*, 7461–7469. [CrossRef]

6.   Guo, X.; Tang, Y.; Wu, M.; Zhang, Z.; Yuan, J. FPGA-Based Hardware-in-the-Loop Real-Time Simulation Implementation for High-Speed Train Electrical Traction System. *IET Electr. Power Appl.* **2020**, *14*, 850–858. [CrossRef]
7.   Kaufhold, M.; Aninger, H. Electrical Stress and Failure Mechanism of the Winding Insulation in PWM-Inverter-Fed Low-Voltage Induction Motors. *IEEE Trans. Ind. Electron.* **2000**, *2*, 396–402. [CrossRef]
8.   Mbaye, A.; Bellomo, J.P. Electrical Stresses Applied to Stator Insulation in Low-Voltage Induction Motors Fed by PWM Drives. *IET Electr. Power Appl.* **1997**, *144*, 191–198. [CrossRef]
9.   Hwang, D.H.; Park, D.Y.; Kim, Y.J.; Lee, Y.H.; Hur, I.G. A Comparison with Insulation System for PWM-Inverter-Fed Induction Motors. In Proceedings of the International Conference on Electrical Machines & Systems, Shenyang, China, 18–20 August 2001. [CrossRef]
10. Otero, M.; Barrera, P.; Bossio, G.R.; Leidhold, R. Stator Inter-turn Faults Diagnosis in Induction Motors Using Zero-sequence Signal Injection. *IET Electr. Power Appl.* **2020**, *14*, 273–2738. [CrossRef]
11. Singh, M.; Shaik, A.G. Incipient Fault Detection in Stator Windings of an Induction Motor Using Stockwell Transform and SVM. *IEEE Trans. Instrum. Meas.* **2020**, *69*, 9496–9504. [CrossRef]
12. Sonje, D.M.; Kundu, P.; Chowdhury, A. A Novel Approach for Sensitive Inter-turn Fault Detection in Induction Motor Under Various Operating Conditions. *Arab. J. Sci. Eng.* **2019**, *44*, 6887–6900. [CrossRef]
13. Namdar, A. A robust principal component analysis-based approach for detection of a stator inter-turn fault in induction motors. *Prot. Control. Mod. Power Syst.* **2022**, *7*, 48. [CrossRef]
14. Mejia-Barron, A.; Tapia-Tinoco, G.; Razo-Hernandez, J.R.; Valtierra-Rodriguez, M.; Granados-Lieberman, D. A neural network-based model for MCSA of ITSC faults in induction motors and its power hardware in the loop simulation. *Comput. Electr. Eng.* **2021**, *93*, 107234. [CrossRef]
15. Tallam, R.; Habetler, T.; Harley, R. Transient model for induction machines with stator winding turn faults. *IEEE Trans. Ind. Appl.* **2002**, *38*, 632–637. [CrossRef]
16. Zhao, Z.; Fan, F.; Wang, W.; Liu, Y.; See, K.Y. Detection of Stator Interturn Short-Circuit Faults in Inverter-Fed Induction Motors by Online Common-Mode Impedance Monitoring. *IEEE Trans. Instrum. Meas.* **2021**, *70*, 3513110. [CrossRef]
17. Duan, F.; Zivanovic, R. Induction Motor Stator Fault Detection by a Condition Monitoring Scheme Based on Parameter Estimation Algorithms. *Electr. Power Compon. Syst.* **2016**, *44*, 1138–1148. [CrossRef]
18. Bazine, I.B.A.; Tnani, S.; Poinot, T.; Champenois, G.; Jelassi, K. On-Line Detection of Stator and Rotor Faults Occurring in In-duction Machine Diagnosis by Parameters Estimation. In Proceedings of the 8th IEEE Symposium on Diagnostics for Electrical Machines, Power Electronics & Drives, Bologna, Italy, 5–8 September 2011. [CrossRef]
19. Abdallah, H.; Benatman, K. Stator winding inter-turn short-circuit detection in induction motors by parameter identification. *IET Electr. Power Appl.* **2017**, *11*, 272–288. [CrossRef]
20. Nguyen, V.; Seshadrinath, J.; Wang, D.; Nadarajan, S.; Vaiyapuri, V. Model-Based Diagnosis and RUL Estimation of Induction Machines Under Interturn Fault. *IEEE Trans. Ind. Appl.* **2017**, *53*, 2690–2701. [CrossRef]
21. Kallesoe, C.S.; Vadstrup, P.; Rasmussen, H.; Izadi-Zamanabadi, R. Observer Based Estimation of Stator Winding Faults in Delta-Connected Induction Motors, a LMI Approach. In Proceedings of the IAS Annual Meeting, Tampa, FL, USA, 8–12 October 2006. [CrossRef]
22. De Angelo, C.H.; Bossio, G.R.; Giaccone, S.J.; Valla, M.I.; Solsona, J.A.; Garcia, G.O. Online Model-Based Stator-Fault Detection and Identification in Induction Motors. *IEEE Trans. Ind. Electron.* **2009**, *56*, 4671–4680. [CrossRef]
23. Guezmil, A.; Berriri, H.; Pusca, R.; Sakly, A.; Romary, R.; Mimouni, M.F. Detecting Inter-Turn Short-Circuit Fault in Induction Machine Using High-Order Sliding Mode Observer: Simulation and Experimental Verification. *J. Control. Autom. Electr. Syst.* **2017**, *28*, 532–540. [CrossRef]
24. Kia, M.Y.; Khedri, M.; Najafi, H.R.; Nejad, M.A.S. Hybrid modelling of doubly fed induction generators with inter-turn stator fault and its detection method using wavelet analysis. *IET Gener. Transm. Distrib.* **2013**, *7*, 982–990. [CrossRef]
25. Kumar, P.S.; Xie, L.; Halick, M.S.M.; Vaiyapuri, V. Stator End-Winding Thermal and Magnetic Sensor Arrays for Online Stator Inter-Turn Fault Detection. *IEEE Sens. J.* **2020**, *21*, 5312–5321. [CrossRef]
26. Chen, P.; Xie, Y.; Hu, S. Electromagnetic Performance and Diagnosis of Induction Motors With Stator Interturn Fault. *IEEE Trans. Ind. Appl.* **2020**, *57*, 1354–1364. [CrossRef]
27. Lee, S.-H.; Wang, Y.-Q.; Song, J.-I. Fourier and wavelet transformations application to fault detection of induction motor with stator current. *J. Cent. S. Univ. Technol.* **2010**, *17*, 93–101. [CrossRef]
28. Liu, H.; Huang, J.; Hou, Z.; Yang, J.; Ye, M. Stator inter-turn fault detection in closed-loop controlled drive based on switching sideband harmonics in CMV. *IET Electr. Power Appl.* **2017**, *11*, 178–186. [CrossRef]
29. Sadeghi, R.; Samet, H.; Ghanbari, T. Detection of Stator Short-Circuit Faults in Induction Motors Using the Concept of Instanta-neous Frequency. *IEEE Trans. Ind. Inform.* **2018**, *15*, 4506–4515. [CrossRef]
30. Vinayak, B.A.; Anand, K.A.; Jagadanand, G. Wavelet-based real-time stator fault detection of inverter-fed induction motor. *IET Electr. Power Appl.* **2019**, *14*, 82–90. [CrossRef]
31. Almounajjed, A.; Sahoo, A.K.; Kumar, M.K. Diagnosis of stator fault severity in induction motor based on discrete wavelet analysis. *Measurement* **2021**, *182*, 109780. [CrossRef]
32. Tian, R.; Chen, F.; Dong, S. Compound Fault Diagnosis of Stator Interturn Short Circuit and Air Gap Eccentricity Based on Random Forest and XGBoost. *Math. Probl. Eng.* **2021**, *42*, 2149048. [CrossRef]

33. Xu, Z.; Hu, C.; Yang, F.; Kuo, S.-H.; Goh, C.-K.; Gupta, A.; Nadarajan, S. Data-Driven Inter-Turn Short Circuit Fault Detection in Induction Machines. *IEEE Access* **2017**, *5*, 25055–25068. [CrossRef]
34. Husari, F.; Seshadrinath, J. Incipient Interturn Fault Detection and Severity Evaluation in Electric Drive System Using Hybrid HCNN-SVM Based Model. *IEEE Trans. Ind. Inform.* **2021**, *18*, 1823–1832. [CrossRef]
35. Rajamany, G.; Srinivasan, S.; Rajamany, K.; Natarajan, R.K. Induction Motor Stator Interturn Short Circuit Fault Detection in Accordance with Line Current Sequence Components Using Artificial Neural Network. *J. Electr. Comput. Eng.* **2019**, *1*, 4825787. [CrossRef]
36. Bensaoucha, S.; Ameur, A.; Bessedik, S.A.; Moati, Y. Artificial Neural Networks Technique to Detect and Locate an Interturn Short-Circuit Fault in Induction Motor. In *Renewable Energy for Smart and Sustainable Cities*; Hatti, M., Ed.; Lecture Notes in Networks and Systems; Springer International Publishing: Cham, Switzerland, 2019; Volume 62, pp. 103–113, ISBN 978-3-030-04788-7.
37. Skowron, M.; Orlowska-Kowalska, T.; Wolkiewicz, M.; Kowalski, C.T. Convolutional Neural Network-Based Stator Current Data-Driven Incipient Stator Fault Diagnosis of Inverter-Fed Induction Motor. *Energies* **2020**, *13*, 1475. [CrossRef]
38. Urresty, J.-C.; Riba, J.-R.; Romeral, L. Application of the ZSVC component to detect stator winding inter-turn faults in PMSMs. *Electr. Power Syst. Res.* **2012**, *89*, 38–44. [CrossRef]
39. Hang, J.; Zhang, J.; Cheng, M.; Huang, J. Online Interturn Fault Diagnosis of Permanent Magnet Synchronous Machine Using Zero-Sequence Components. *IEEE Trans. Power Electron.* **2015**, *30*, 6731–6741. [CrossRef]
40. Cash, M.; Habetler, T.; Kliman, G. Insulation failure prediction in AC machines using line-neutral voltages. *IEEE Trans. Ind. Appl.* **1998**, *34*, 1234–1239. [CrossRef]
41. Cash, M.A.; Habetler, T.G. Insulation failure prediction in inverter-fed induction machines using line-neutral voltages. In Proceedings of the IAS Annual Meeting, New Orleans, LA, USA, 5–9 October 1997. [CrossRef]
42. Garcia, P.; Briz, F.; Degner, M.W.; Diez, A.B. Diagnostics of Induction Machines Using the Zero Sequence Voltage. In Proceedings of the IAS Annual Meeting, Seattle, WA, USA, 3–7 October 2004. [CrossRef]
43. Wang, Z.H.; Huang, X.D. Principle of Phase Measurement and Its Application Based on All-Phase Spectral Analysis. *J. Data Acquis Process* **2009**, *24*, 777–782. [CrossRef]
44. Huang, X.D.; Wang, Z.H. Anti-noise Performance of All-phase FFT Phase Measuring Method. *J. Data Acquis Process* **2011**, *26*, 286–291. [CrossRef]
45. Huang, K.H.; Wang, D.M.; Zhu, Z.Y.; Wei, H.F.; Jiang, W.U. Power System Harmonic Detection Algorithm Based on Co-sin-Window and Interpolated FFT and APFFT. *Comput. Technol. Dev.* **2011**, *21*, 223–230.
46. Huang, X.; Wang, Y.; Jin, X.; Lü, W. No-Windowed ApFFT/FFT Phase Difference Frequency Estimator Based on Frequency-Shift & Compensation. *J. Electron. Inf.* **2016**, *38*, 124–131. [CrossRef]
47. Qi, G.Q.; Jia, X.L. High-Accuray Frequency and Phase Estimation of single-Tone Based on Phase of DFT. *Acta Electron. Sin.* **2001**, *9*, 1164–1167. [CrossRef]
48. Li, X.F.; Li, L.; Kou, K.; Wu, T.F.; Yang, Y. Analysis and Improvement of Time-Shift Phase Difference Spectral Correction Based on All-Phase FFT. *J. Tianjin Univ. Sci. Technol.* **2016**, *49*, 1290–1295.
49. Li, S. Multi-Sensor Fusion by CWT-PARAFAC-IPSO-SVM for Intelligent Mechanical Fault Diagnosis. *Sensors* **2022**, *22*, 367. [CrossRef]
50. Todkar, S.S.; Baltazart, V.; Ihamouten, A.; Dérobert, X.; Guilbert, D. One-class SVM based outlier detection strategy to detect thin interlayer debondings within pavement structures using Ground Penetrating Radar data. *J. Appl. Geophys.* **2021**, *192*, 104392. [CrossRef]
51. Guan, S.; Wang, X.; Hua, L.; Li, L. Quantitative ultrasonic testing for near-surface defects of large ring forgings using feature extraction and GA-SVM. *Appl. Acoust.* **2020**, *173*, 107714. [CrossRef]
52. Zhao, Y.; Wei, G. Using an Improved PSO-SVM Model to Recognize and Classify the Image Signals. *Complexity* **2021**, *21*, 8328532. [CrossRef]

Review

Physical Variable Measurement Techniques for Fault Detection in Electric Motors

Sarahi Aguayo-Tapia, Gerardo Avalos-Almazan, Jose de Jesus Rangel-Magdaleno * and Juan Manuel Ramirez-Cortes

Digital Systems Group, Electronics Department, Instituto Nacional de Astrofísica, Óptica y Electrónica, Luis Enrique Erro #1, Tonantzintla, Puebla 72840, Mexico; sarahi.aguayo@inaoe.mx (S.A.-T.); gerardo.avalos@inaoe.mx (G.A.-A.); jmram@inaoep.mx (J.M.R.-C.)
* Correspondence: jrangel@inaoe.mx

Abstract: Induction motors are widely used worldwide for domestic and industrial applications. Fault detection and classification techniques based on signal analysis have increased in popularity due to the growing use of induction motors in new technologies such as electric vehicles, automatic control, maintenance systems, and the inclusion of renewable energy sources in electrical systems, among others. Hence, monitoring, fault detection, and classification are topics of interest for researchers, given that the presence of a fault can lead to catastrophic consequences concerning technical and financial aspects. To detect a fault in an induction motor, several techniques based on different physical variables, such as vibrations, current signals, stray flux, and thermographic images, have been studied. This paper reviews recent investigations into physical variables, instruments, and techniques used in the analysis of faults in induction motors, aiming to provide an overview on the pros and cons of using a certain type of physical variable for fault detection. A discussion about the detection accuracy and complexity of the signals analysis is presented, comparing the results reported in recent years. This work finds that current and vibration are the most popular signals employed to detect faults in induction motors. However, stray flux signal analysis is presented as a promising alternative to detect faults under certain operating conditions where other methods, such as current analysis, may fail.

Keywords: fault detection; fault classification; induction motors; measurement techniques; physical variables; signal analysis

Citation: Aguayo-Tapia, S.; Avalos-Almazan, G.; Rangel-Magdaleno, J.d.J.; Ramirez-Cortes, J.M. Physical Variable Measurement Techniques for Fault Detection in Electric Motors. *Energies* **2023**, *16*, 4780. https://doi.org/10.3390/en16124780

Academic Editors: Daniel Morinigo-Sotelo, Rene Romero-Troncoso and Joan Pons-Llinares

Received: 4 May 2023
Revised: 4 June 2023
Accepted: 15 June 2023
Published: 18 June 2023

1. Introduction

Induction motors (IMs) are the principal source of power in the manufacturing industry due to their sturdiness, energetic efficiency, and repair and maintenance costs.

IMs are prone to present faults associated with thermal, mechanical, and magnetic stresses or to expected causes, such as natural wear of bearings [1]. Bearing damage is the type of fault that appears more frequently. However, other faults may appear due to unbalanced loads, unbalanced power supply, and high operational frequencies, among others. Typically, the faults with a higher frequency of appearance are: bearing faults, stator winding faults, and rotor broken bar faults [2,3].

Researchers have documented advances in IMs' condition monitoring and fault detection in this context. Atta et al. [4] summarized recent techniques for broken bar faults under starting and steady-state conditions using current, voltage, flux, vibration, and acoustic signatures. In that work, the authors subgroup the methods based on resistance estimation, parameter estimation, digital estimation, time-domain, frequency domain, time–frequency domain, and data-driven approaches. The main drawback of this work is that it is highly specialized, given that it focused only on one type of fault, in this case, broken bar fault detection. However, it provides a good panorama of the evolution of broken bar fault detection techniques in the last 8 years. The authors pointed out that some issues still require attention, such as methods for multiple-fault detection, model-based methods, transfer learning in driven-based methods, and assessing the performance of time–frequency methods under short-start duration, among others. Gundewar et al. [5] reviewed recent techniques used for fault diagnosis in IM according to the type of fault: bearing fault (BF), rotor bar faults, air gap eccentricity, and stator faults. Moreover, they highlighted the use of artificial intelligence (AI) techniques for condition monitoring, such as support vector machines (SVMs), fuzzy logic, adaptive neuro-fuzzy inference system (ANFIS), and genetic algorithm (GA). They provided an extensive overview of different types of IM faults. Nevertheless, the techniques for fault detection are limited to AI-based, which have been proven to have high detection accuracies but at the cost of high computational resources required for its implementation. A review on IMs fault detection based on current analysis was presented by Yakhni et al. [6], focusing on applications with variable speeds. The authors summarized the methods based on current analysis for detecting particular faults: broken rotor bar, bearing, eccentricity, stator fault, inter-turn short circuit, and unbalanced supply. This paper contributes to reviews of current-based analysis. However, it only focuses on two approaches: the adaptive observer and the adaptive notch filtering method.

This paper presents a revision of the recent works and tendencies in the types of physical variables employed for fault detection. Although previous works have discussed advances in fault detection and condition monitoring, these are mainly centered on the type of fault or the methods themselves, not so in detail regarding the type of physical variable.

2. Common Types of Faults in Induction Motors

Different types of physical variables are measured and analyzed to detect the presence of a fault in IM. Induction motors are prone to present faults associated with electrical, mechanical, or environmental aspects. Electrical faults can be induced, among others, by supply current unbalance and overload, and faults such as short circuits are produced. Mechanical faults, however, are related to broken bars and bearing damage. Certain environmental conditions, such as humidity and very high or low temperatures, may affect the IM and progress into electrical or mechanical faults. Some of the faults that appear in IMs more commonly are presented next.

Broken Rotor Bars

Damage in the rotor bars is difficult to detect since the induced spectral components lie near the supply frequency. Moreover, as the severity of the fault increases, the spectral components are greater in amplitude. As the load of the motor increases, the fault components are further from the fundamental frequency. Hence, incipient faults, commonly less than half broken bars, are more difficult to detect; the load condition also affects the fault detection accuracy.

The fault is created synthetically by drilling the bar to study broken rotor bars, as illustrated in Figure 1. Most works that study the presence of broken bars in IMs analyze 1 and 2 BRB; few works are able to detect less than 1 BRB, given that it usually requires complex algorithms for signal processing.

Broken rotor bars are commonly detected by means of motor current signature analysis, where the current spectrum is processed and analyzed, searching for changes in the amplitude of certain frequency spikes. These components usually appear near the fundamental frequency, as the following equations indicate this:

$$f_{bb} = [\frac{h}{p}(1-s) \pm s]f_s \qquad (1)$$

where f_{bb} are the spurious frequencies, p represents the number of pairs of poles, s is the slip, f_s represents the frequency of the grid, and h is an index that represents the number of harmonics.

In a similar way, the localization of spurious frequencies can be performed by using vibration analysis, following the next equation:

$$f_{bk\pm} = f_s[1 \pm 2ks] \tag{2}$$

where $f_{bk\pm}$ represents the spurious frequency, k represents an integer index, f_s is the frequency of the grid, and s represents the slip.

Figure 1. Rotor bar drilled to emulate a one broken bar fault.

Bearing Damage

It is well known that bearings in an IM are fundamental elements since this allows the rotor to rotate on its own axis. Due to its mechanical nature, bearings are under mechanical and electrical stresses. A common IM bearing is conformed by static and mobile elements, as shown in Figure 2.

The seals maintain the lubricant inside the bearing and keep the contaminants out; the mechanic load is transferred from a foothold to a rolling point, i.e., from the inner raceway to the outer raceway through the rolling elements (balls or rollers), and the cage maintains the rolling elements in its position.

Figure 2. Bearing parts.

Bearing faults are mostly identified using vibration signals [7,8]. Spurious frequencies based on vibration analysis can be localized according to the next equations [9]:

$$f_o = \frac{N_b}{2} f_r \left[1 - \frac{Db}{Dc} cos(\beta)\right] \tag{3}$$

$$f_i = \frac{N_b}{2} f_r [1 + \frac{Db}{Dc} cos(\beta)] \tag{4}$$

$$f_b = \frac{D_b}{D_c} f_r [1 - \frac{Db^2}{Dc^2} cos(\beta)^2] \tag{5}$$

where f_o, f_i, and f_b represent the spurious frequencies for outer race damage, inner race damage, and bearing ball damage, respectively; N_b represents the number of balls; β represents the angle between the ball and the defect; D_b is the diameter of the ball; and D_c is the pass diameter.

In this way, previous works have studied bearing faults through vibration analysis, expecting the appearance of the fault's frequencies, as stated by the aforementioned equations.

Short Circuit

The most recurrent faults related to a shortcut are the stator winding and rotor short-cuts. A short circuit is one of the most difficult faults to detect; protection systems commonly detect the shortcut when damage is critical. A grounded shortcut results in severe damage to the machine's integrity.

There are several approaches used for shortcut detection. The most used are as follows:

- Analysis of the magnetomotive forces;
- Winding function approach;
- Dynamic mesh reluctance approach;
- Finite element.

If a stator fault exists, the current through the shorted winding will cause variations in the normal magnetic field and variations in the axial flux according to the next equation [10]:

$$f_{ss} = f_s [(k \pm n \frac{1-s}{\rho})] \tag{6}$$

where f_s is the grid frequency, n is the order of the shorted coil space harmonic, k is the harmonic order, s represents the slip of the machine, and ρ is the number of pole pairs.

2.1. Effects of Vibrations in Induction Machines

Faults in induction machines may cause negative effects in the normal operation of the motor regardless of the cause; some of the principal vibration sources are voltage/current imbalances, rotor bar defects, bearing defects, voltage distortion of the power grid, and speed driver issues, among others. There are two main sources of vibrations: mechanical and electromagnetic origin.

Vibrations of mechanical origin are mainly caused by rotor imbalance, shaft bow, and misalignment or wear in mechanical elements.

Electromagnetic vibrations provoke for radial electromagnetic forces and tangential electromagnetic forces. Under normal operation conditions, these forces do not represent significant vibrations. Excessive electromagnetic motor vibrations are often a result of a resonance condition on the structure of the motor as a unit or even on the motor components [11]. Vibrations may cause or accentuate problems, such as misalignment and eccentricity, among others. These problems are discussed next.

2.2. Misalignment

Misalignment can be caused by the deviation of the coupling axis between the load and motor shafts. Excessive vibrations will provoke asymmetries in the motor frame, shaft, and bearings in the IM. If the misalignment fault is not detected early, the consequences could be bearing damages, eccentricity, and gearbox failures.

The two types of misalignment issues in IMs are parallel and angular faults. A parallel fault occurs when there is a height difference between the shaft centers of the load and the IM, and an angular fault occurs when an inclination exists between the load and the IM.

The combination of parallel and angular misalignment is called mixed misalignment. The coupling differences between the shafts emerge in the frequency spectrum of current as follows [12]:

$$f_{mc} = [1 \pm k(1-s)/p]f_s \tag{7}$$

where f_{mc} represents the spurious frequency of the misalignment signal frequency, k is an integer operator, p represents the number of pole pairs, f_s is the supply frequency, and s is the slip of the motor. Similarly, the spurious frequency can be obtained for the vibration spectrum as follows,

$$f_{mv} = kf_r \tag{8}$$

where f_r is the rotational frequency.

2.3. Eccentricity Air Gap

Eccentricity can be classified as dynamic and static. Dynamic eccentricity occurs when the center of the rotor is not at the center of rotation, and the position of the minimum radial air gap rotates with the rotor. Static eccentricity occurs when the position of the minimum radial air gap is fixed. Both static and dynamic eccentricity produce a polar pair, $p \pm 1$, in the spatial distribution of the magnetic field in the air gap, where p represents the number of the fundamental polar pair [13,14].

An eccentricity fault provokes spurious components in the current spectrum according to the following equation:

$$f_h = (2(kR \pm n_d)(\frac{1-s}{p}) \pm v)f \tag{9}$$

where n_d represents the eccentricity order equal to 0 for statistic eccentricity and integer coefficient $1, 2, 3 \ldots$ for dynamic eccentricity; f is the frequency of the power supply; s represents the slip; p is the number of poles; R represents the number of rotor bars; k is a positive integer coefficient; and, finally, v represents the harmonic order of the stator supply voltage [15].

In the next section, fault detection techniques are discussed.

3. Analysis of Faults Based on Physical Variables

According to the features of the fault detection techniques, these can be classified into model-based, signal-based, and data-based techniques, as illustrated in Figure 3. Depending on the type of physical variable, some of these methods are more adequate than others for fault detection.

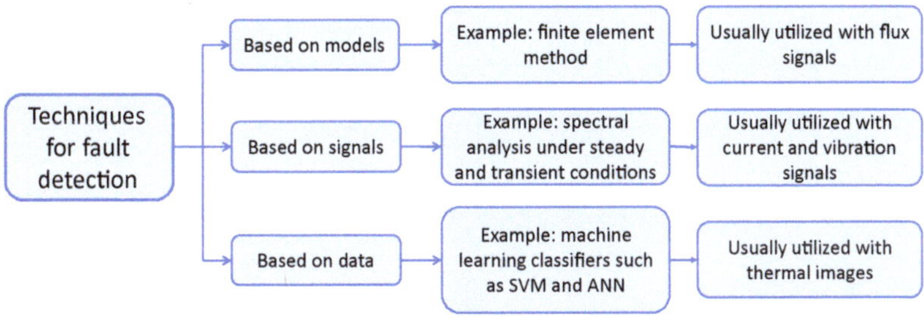

Figure 3. A classification of fault detection methods.

For instance, a method based on models is proposed using flux signals, where a finite element method (FEM) simulation is implemented, as illustrated in Figure 4. These methods usually imply converting the flux signals into the DQ frame.

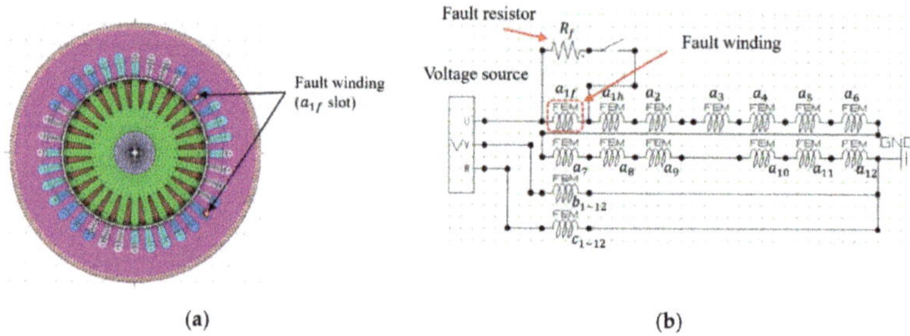

Figure 4. FEM simulation for a fault detection method based on flux signals: (**a**) mesh model, (**b**) stator circuit model. Source: [16].

On the other hand, an example of a technique based on signals may comprehend selective filtering of current or vibration signals and statistical analysis, such as the works in [17,18]. The filtering of the signal that is illustrated in Figure 5 depicts a narrow band-width Taylor–Fourier filter over a current signal, and Figure 6 illustrates the decision making according to the three standard deviation criteria for fault detection based on statistical distributions of the amplitude of the filtered signals.

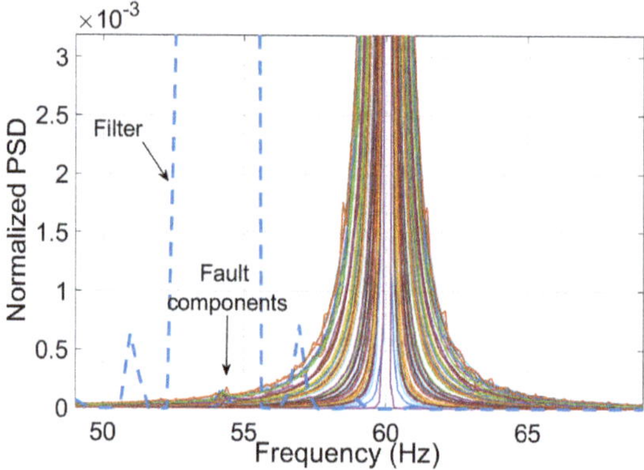

Figure 5. Filtering of the current signal. Source: [18].

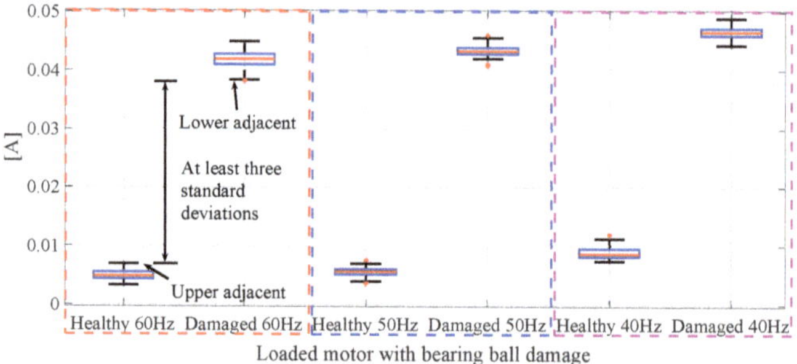

Figure 6. Fault detection criteria based on statistical distributions.

An example of a method based on data is presented in Figure 7, where an architecture based on a convolutional neural network is applied to detect electrical faults by using thermal images. Thermal images are commonly processed, as presented in [19], with AI methods.

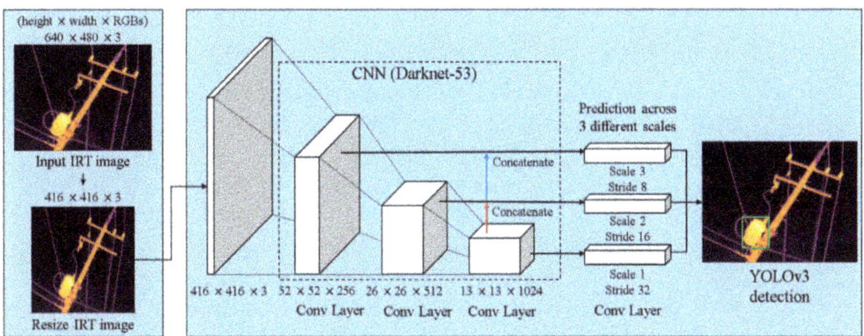

Figure 7. Scheme of a fault detection method based on data. Source: [19].

Next, details regarding detecting faults based on certain types of signals are given.

4. Methodologies for IMs Diagnostic

Fault detection and classification methodologies depend mainly on the implemented algorithm. The elements to traduce the physical variables to processable information are known as sensors. Sensors are indispensable for measuring physical variables and looking for valuable information for detection and classification. The most used sensors are current and voltage sensors, accelerometers, encoders, thermographic cameras, acoustic sensors, and antennae, among others [20].

Inherent to the sensor's use, data acquisition systems must store the physical variables to be processed. Those systems can be digital or analog; commonly, aiming to avoid complications, data acquisition boards are used. Other possibilities include embedded systems such as microcontrollers or field programmable gate arrays (FPGAs). Commonly, the sampling frequencies can be fixed from a range going from 1 (Khz) to 100 (Mhz), and the data acquisition systems can include filters to avoid aliasing or noise.

A variable analysis is commonly performed in the frequency domain. One of the most used techniques for signal processing is Fourier transform, and recently, techniques based on AI for characterization and classification have been developed. Regardless of the physical variable that is analyzed, machine learning methods have stood out as one of the best options for fault classification and pattern recognition, at the cost of being the most complex algorithms and requiring high computational resources for its implementation. Some of these methods are mentioned next:

- Convolutional neural network (CNN): This is a neural network conformed by multiple filtering and classification stages, where the filtering layer extracts the features from the inputs (current, vibration, or another signal), and the classification stage is a multi-layer perceptron that deals with the classification of the faults [21]. The advantage of using CNN is that it can provide good results with raw data; in other words, usually, no pre-processing is required, and previous feature extraction is unnecessary.
- Support vector machine (SVM) and k-nearest neighbor (KNN): SVM establishes a decision boundary in a way that the location between classes is as far as possible; it is fundamentally a binary classification where a hyperplane separates the classes. On the other hand, KNN requires calculating a Euclidean distance between samples to determine similarities between these samples. Then, according to the similarities, a sample is classified into a certain class. KNN is one of the simplest machine learning methods for classification problems. These methods have been implemented using both current and vibration signals [22–24].
- Deep learning: This is a type of machine learning based on artificial neural networks with multiple layers. These neural networks attempt to simulate how the human brain works. Recently, several techniques based on deep learning have been proposed. Among the most interesting methodologies are those that are based on the recognition of thermal images because those methods do not disturb the normal operation of the

system; these kinds of techniques are based on the analysis of the thermal behavior of the faulty region in an induction machine [25].

In the next sections, physical variables and their analysis techniques are listed.

4.1. Current

Motor current signal analysis (MCSA) refers to the process used to obtain information about the dynamic of the electric machine by signal processing from the stator currents. The current signals in the time domain are usually obtained through current sensors with resistive shunts at their outputs [26]. Frequently, researchers obtain the current samples from different speeds, frequencies, and load conditions to acquire behavioral information on the dynamics under different conditions to apply the detection and classification techniques in practical scenarios, such as industrial applications.

Motor faults affect the spectrum of the stator current signals; several techniques for fault detection in induction motors based on MCSA have been proposed for detecting both electrical and mechanical issues, and commonly, these techniques are based on algorithms for spectrum analysis to find fault signatures, aiming to detect and classify the location and severity of the fault [27–37].

Current-based analysis for fault detection usually involves using a Hall-effect sensor concerning the signal acquisition; these are commonly included in the instrumentation installed in the IMs for control and protection purposes. Therefore, no additional expenses are required, nor is invasive sensing needed. Current signals are studied for fault detection, mostly in the frequency domain, to identify components introduced by the fault, such as broken rotor bars (BRBs) and inter-turn short circuits (ITSCs).

In 2016, Romero et al. [38] proposed a method to detect 1 BRB and eccentricity produced by a motor-load misalignment relying on complete ensemble empirical mode decomposition (CEEMD) and multiple signal classification (MUSIC). In that work, a time–frequency analysis is conducted to identify the faults under the start-up and steady-state regimes of the motor. The authors presented an experimental setup where the current signals were acquired with a Honeywell CSNE151 Hall-effect current sensor. Then, the proposed methodology was compared with others such as STFT, Wavelet and STFT, EMD and STFT, MUSIC, and Wigner–Ville. The results highlighted that CEEMD and MUSIC exhibited better detectability than the other methods under both start-up and steady-state conditions.

In 2021, Yang et al. [37] explored the combination of modified ensemble empirical mode decomposition (MEEMD) energy entropy and ANN to analyze stator current signals. In that work, current signals were decomposed with MEEMD, and a cross-correlation criterion selected the components with more information. Then, the energy entropy of these signals was calculated and stored in a vector. Finally, the vectors fed the ANN to identify BRB and air-gap eccentricity. The work reported a detection accuracy of 99% and an improvement respecting EEMD with ANN and MEEMD with SVM. The authors used simulated data, which can be considered a drawback given that data collected from an experiment usually includes some of the effects that are more likely to be present in the industry.

Various processing methods combined with AI classifiers have been reported in the last few years for fault detection of different types. For instance, in [39], faulty bearings, different levels of short circuits in the stator coils, and different levels of BRB are detected using a multi-agent system based on AI. In other work, for detecting winding insulation burn, bearing damage, and BRB, the authors in [40] proposed using spiking neural networks to analyze the current signals of the IM.

Hence, it is noticeable that the current analysis is very popular for condition monitoring and fault detection in IMs. Table 1 summarizes different works based on current analysis. Each work is divided by year, the type of sensor employed for the signal acquisition, the type of fault that is detected, and the method for the analysis.

Table 1. Works based on current analysis.

Year	Type of Sensor	Type of Fault	Method
2017 [41]	Current clamp model i200s from Fluke	Half BRB, 1 BRB, and 2 BRBs	Homogeneity
2017 [42]	Digital signal processor board dSpace 1104	1 BRB for different loads	Sliding window discrete Fourier transform
2017 [43]	LA-55P current sensor	6 types of stator faults, 3 types of rotor faults, 3 types of unbalance voltages, and outer raceways bearing fault	Fast Fourier transform (FFT) and multiple Park's vector (MPV) approaches
2018 [44]	Software	1 BRB for full load and half load	MUSIC, least-squares magnitude estimation, niche bare-bones particle swarm optimization, and singular value decomposition (SVD)
2019 [45]	Data acquisition system	Bearing axis deviation, stator, and rotor friction; rotor end ring break; and poor insulation	Recurrence plots and convolutional neural networks (CNNs)
2021 [46]	Current sensor	1 BRB, 2 BRBs, and 4 levels of ITSC	Discrete wavelet transform (DWT), ANFIS, and Clark's transformation
2021 [47]	Software	Stator winding faults: phase to phase and phase to ground	FFT, Short-Time Fourier Transform (STFT), Continuous Wavelet Transform (CWT) and Long Short-Term Memory (LSTM)
2021 [48]	Software	1-1 broken diametrically opposite rotor bars	Finite element method and multilayer perceptron
2021 [49]	Current sensor	1 BRB, 2 BRBs, and bearing damage	Principal component analysis (PCA), radial basis function neural networks, and probabilistic neural networks
2021 [50]	Current sensor	2 BRBs	MUSIC
2022 [51]	NI 9246 module and cDAQ-9178	Rotor misalignment, BRB, and ITSC	FFT, power spectral density (PSD), and autocorrelation function
2022 [52]	Hall-effect current sensor	BRB and ITSC	FFT and motor current normalized residual harmonic analysis
2022 [53]	Signal extractor NI PX1-1033	Broken bearings, BRB and ITSC	Symmetrical uncertainty, genetic algorithm, Hilbert–Huang transform, and SVM

In Figure 8, aspects related to the analysis of current signals are summarized: the preferred instrument for acquisition, method for analysis, and pros and cons of using this physical variable.

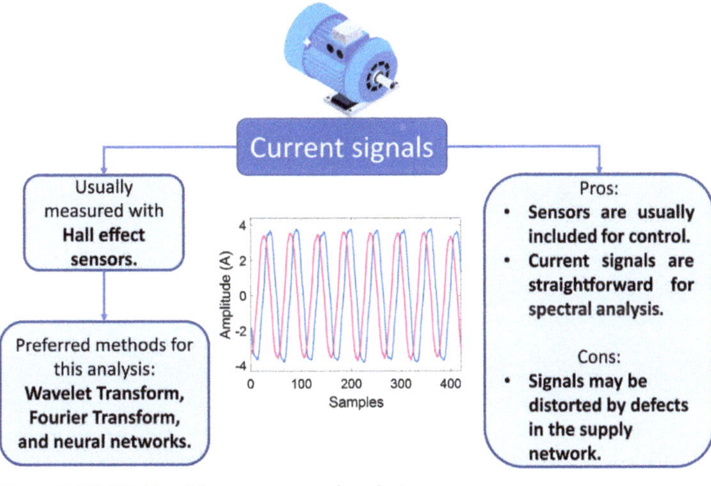

Figure 8. Highlights of the current-based analysis.

4.2. Vibrations

IMs inevitably produce vibrations during operation because of their rotatory nature. Vibrations may cause imbalance and misalignment in the electromechanical system, and those issues are accentuated when a defect exists in the motor. Vibrations directly affect the air-gap eccentricity, which influences the uniformity of the distance between the rotor and the stator, indirectly affecting the stator currents.

Fault signature depends mainly on the geometry of the damaged element. The principal disadvantage of vibration analysis on an IM is system isolation since vibrations of the surrounding elements introduce noise to the measurements. Devices such as accelerometers are commonly used to sense the induction machine's three-axis movements with respect to time [54].

Vibration analysis has been widely studied for fault detection, mainly because most of the faults that can appear in an IM will have an effect on the vibration measurements [55,56]. In 2016, Sun et al. [57] employed a deep neural network (DNN) to classify BRB, bowed rotor, inner race bearing fault, ITSC, and unbalance rotor. This approach utilized a sparse auto-encoder based on the measured vibration data for feature learning. Moreover, the authors compared the DNN performance with a neural network, SVM, and linear regression; for the five types of faults, DNN presented the best accuracy results.

In 2019, Lee et al. [58] measured vibration data from an IM with rotor fault and bearing fault and used a CNN for fault classification. Vibration data were measured by means of a vibration sensor and NI-9231. Classification accuracies of 98–100% were reported in this work, highlighting that the vibrations signals did not require a frequency domain transformation, as is commonly required for a current-based analysis.

Other recent works have frequently proposed methods, including neural networks such as CNN and deep convolutional neural networks (DCNNs), for bearing fault classification [59–61]. Some of the pros and cons of vibration-based analysis are summarized in Figure 9.

Recent works on fault detection based on vibration analysis are listed in Table 2.

Table 2. Works based on vibrations analysis.

Year	Type of Sensor	Type of Fault	Method
2016 [62]	Software	Unbalance and misalignment motor faults	Orbital analysis, Steinbuch Lernmatrix for classification and binarization process
2017 [63]	Accelerometer of an Android mobile phone	2 BRBs and inner raceway bearing fault	FFT and motor speed estimation
2017 [23]	Accelerometer 604B31	3 turns shorted stator winding, unbalance rotor, inner race bearing fault defect, 3 BRBs and bowed rotor	Back-propagation neural network and SVM
2019 [58]	Vibration sensor and NI-9234	1 BRB and bearing fault	CNN
2019 [64]	Spectrum analyzer	Rotor bending, inner ring bearing fault and BRB	FFT and cerebellar model articulation controller
2021 [61]	Single antenna's nearfield effect	Inner and outer race bearing damage, unbalance	DCNN and FFT
2021 [65]	Accelerometer	BRB, rotor deflection, bearing fault, stator winding fault and rotor unbalance	Transfer PCA
2021 [60]	Accelerometer	Short circuit of 2, 4, and 8 turns; air-gap eccentricity; BRB, broken bearing cage; bearing abrasion fault; bearing ball fault; and inner and outer race bearing damage	Nearest centroid classifier and CNN
2022 [66]	IMI 608A11 accelerometer	Outer and inner race bearing damage	Electromagnetic dynamic coupled modeling method
2022 [67]	Piezoelectric accelerometer sensor and PCI-1711	BRB and cavitation fault	Ensemble framework, fuzzy rough active learning, and drift detection
2022 [24]	Software	1 BRB, 2 BRBs, 3 BRBs, and 4 BRBs	Random forest, decision tree, k-nearest neighbor (KNN), STFT, and CNN
2022 [68]	Accelerometer	1 BRB, 2 BRBs, and outer race of rolling bearing fault	Cyclic modulation spectrum, fast spectral correlation, Teager–Kaiser energy operator, and STFT

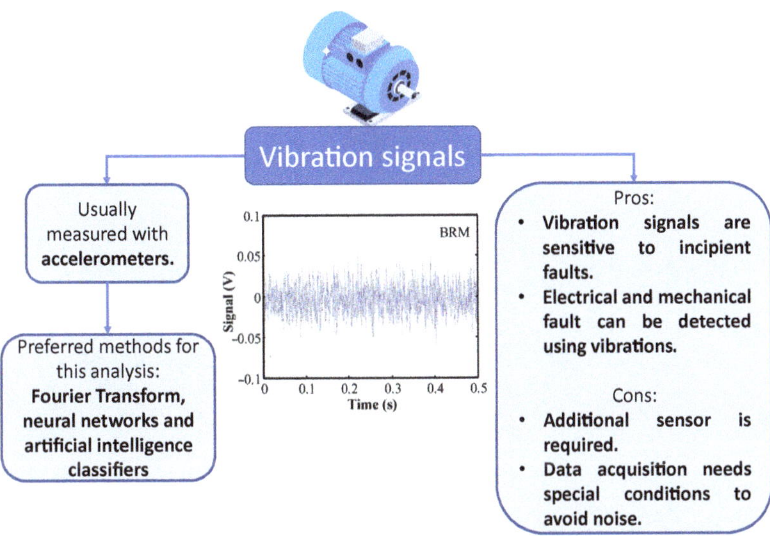

Figure 9. Highlights of the vibration-based analysis. Vibration signal taken from [65].

4.3. Temperature

Thermal analysis is a process based on a nonintrusive and noncontact technique for monitoring systems. In recent years, this kind of analysis has grown in popularity due to the large amount of information provided by thermal images. For instance, when a bearing fault occurs, the friction operation coefficient upon operation increases, which leads to an increase in the temperature of the IM [69].

This kind of analysis is commonly implemented based on complicated computational processes due to image processing. Fault detection based on thermal images requires the use of special cameras; this type of instrument is more unusual than, for example, a current clamp.

Despite the complications of thermal analysis, some works have been committed to proposing new methods for fault detection, such as the one presented by Mahami et al. [70] in 2021, where an infrared thermography technique was applied to detect eight different short circuit faults in the stator winding; a bag-of-visual-words model was used for feature extraction, and the fault patterns were identified using an extremely randomized tree. In that work, the classifier is compared with KNN, SVM, least-squares SVM, random forest, deep-rule-based, and self-organizing fuzzy logic classifier approaches, and the results prove that the extremely randomized tree shows the best accuracy. Thermal analysis based on AI methods is an effective option for fault detection in IMs. Other works also rely on the use of advanced techniques for pattern classification.

In Figure 10, highlights of thermal analysis are given, and in Table 3, recent works are listed. Considerably fewer works are committed to thermal analysis than current or vibration analysis.

4.4. Flux

Recent methodologies for monitoring systems based on magnetic flux have been gaining attention mainly because some manufacturers include magnetic sensors embedded in their electric motors.

The magnetic-flux-based methodologies can be grouped into stray flux analysis, where the magnetic flux outside the machine is studied, and the air-gap flux is analyzed. In general, flux analysis is noninvasive, sensors are low cost, and their installation is simple.

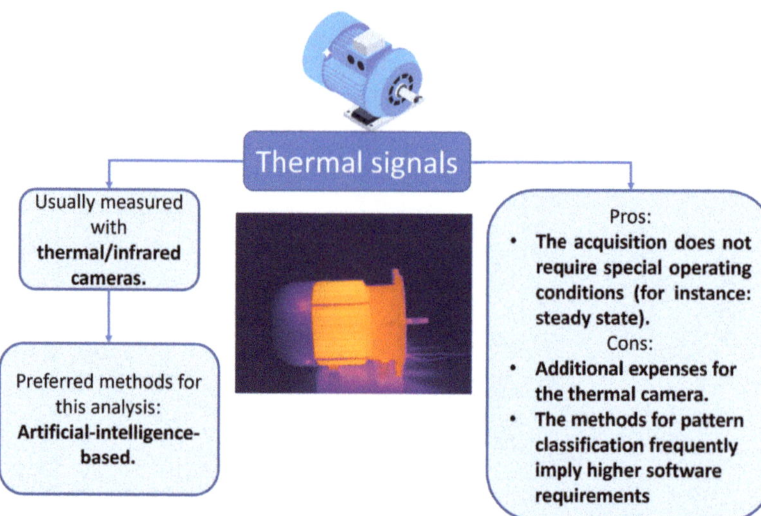

Figure 10. Highlights of the thermal-based analysis. Thermal image taken from [70].

Table 3. Works based on temperature analysis.

Year	Type of Sensor	Type of Fault	Method
2021 [70]	Dali-tech T4/T8 Infrared camera	8 types of stator short-circuit faults	Bag-of-visual-words, speed-up robust features, and extremely randomized tree
2022 [71]	Infrared thermal image camera Dali-tech T4/T8	8 forms of short-circuit failures in the stator windings, stuck rotor fault, and cooling fan failure	Coordinate attention feature extraction and CAPNet training
2022 [72]	Thermal camera	Inner and outer race bearing damage, ball bearing fault, 1 BRB, 5 BRBs, 8 BRBs, inner bearing plus 1 BRB, and outer bearing plus 5 BRBs	Transfer learning model based on a VGG-19 CNN
2022 [73]	Infrared micro-camera model FLIR LEPTON 3	Misalignment, unbalance, 1 BRB, 2 BRBs, outer race bearing faults, and gearbox wearing	Otsu algorithm, scale-invariant feature transform, PCA, and artificial neural network

In 2016, Mirzaeva et al. [74] proposed an online fault diagnostic system based on internal main air-gap flux density measurements using Hall-effect sensors; stator turn-to-turn shorts, rotor bar damage, and dynamic and static eccentricity were detected with this system. Using this method, faults were detected, and their location and severity were obtained. Moreover, the flux density measurements provided time and space dimensions, offering more possibilities for condition monitoring.

In 2019, spectrum analysis of the radial stray flux was proposed by Park et al. [75] to detect one BRB and two BRBs. This paper demonstrates the reliability of this proposal over the current analysis under the influence of rotor axial air ducts, a condition that produces magnetic asymmetries and may lead current-based detection to fail. Another advantage of this method is the low-cost flux coil and the sensitivity and reliability of this flux-based fault detection.

Minervini et al. [76] used current and flux signals to classify bearing faults by using a pre-trained DCNN. In that work, two severity levels of generalized roughness and localized defects in the outer ring of the bearing were analyzed. It is important to highlight that, in this case, the flux signal provided 100% accuracy, contrasting with an accuracy of 80% when using current signals. The authors suggested that this difference may be attributed to the fact that the flux sensor could be installed very close to the faulted bearing, given the size of the motor.

Overall, the flux analysis provides good results for IM fault detection, whereas another kind of methodology based on another physical phenomenon exhibits false positives. The drawback of flux-based methods is the restriction for remote monitoring, unlike the current signals that are usually the base of condition monitoring of IMs.

In Figure 11, some highlights of the flux analysis are given, and recent works based on this physical variable are presented in Table 4.

Table 4. Works based on flux analysis.

Year	Type of Sensor	Type of Fault	Method
2017 [74]	36 Hall-effect flux sensor	ITSC, rotor bar damage, and static and dynamic eccentricity	Magnitude analysis of air-gap flux density
2019 [75]	320 turns Helmholtz coil with 121 cm inner diameter and 155 cm outer diameter	1 BRB and 2 BRBs	Spectral analysis
2022 [16]	Software and current sensor	ITSC	Clarke transformation
2022 [77]	Five-turn search coil	1 and 2 BRBs adjacent and non-adjacent, and 3 BRBs	STFT time–frequency analysis
2023 [78]	Coil-based sensor	1 BRB and 2 BRBs	Persistence spectrum, data augmentation techniques, and CNN

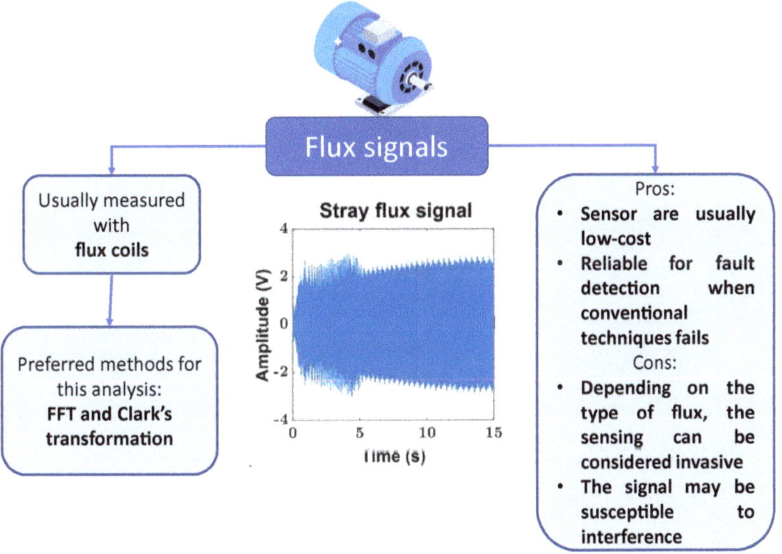

Figure 11. Highlights of the flux-based analysis. Flux signal taken from [79].

4.5. Combined Analysis

Aiming to obtain better results, researchers have explored two or even three modal signal analyses to take advantage of the properties of the different variables from the IM.

It is noticeable from Table 5 that most works combine current and vibration signals, taking advantage of the wide knowledge built around analyzing these variables. For instance, in 2017, Cruz et al. [80] processed current and vibration signals in the time–frequency domain by using short-time Fourier transform and wavelet multiresolution analysis. With this method, authors detected failures in the motor helix and unbalanced loads in the motor shaft and classified them using fuzzy logic.

AI methods were employed by Zawad et al. in 2019 [81] for detecting unbalanced shaft rotation; bearing fault; unbalanced voltage; one, two, and three BRBs; and four combinations of these faults. The authors proposed using matching pursuit and DWT for feature extraction from current and vibration signals. The signals were acquired from an experimental array, where an eight-channel power quality analyzer was employed to measure three-phase currents, and an accelerometer 356A32 was utilized for the vibration measurements. For comparative purposes, SVM, KNN, and ensemble were used, along with 17 different classifiers available in the Classification Learner toolbox of MATLAB. Within these options, fine KNN, bagged trees, subspace KNN, fine Gaussian SVM, and weighted KNN performed better, exhibiting accuracies near 100%. The authors observed that the number of features for the classification affects detection accuracy; in this work, a higher number of features provided the best results. This method is useful for comparative purposes.

In [82], a current and vibration analysis is performed based on different algorithms, such as Matching pursuit, DWT, invasive weed optimization algorithm, K-Nearest neighbor, SVM, random forest, genetic algorithm, receiver operation characteristic curve, and statistical analysis; these works compare these algorithms to detect 1 BRB, 5-BRB, 8-BRB, ball bearing fault, inner and outer bearing fault, and stator fault. Other works combine analysis of current, speed, and flux, such as the one presented in [83], or speed, current, and torque, such as the one in [84]. Another combination of physical variables is explored in [85], where authors analyze temperature and flux signals, and in [79], where current and flux signals are employed. Some of the latest works reported for fault detection based on analyzing more than one physical variable are presented in Table 5.

Table 5. Works based on combined analysis.

Year	Physicial Variables	Type of Sensor	Type of Fault	Method
2016 [86]	Voltage and current	PCI data acquisition board	2 BRBs	Clark transformation and FFT
2017 [80]	Vibration and current	Data acquisition system NI USB-6211	Unbalanced load in the motor shaft and in the motor helix, 1 BRB, 3 BRBs, 5 BRBs, and 7 BRBs	Short-time Fourier transform, wavelet multiresolution analysis, and fuzzy logic classifier
2018 [87]	Current, voltage, and temperature	Nicolet Odyssey data acquisition system	1 BRB, 2 BRBs, 4 BRBs, damaged capacitor bank, and damaged bearings	SVM, fuzzy membership functions, fuzzy information fusion, and cross-validation
2019 [88]	Current and vibration	Accelerometer LIS3L02AS4 and Fluke current clamp i200s	Bearing fault and mechanical unbalance	Statistical analysis and classification tree
2019 [81]	Current and vibration	8-channel power quality analyzer PQPro by CANDURA and accelerometer 356A32	1 BRB; 2 BRBs; 3 BRBs; unbalanced shaft rotation (USR); bearing fault; combined bearing fault and USR; combined bearing fault and 1 BRB; combined bearing fault, USR, and unbalanced voltage (UV); and combined UV and 3 BRBs	Matching pursuit, DWT, SVM, KNN, and ensemble
2019 [89]	Current and vibration	Current probe Fluke i200s and accelerometer PCB 352460	BRB, built-in bowed rotor, and faulted bearing	Hilbert transform, CNN, and LSTM
2021 [76]	Current and flux	Stray flux sensor and Hall-effect sensor	2 types of bearing faults: generalized roughness and single-point defect, and the simultaneous presence of torque oscillation and bearing fault	CWT and STFT
2022 [90]	Vibration and current	Accelerometer and data acquisition board	3 BRBs, rotor bent in the center, and inner race bearing defect in the shaft end	Hilbert transform, recurrent neural network, and lightweight multisensory fusion model
2022 [16]	Flux and current	Software	ITSC	Fourier transform and Runge–Kutta 4th method
2022 [91]	Giant magnetoresistance (GMR), temperature, current, voltage, and vibration	Current sensor, SCR013; voltage sensor, ZMPT101B; vibration sensor, ADXL335; GMR sensor, AA002-02; and temperature sensor, LM335	Shaft bend, rotor burn, BRB, 2 mm holes on the lamination sheet of the rotor	Polynomial regression, multiple linear regression, logistic regression, and polynomial chirplet transform

5. Future Trends

The evolution of fault detection techniques has brought interest to study methods based on less conventional types of physical variables. Even though current and vibration analyses remain the preferred types of signals for fault detection, other signals, such as flux and thermal images, have gained attention in recent years [76,84,85].

Works that propose the use of combined analysis, such as current and vibrations, are also very popular; usually, these take advantage of each physical variable and extract more information than using only one variable but with a more complex experimental array. These works commonly compare the results for various faults and detection methods.

Most of the works studied in this paper use AI-based methods, which reveal that no matter the physical variable employed for the analysis, these methods provide good results with accuracies superior than 90%.

In summary, current and vibration signals are commonly analyzed by their frequency spectrum using techniques, such as Fourier and wavelet transform, but their disadvantage is that both of them are prone to distortion, given the sensitivity of the sensor, for example, current signals by means of electrical noise and vibration signals by the vibrations or surrounding motors.

Some of the trends that have been observed during the elaboration of this work are listed below:

- Current and vibration signals remain the leading variable for fault detection and classification, mainly because these signals provide great information and can be analyzed alone or combined using straightforward methods. However, without preprocessing method, these signals rarely achieve the detection of incipient faults. Thus, developing processing techniques (or improving the existing ones) of low complexity and fast response are desirable, aiming to implement online fault detection systems.
- AI methods are extensively used for fault detection and can be used for any signal. Nevertheless, some types of variables rely on these methods, such as pattern classification for thermal images. In this regard, an interesting contribution may be the development of less complex algorithms, such as a pattern classifier based on memristive networks.
- Flux signals are presented in different works as a reliable physical variable for detecting faults. However, some authors have highlighted the lack of online monitoring. Concerning this matter, future works may be oriented to surpass this disadvantage.

It is expected that this work could provide useful information for researchers to prefer a certain type of physical variable in the study of new methods for fault detection.

Finally, Table 6 summarizes a comparison of the characteristics of the different physical variables treated in this work.

Table 6. Comparison of methods for fault detection based on different physical variables.

	Current	Vibrations	Thermal	Flux
Type of sensor	Hall-effect	Accelerometer	Infrared camera	Coil sensor
Straightforward sensing	Yes	No	Yes	No
Low-cost sensing	Yes	Yes	No	Yes
Popular methods for signal processing	FFT and wavelet transform	FFT, wavelet transform, and neural networks	AI-based method	Spectral analysis and Clarke transformation
Types of faults detected	BRB (1/2, 1, and 2), bearing faults (axis deviation, misalignment, and inner and outer race), unbalanced voltage, ITSC (4 levels), rotor faults (misalignment, friction, and end ring break), poor insulation, and stator faults (6 types)	BRB (1, 2, 3, and 4), bearing faults (inner and outer race, abrasion and balls), air-gap eccentricity, ITSC (2, 3, 4, and 8 turns) and rotor faults (unbalanced, bowed rotor, and bending)	ITSC (8 types), bearing faults (inner and outer race and ball), gearbox wearing, cooling fan failure, and stuck rotor fault	ITSC, BRB (1, 2, and 3), rotor bar damage, and static and dynamic eccentricity

6. Conclusions

Fault detection in IMs is a topic that has been studied using different physical variables of the motor, such as current, vibrations, flux, and temperature. Among them, current analysis remains the most popular one, mainly because of the non-invasive sensing and considering that the IMs have the current sensor already installed for control purposes [18,92–117]. A promising physical variable that has been employed for fault detection is flux analysis, which exhibits good results, even in cases where the MCSA fails; thus, it is expected that works based on flux analysis will grow in the next years.

24. Misra, S.; Kumar, S.; Sayyad, S.; Bongale, A.; Jadhav, P.; Kotecha, K.; Abraham, A.; Gabralla, L.A. Fault Detection in Induction Motor Using Time Domain and Spectral Imaging-Based Transfer Learning Approach on Vibration Data. *Sensors* **2022**, *22*, 8210. [CrossRef]

25. Sabah, A.N.; Jaffery, Z.A. Fault Detection of Induction Motor Using Thermal Imaging. In Proceedings of the 2022 IEEE IAS Global Conference on Emerging Technologies (GlobConET), Arad, Romania, 20–22 May 2022; pp. 84–90. [CrossRef]

26. Miljković, D. Brief Review of Motor Current Signature Analysis. *CrSNDT J.* **2015**, *5*, 14–26.

27. Asad, B.; Vaimann, T.; Belahcen, A.; Kallaste, A.; Rassõlkin, A.; Ghafarokhi, P.S.; Kudelina, K. Transient Modeling and Recovery of Non-Stationary Fault Signature for Condition Monitoring of Induction Motors. *Appl. Sci.* **2021**, *11*, 2806. [CrossRef]

28. de la Barrera, P.M.; Otero, M.; Schallschmidt, T.; Bossio, G.R.; Leidhold, R. Active Broken Rotor Bar Diagnosis in Induction Motor Drives. *IEEE Trans. Ind. Electron.* **2021**, *68*, 7556–7566. [CrossRef]

29. Fernandez-Cavero, V.; García-Escudero, L.A.; Pons-Llinares, J.; Fernández-Temprano, M.A.; Duque-Perez, O.; Morinigo-Sotelo, D. Diagnosis of Broken Rotor Bars during the Startup of Inverter-Fed Induction Motors Using the Dragon Transform and Functional ANOVA. *Appl. Sci.* **2021**, *11*, 3769. [CrossRef]

30. Fernandez-Cavero, V.; Pons-Llinares, J.; Duque-Perez, O.; Morinigo-Sotelo, D. Detection of Broken Rotor Bars in Nonlinear Startups of Inverter-Fed Induction Motors. *IEEE Trans. Ind. Appl.* **2021**, *57*, 2559–2568. [CrossRef]

31. Fernandez-Cavero, V.; Pons-Llinares, J.; Duque-Perez, O.; Morinigo-Sotelo, D. Detection and quantification of bar breakage harmonics evolutions in inverter-fed motors through the dragon transform. *ISA Trans.* **2021**, *109*, 352–367. [CrossRef]

32. Farag, K.; Shawier, A.; Abdel-Khalik, A.S.; Ahmed, M.M.; Ahmed, S. Applicability Analysis of Indices-Based Fault Detection Technique of Six-Phase Induction Motor. *Energies* **2021**, *14*, 5905. [CrossRef]

33. Ferrucho-Alvarez, E.R.; Martinez-Herrera, A.L.; Cabal-Yepez, E.; Rodriguez-Donate, C.; Lopez-Ramirez, M.; Mata-Chavez, R.I. Broken Rotor Bar Detection in Induction Motors through Contrast Estimation. *Sensors* **2021**, *21*, 7446. [CrossRef] [PubMed]

34. Fu, Q.; Guo, Q.; Zhang, W.; Cui, W.; Zhao, L. An analytical model for squirrel cage induction machine with broken rotor bars derived based on the multiple coupled circuit theory and the winding function approach. *Int. J. Circuit Theory Appl.* **2021**, *49*, 1633–1658. [CrossRef]

35. Jorkesh, S.; Poshtan, J. Fault diagnosis of an induction motor using data fusion based on neural networks. *IET Sci. Meas. Technol.* **2021**, *15*, 681–689. [CrossRef]

36. Park, C.H.; Kim, H.; Lee, J.; Ahn, G.; Youn, M.; Youn, B.D. A feature inherited hierarchical convolutional neural network (FI-HCNN) for motor fault severity estimation using stator current signals. *Int. J. Precis. Eng. Manuf.-Green Technol.* **2021**, *8*, 1253–1266. [CrossRef]

37. Yang, Z.; Kong, C.; Wang, Y.; Rong, X.; Wei, L. Fault diagnosis of mine asynchronous motor based on MEEMD energy entropy and ANN. *Comput. Electr. Eng.* **2021**, *92*, 107070. [CrossRef]

38. Romero-Troncoso, R.; Garcia-Perez, A.; Morinigo-Sotelo, D.; Duque-Perez, O.; Osornio-Rios, R.; Ibarra-Manzano, M. Rotor unbalance and broken rotor bar detection in inverter-fed induction motors at start-up and steady-state regimes by high-resolution spectral analysis. *Electr. Power Syst. Res.* **2016**, *133*, 142–148. [CrossRef]

39. Palácios, R.H.C.; da Silva, I.N.; Goedtel, A.; Godoy, W.F. A novel multi-agent approach to identify faults in line connected three-phase induction motors. *Appl. Soft Comput.* **2016**, *45*, 1–10. [CrossRef]

40. Yin, X.; Liu, X.; Sun, M.; Dong, J.; Zhang, G. Fuzzy Reasoning Numerical Spiking Neural P Systems for Induction Motor Fault Diagnosis. *Entropy* **2022**, *24*, 1385. [CrossRef]

41. Lizarraga-Morales, R.A.; Rodriguez-Donate, C.; Cabal-Yepez, E.; Lopez-Ramirez, M.; Ledesma-Carrillo, L.M.; Ferrucho-Alvarez, E.R. Novel FPGA-based Methodology for Early Broken Rotor Bar Detection and Classification through Homogeneity Estimation. *IEEE Trans. Instrum. Meas.* **2017**, *66*, 1760–1769. [CrossRef]

42. Moussa, M.A.; Boucherma, M.; Khezzar, A. A Detection Method for Induction Motor Bar Fault Using Sidelobes Leakage Phenomenon of the Sliding Discrete Fourier Transform. *IEEE Trans. Power Electron.* **2017**, *32*, 5560–5572. [CrossRef]

43. Vilhekar, T.G.; Ballal, M.S.; Suryawanshi, H.M. Application of Multiple Parks Vector Approach for Detection of Multiple Faults in Induction Motors. *J. Power Electron.* **2017**, *17*, 972–982.

44. Lu, J.; Wang, P.; Duan, S.; Shi, L.; Han, L. Detection of broken rotor bars fault in induction motors by using an improved MUSIC and least-squares amplitude estimation. *Math. Probl. Eng.* **2018**, *2018*, 5942890. [CrossRef]

45. Hsueh, Y.; Ittangihala, V.R.; Wu, W.B.; Chang, H.C.; Kuo, C.C. Condition Monitor System for Rotation Machine by CNN with Recurrence Plot. *Energies* **2019**, *12*, 3221. [CrossRef]

46. Mohamed, M.A.; Hassan, M.A.M.; Albalawi, F.; Ghoneim, S.S.M.; Ali, Z.M.; Dardeer, M. Diagnostic Modelling for Induction Motor Faults via ANFIS Algorithm and DWT-Based Feature Extraction. *Appl. Sci.* **2021**, *11*, 9115. [CrossRef]

47. Hussain, M.; Soother, D.K.; Kalwar, I.H.; Memon, T.D.; Memon, Z.A.; Nisar, K.; Chowdhry, B.S. Stator Winding Fault Detection and Classification in Three-Phase Induction Motor. *Intell. Autom. Soft Comput.* **2021**, *29*, 869–883. [CrossRef]

48. Lopes, T.D.; Raizer, A.; Valente Júnior, W. The Use of Digital Twins in Finite Element for the Study of Induction Motors Faults. *Sensors* **2021**, *21*, 7833. [CrossRef] [PubMed]

49. Marmouch, S.; Aroui, T.; Koubaa, Y. Statistical Neural Networks for Induction Machine Fault Diagnosis and Features Processing Based on Principal Component Analysis. *IEEJ Trans. Electr. Electron. Eng.* **2021**, *16*, 307–314. [CrossRef]

50. Dehina, W.; Boumehraz, M.; Kratz, F. Detectability of rotor failure for induction motors through stator current based on advanced signal processing approaches. *Int. J. Dyn. Control* **2021**, *9*, 1381–1395. [CrossRef]

51. Okwuosa, C.N.; Akpudo, U.E.; Hur, J.W. A Cost-Efficient MCSA-Based Fault Diagnostic Framework for SCIM at Low-Load Conditions. *Algorithms* **2022**, *15*, 212. [CrossRef]
52. Allal, A.; Khechekhouche, A. Diagnosis of induction motor faults using the motor current normalized residual harmonic analysis method. *Int. J. Electr. Power Energy Syst.* **2022**, *141*, 108219. [CrossRef]
53. Lee, C.Y.; Wen, M.S.; Zhuo, G.L.; Le, T.A. Application of ANN in Induction-Motor Fault-Detection System Established with MRA and CFFS. *Mathematics* **2022**, *10*, 2250. [CrossRef]
54. Ghazali, M.; Rahiman, W. Vibration Analysis for Machine Monitoring and Diagnosis: A Systematic Review. *Shock Vib.* **2021**, *2021*, 9469318. [CrossRef]
55. Sharma, A.; Jia, Z. Review on engine vibration fault analysis based on data mining. *J. Vibroeng.* **2021**, *23*, 1433–1445. [CrossRef]
56. Zhou, Z.; Sun, J.; Cai, W.; Liu, W. Test Investigation and Rule Analysis of Bearing Fault Diagnosis in Induction Motors. *Energies* **2023**, *16*, 699. [CrossRef]
57. Sun, W.; Shao, S.; Zhao, R.; Yan, R.; Zhang, X.; Chen, X. A sparse auto-encoder-based deep neural network approach for induction motor faults classification. *Measurement* **2016**, *89*, 171–178. [CrossRef]
58. Lee, J.H.; Pack, J.H.; Lee, I.S. Fault Diagnosis of Induction Motor Using Convolutional Neural Network. *Appl. Sci.* **2019**, *9*, 2950. [CrossRef]
59. Irgat, E.; Çinar, E.; Ünsal, A.; Yazıcı, A. An IoT-Based Monitoring System for Induction Motor Faults Utilizing Deep Learning Models. *J. Vib. Eng. Technol.* **2022** . [CrossRef]
60. Jiménez-Guarneros, M.; Grande-Barreto, J.; Rangel-Magdaleno, J.D.J. Multiclass incremental learning for fault diagnosis in induction motors using fine-tuning with a memory of exemplars and nearest centroid classifier. *Shock Vib.* **2021**, *2021*, 6627740. [CrossRef]
61. Dutta, S.; Basu, B.; Talukdar, F.A. Classification of Induction Motor Fault and Imbalance Based on Vibration Signal Using Single Antenna's Reactive Near Field. *IEEE Trans. Instrum. Meas.* **2021**, *70*, 1–9. [CrossRef]
62. Juan Carbajal-Hernández, J.; Sánchez-Fernández, L.P.; Hernández-Bautista, I.; de J. Medel-Juárez, J.; Sánchez-Pérez, L.A. Classification of unbalance and misalignment in induction motors using orbital analysis and associative memories. *Neurocomputing* **2016**, *175*, 838–850. [CrossRef]
63. Naha, A.; Thammayyabbabu, K.R.; Samanta, A.K.; Routray, A.; Deb, A.K. Mobile Application to Detect Induction Motor Faults. *IEEE Embed. Syst. Lett.* **2017**, *9*, 117–120. [CrossRef]
64. Chen, P.Y.; Chao, K.H.; Tseng, Y.C. A Motor Fault Diagnosis System Based on Cerebellar Model Articulation Controller. *IEEE Access* **2019**, *7*, 120326–120336. [CrossRef]
65. Ruqiang, Y.; Fei, S.; Mengjie, Z. Induction Motor Fault Diagnosis Based on Transfer Principal Component Analysis. *Chin. J. Electron.* **2021**, *30*, 18–25 [CrossRef]
66. Huang, L.; Shen, G.; Hu, N.; Chen, L.; Yang, Y. Coupled Electromagnetic-Dynamic Modeling and Bearing Fault Characteristics of Induction Motors considering Unbalanced Magnetic Pull. *Entropy* **2022**, *24*, 1386. [CrossRef]
67. Hosseinpour, Z.; Arefi, M.M.; Mozafari, N.; Luo, H.; Yin, S. An Ensemble-Based Fuzzy Rough Active Learning Approach for Broken Rotor Bar Detection in Nonstationary Environment. *IEEE Trans. Instrum. Meas.* **2022**, *71*, 2511808. [CrossRef]
68. Wang, Z.; Yang, J.; Li, H.; Zhen, D.; Gu, F.; Ball, A. Improved cyclostationary analysis method based on TKEO and its application on the faults diagnosis of induction motors. *ISA Trans.* **2022**, *128*, 513–530. [CrossRef] [PubMed]
69. Kumar, R.R.; Andriollo, M.; Cirrincione, G.; Cirrincione, M.; Tortella, A. A Comprehensive Review of Conventional and Intelligence-Based Approaches for the Fault Diagnosis and Condition Monitoring of Induction Motors. *Energies* **2022**, *15*, 8938. [CrossRef]
70. Mahami, A.; Rahmoune, C.; Bettahar, T.; Benazzouz, D. Induction motor condition monitoring using infrared thermography imaging and ensemble learning techniques. *Adv. Mech. Eng.* **2021**, *13*, 16878140211060956. [CrossRef]
71. Xu, Z.; Tang, G.; Pang, B. An Infrared Thermal Image Few-Shot Learning Method Based on CAPNet and Its Application to Induction Motor Fault Diagnosis. *IEEE Sens. J.* **2022**, *22*, 16440–16450. [CrossRef]
72. Ibrahim, A.; Anayi, F.; Packianather, M. New Transfer Learning Approach Based on a CNN for Fault Diagnosis. *Eng. Proc.* **2022**, *24*, 16. [CrossRef]
73. Alfredo Osornio-Rios, R.; Yosimar Jaen-Cuellar, A.; Ivan Alvarado-Hernandez, A.; Zamudio-Ramirez, I.; Armando Cruz-Albarran, I.; Alfonso Antonino-Daviu, J. Fault detection and classification in kinematic chains by means of PCA extraction-reduction of features from thermographic images. *Measurement* **2022**, *197*, 111340. . [CrossRef]
74. Mirzaeva, G.; Saad, K.I.; Jahromi, M.G. Comprehensive Diagnostics of Induction Motor Faults Based on Measurement of Space and Time Dependencies of Air Gap Flux. *IEEE Trans. Ind. Appl.* **2017**, *53*, 2657–2666. [CrossRef]
75. Park, Y.; Yang, C.; Kim, J.; Kim, H.; Lee, S.B.; Gyftakis, K.N.; Panagiotou, P.A.; Kia, S.H.; Capolino, G.A. Stray Flux Monitoring for Reliable Detection of Rotor Faults Under the Influence of Rotor Axial Air Ducts. *IEEE Trans. Ind. Electron.* **2019**, *66*, 7561–7570. [CrossRef]
76. Minervini, M.; Mognaschi, M.E.; Di Barba, P.; Frosini, L. Convolutional Neural Networks for Automated Rolling Bearing Diagnostics in Induction Motors Based on Electromagnetic Signals. *Appl. Sci.* **2021**, *11*, 7878. [CrossRef]
77. Rafaq, M.S.; Faizan Shaikh, M.; Park, Y.; Lee, S.B. Reliable Airgap Search Coil Based Detection of Induction Motor Rotor Faults Under False Negative Motor Current Signature Analysis Indications. *IEEE Trans. Ind. Inform.* **2022**, *18*, 3276–3285. [CrossRef]

78. Biot-Monterde, V.; Navarro-Navarro, A.; Zamudio-Ramirez, I.; Antonino-Daviu, J.A.; Osornio-Rios, R.A. Automatic Classification of Rotor Faults in Soft-Started Induction Motors, Based on Persistence Spectrum and Convolutional Neural Network Applied to Stray-Flux Signals. *Sensors* **2023**, *23*, 316. [CrossRef] [PubMed]

79. Navarro-Navarro, A.; Zamudio-Ramirez, I.; Biot-Monterde, V.; Osornio-Rios, R.A.; Antonino-Daviu, J.A. Current and Stray Flux Combined Analysis for the Automatic Detection of Rotor Faults in Soft-Started Induction Motors. *Energies* **2022**, *15*, 2511. [CrossRef]

80. Guerra de Araujo Cruz, A.; Delgado Gomes, R.; Antonio Belo, F.; Cavalcante Lima Filho, A. A Hybrid System Based on Fuzzy Logic to Failure Diagnosis in Induction Motors. *IEEE Lat. Am. Trans.* **2017**, *15*, 1480–1489. . [CrossRef]

81. Ali, M.Z.; Shabbir, M.N.S.K.; Liang, X.; Zhang, Y.; Hu, T. Machine Learning-Based Fault Diagnosis for Single- and Multi-Faults in Induction Motors Using Measured Stator Currents and Vibration Signals. *IEEE Trans. Ind. Appl.* **2019**, *55*, 2378–2391. [CrossRef]

82. Ibrahim, A.; Anayi, F.; Packianather, M.; Alomari, O.A. New Hybrid Invasive Weed Optimization and Machine Learning Approach for Fault Detection. *Energies* **2022**, *15*, 1488. [CrossRef]

83. Ameid, T.; Talhaoui, H.; Azzoug, Y.; Chebaani, M.; Laidoudi, A. Rotor fault detection using hybrid signal processing approach for sensorless Backstepping control driven induction motor at low-speed operation. *Int. Trans. Electr. Energy Syst.* **2021**, *31*, e13150. [CrossRef]

84. Bejaoui, I.; Bruneo, D.; Xibilia, M.G. Remaining Useful Life Prediction of Broken Rotor Bar Based on Data-Driven and Degradation Model. *Appl. Sci.* **2021**, *11*, 7175. [CrossRef]

85. Najafi, A.; Çavuş, B. Optimized Thermal Modeling and Broken Rotor Bar Fault Detection of BDFIG with Discrete Wavelet Transform. *Electr. Power Components Syst.* **2022**, *50*, 776–787. [CrossRef]

86. Goktas, T.; Arkan, M. Discerning broken rotor bar failure from low-frequency load torque oscillation in DTC induction motor drives. *Trans. Inst. Meas. Control* **2018**, *40*, 279–286. [CrossRef]

87. Breuneval, R.; Clerc, G.; Nahid-Mobarakeh, B.; Mansouri, B. Classification with automatic detection of unknown classes based on SVM and fuzzy MBF: Application to motor diagnosis. *AIMS Electron. Electr. Eng.* **2018**, *2*, 59–84. [CrossRef]

88. Contreras-Hernandez, J.L.; Almanza-Ojeda, D.L.; Ledesma-Orozco, S.; Garcia-Perez, A.; Romero-Troncoso, R.J.; Ibarra-Manzano, M.A. Quaternion Signal Analysis Algorithm for Induction Motor Fault Detection. *IEEE Trans. Ind. Electron.* **2019**, *66*, 8843–8850. [CrossRef]

89. Wang, J.; Fu, P.; Zhang, L.; Gao, R.X.; Zhao, R. Multilevel Information Fusion for Induction Motor Fault Diagnosis. *IEEE/ASME Trans. Mechatronics* **2019**, *24*, 2139–2150. [CrossRef]

90. Wang, J.; Fu, P.; Ji, S.; Li, Y.; Gao, R.X. A Light Weight Multisensory Fusion Model for Induction Motor Fault Diagnosis. *IEEE/ASME Trans. Mechatronics* **2022**, *27*, 4932–4941. [CrossRef]

91. Kumar, J.A.; Swaroopan, N.J.; Shanker, N. Prediction of Rotor Slot Size Variations in Induction Motor Using Polynomial Chirplet Transform and Regression Algorithms. *Arab. J. Sci. Eng.* **2023**, *48*, 6099–6109. [CrossRef]

92. Gholaminejad, A.; Jorkesh, S.; Poshtan, J. *A Comparative Case Study between Shallow and Deep Neural Networks in Induction Motor's Fault Diagnosis*; IET Science, Measurement & Technology: Hoboken, NJ, USA, 2023. . [CrossRef]

93. Jawadekar, A.; Paraskar, S.; Jadhav, S.; Dhole, G. Artificial neural network-based induction motor fault classifier using continuous wavelet transform. *Syst. Sci. Control Eng.* **2014**, *2*, 684–690.

94. Yang, C.; Kang, T.J.; Lee, S.B.; Yoo, J.Y.; Bellini, A.; Zarri, L.; Filippetti, F. Screening of False Induction Motor Fault Alarms Produced by Axial Air Ducts Based on the Space-Harmonic-Induced Current Components. *IEEE Trans. Ind. Electron.* **2015**, *62*, 1803–1813. [CrossRef]

95. Ferracuti, F.; Giantomassi, A.; Iarlori, S.; Ippoliti, G.; Longhi, S. Electric motor defects diagnosis based on kernel density estimation and Kullback–Leibler divergence in quality control scenario. *Eng. Appl. Artif. Intell.* **2015**, *44*, 25–32. [CrossRef]

96. Abu Elhaija, W.; Abu Al-Haija, Q. A novel dataset and lightweight detection system for broken bars induction motors using optimizable neural networks. *Intell. Syst. Appl.* **2023**, *17*, 200167. [CrossRef]

97. Rouabah, B.; Toubakh, H.; Kafi, M.R.; Sayed-Mouchaweh, M. Adaptive data-driven fault-tolerant control strategy for optimal power extraction in presence of broken rotor bars in wind turbine. *ISA Trans.* **2022**, *130*, 92–103. [CrossRef]

98. Sabir, H.; Ouassaid, M.; Ngote, N. An experimental method for diagnostic of incipient broken rotor bar fault in induction machines. *Heliyon* **2022**, *8*, e09136. [CrossRef] [PubMed]

99. Kumar, R.S.; Raj, I.G.C.; Suresh, K.P.; Leninpugalhanthi, P.; Suresh, M.; Panchal, H.; Meenakumari, R.; Sadasivuni, K.K. A method for broken bar fault diagnosis in three phase induction motor drive system using Artificial Neural Networks. *Int. J. Ambient Energy* **2022**, *43*, 5138–5144.

100. Talhaoui, H.; Ameid, T.; Aissa, O.; Kessal, A. Wavelet packet and fuzzy logic theory for automatic fault detection in induction motor. *Soft Comput.* **2022**, *26*, 11935–11949. [CrossRef] [PubMed]

101. Talhaoui, H.; Ameid, T.; Kessal, A. Energy eigenvalues and neural network analysis for broken bars fault diagnosis in induction machine under variable load: Experimental study. *J. Ambient Intell. Humaniz. Comput.* **2022**, *13*, 2651–2665. [CrossRef]

102. Trujillo Guajardo, L.A.; Platas Garza, M.A.; Rodríguez Maldonado, J.; González Vázquez, M.A.; Rodríguez Alfaro, L.H.; Salinas Salinas, F. Prony Method Estimation for Motor Current Signal Analysis Diagnostics in Rotor Cage Induction Motors. *Energies* **2022**, *15*, 3513. [CrossRef]

103. Wang, W.; Song, X.; Liu, G.; Chen, Q.; Zhao, W.; Zhu, H. Induction Motor Broken Rotor Bar Fault Diagnosis Based on Third-Order Energy Operator Demodulated Current Signal. *IEEE Trans. Energy Convers.* **2022**, *37*, 1052–1059. [CrossRef]

104. Lopez-Ramirez, M.; Rodriguez-Donate, C.; Ledesma-Carrillo, L.M.; Villalobos-Pina, F.J.; Munoz-Minjares, J.U.; Cabal-Yepez, E. Walsh–Hadamard Domain-Based Intelligent Online Fault Diagnosis of Broken Rotor Bars in Induction Motors. *IEEE Trans. Instrum. Meas.* **2022**, *71*, 3505511. [CrossRef]
105. Messaoudi, M.; Flah, A.; Alotaibi, A.A.; Althobaiti, A.; Sbita, L.; Ziad El-Bayeh, C. Diagnosis and Fault Detection of Rotor Bars in Squirrel Cage Induction Motors Using Combined Park's Vector and Extended Park's Vector Approaches. *Electronics* **2022**, *11*, 380. [CrossRef]
106. Senthil Kumar, R.; Gerald Christopher Raj, I.; Alhamrouni, I.; Saravanan, S.; Prabaharan, N.; Ishwarya, S.; Gokdag, M.; Salem, M. A combined HT and ANN based early broken bar fault diagnosis approach for IFOC fed induction motor drive. *Alex. Eng. J.* **2023**, *66*, 15–30. [CrossRef]
107. Valtierra-Rodriguez, M.; Rivera-Guillen, J.R.; De Santiago-Perez, J.J.; Perez-Soto, G.I.; Amezquita-Sanchez, J.P. Expert System Based on Autoencoders for Detection of Broken Rotor Bars in Induction Motors Employing Start-Up and Steady-State Regimes. *Machines* **2023**, *11*, 156. [CrossRef]
108. Wang, P.; Wang, K.; Chen, L. Broken Rotor Bars Detection in Inverter-fed Induction Motors Under Continuous Switching of Different Speed Modes. *IEEE Trans. Ind. Electron.* **2023**, 1–10. [CrossRef]
109. Yatsugi, K.; Pandarakone, S.E.; Mizuno, Y.; Nakamura, H. Common Diagnosis Approach to Three-Class Induction Motor Faults Using Stator Current Feature and Support Vector Machine. *IEEE Access* **2023**, *11*, 24945–24952. [CrossRef]
110. Zhu, H.; Jia, Z.; Song, X.; Sun, W. An approach to detect broken rotor bars based on instantaneous frequency of the fault characteristic harmonic during the start-up transient. *Int. J. Adv. Manuf. Technol.* **2023**, *124*, 4107–4119. [CrossRef]
111. Chehaidia, S.E.; Cherif, H.; Alraddadi, M.; Mosaad, M.I.; Bouchelaghem, A.M. Experimental Diagnosis of Broken Rotor Bar Faults in Induction Motors at Low Slip via Hilbert Envelope and Optimized Subtractive Clustering Adaptive Neuro-Fuzzy Inference System. *Energies* **2022**, *15*, 6746. [CrossRef]
112. Gu, B.G. Development of Broken Rotor Bar Fault Diagnosis Method with Sum of Weighted Fourier Series Coefficients Square. *Energies* **2022**, *15*, 8735. [CrossRef]
113. Jafari, H.; Poshtan, J.; Shamaghdari, S. Stochastic Event-Triggered Fault Detection and Isolation Based on Kalman Filter. *IEEE Trans. Cybern.* **2022**, *52*, 12329–12339. [CrossRef] [PubMed]
114. Jerkan, D.G.; Reljić, D.; Todorović, I.; Isakov, I.; Porobić, V.; Dujić, D. Detection of Broken Rotor Bars in a Cage Induction Machine Using DC Injection Braking. *IEEE Access* **2022**, *10*, 49585–49598. [CrossRef]
115. Jia, Z.; Wang, Y.; Zhu, H.; Song, X. A Novel Method for Diagnosis of Broken Bars Based on Optimum Resolution of Prescient Direction Algorithm. *Wirel. Commun. Mob. Comput.* **2022**, *2022*, 5772168. [CrossRef]
116. De Santiago-Perez, J.J.; Valtierra-Rodriguez, M.; Amezquita-Sanchez, J.P.; Perez-Soto, G.I.; Trejo-Hernandez, M.; Rivera-Guillen, J.R. Fourier-Based Adaptive Signal Decomposition Method Applied to Fault Detection in Induction Motors. *Machines* **2022**, *10*, 757. [CrossRef]
117. Agah, G.R.; Rahideh, A.; Khodadadzadeh, H.; Khoshnazar, S.M.; Hedayatikia, S. Broken Rotor Bar and Rotor Eccentricity Fault Detection in Induction Motors Using a Combination of Discrete Wavelet Transform and Teager–Kaiser Energy Operator. *IEEE Trans. Energy Convers.* **2022**, *37*, 2199–2206. [CrossRef]

Article

Fault Detection of Induction Motors with Combined Modeling- and Machine-Learning-Based Framework

Moritz Benninger [1,*], **Marcus Liebschner** [1] **and Christian Kreischer** [2]

1   Faculty of Electronics and Computer Science, University of Applied Sciences Aalen, 73430 Aalen, Germany; marcus.liebschner@hs-aalen.de
2   Chair for Electrical Machines and Drive Systems, Helmut Schmidt University, 22043 Hamburg, Germany; christian.kreischer@hsu-hh.de
*   Correspondence: moritz.benninger@hs-aalen.de

Abstract: This paper deals with the early detection of fault conditions in induction motors using a combined model- and machine-learning-based approach with flexible adaptation to individual motors. The method is based on analytical modeling in the form of a multiple coupled circuit model and a feedforward neural network. In addition, the differential evolution algorithm independently identifies the parameters of the motor for the multiple coupled circuit model based on easily obtained measurement data from a healthy state. With the identified parameters, the multiple coupled circuit model is used to perform dynamic simulations of the various fault cases of the specific induction motor. The simulation data set of the stator currents is used to train the neural network for classification of different stator, rotor, mechanical, and voltage supply faults. Finally, the combined method is successfully validated with measured data of faults in an induction motor, proving the transferability of the simulation-trained neural network to a real environment. Neglecting bearing faults, the fault cases from the validation data are classified with an accuracy of 94.81%.

Keywords: induction motors; fault detection; machine learning; supervised learning; multiple coupled circuit model; parameter identification

Citation: Benninger, M.; Liebschner, M.; Kreischer, C. Fault Detection of Induction Motors with Combined Modeling- and Machine-Learning-Based Framework. *Energies* 2023, 16, 3429. https://doi.org/10.3390/en16083429

Academic Editors: Daniel Morinigo-Sotelo, Joan Pons-Llinares and Rene Romero-Troncoso

Received: 23 March 2023
Revised: 6 April 2023
Accepted: 11 April 2023
Published: 13 April 2023

1. Introduction

Squirrel cage induction motors are often an essential part of industrial processes and are widely used in various industries due to their robust characteristics. Failures and repairs of individual machines or complete systems can quickly lead to high costs in the industrial environment, as well as a large demand for additional manpower and time. Continuous monitoring and early detection of electrical drive fault conditions offer a potential solution to such problems. These tools can help ensure reliable and predictable machine operation. For the above reasons, the early detection and diagnosis of fault conditions in induction motors and other types of electric drives have been a highly regarded research topic, as evidenced by various review publications [1–3]. According to Gao et al., fault diagnosis methods are generally classified into model-based, signal-based, and data-based approaches [4]: Model-based methods use models for fault diagnosis by monitoring the correlation between the real systems and the models. Signal-based methods utilize measured signals that reflect the fault cases. A diagnostic decision is made based on the extracted features and previous experience about the features in the healthy and faulty states. Data-based approaches use only existing data sets for fault detection. No prior knowledge is required.

Traditionally, signal-based approaches have played a major role in induction machine fault detection [5]. A classic method is the motor current signature analysis (MCSA), whose main objective is a high-resolution Fourier analysis of the stator currents in order to identify specific frequency components [6,7]. The stator currents are also considered in approaches using the Park vector, which transforms the three-phase current into a two-dimensional representation. When a fault condition occurs, the shape of the representation changes compared to the healthy state, allowing the fault condition to be detected [8,9]. In addition, more sophisticated signal-based methods exist, such as the discrete wavelet transform (DWT), which has the advantage of providing powerful frequency analysis of non-stationary signals [10–12]. Similar methods are the Wigner–Ville distribution (WVD) [13] or the Hilbert–Huang transform (HHT) [14]. These methods allow analysis in the combined time and frequency domain and the detection of specific fault characteristics in this domain. As in most of the approaches described so far, stator currents are primarily used as the signals to be evaluated [15]. Alternatively, vibration measurements are also used [16,17], with each quantity having its own advantages and disadvantages [18]. Other approaches use an additional sensor to measure the magnetic flux [19–21] or a thermography camera [22,23] for fault detection, but like the vibration sensor, this involves additional effort and cost. Current sensors, on the other hand, are usually present in all motors. In general, the signal-based approaches can be applied to all fault cases, such as short circuits in the stator [24,25], broken cage bars or end rings [26,27], eccentricities [17,28], or bearing faults [29,30]. The disadvantages of these signal-based methods are that they require precise prior knowledge of the specific fault characteristics, and the detection must often be manually adapted to the particular motor.

Due to major advances in the field of artificial intelligence (AI) and machine learning (ML), data-based approaches for induction machine fault diagnosis have been increasingly developed [31,32]. The advantage over signal-based methods is that now no specific prior knowledge of the fault characteristics and no manual analysis of the signals is required [33]. All that is required is the acquisition of a sufficient amount of data, and the corresponding algorithms learn the necessary fault characteristics independently. In the field of machine learning, there are different methods available, such as the support vector machine (SVM) [34–36], the k-nearest neighbors (kNN) algorithm [35,36], and different types of neural networks, such as regular feedforward neural networks (FFNN) [37–39], convolutional neural networks (CNN) [40,41], recurrent neural networks (RNN) [42,43], autoencoders (AE) [44,45], or deep belief networks (DBN) [45]. The publications listed so far deal with several fault cases, but individual approaches also focus on specific fault types, such as stator short circuits [46,47], broken rotor bars [48–50], or bearing faults [45,51,52]. A combination of signal-based and data-based methods is also common, with the signal-based approaches preprocessing the data to extract known fault features that the algorithms use for diagnosis [14,22,38,48]. However, the disadvantage of data-based methods is that enough data about the different healthy and faulty states must be available. In an industrial environment, detailed data acquisition for motors, especially for fault conditions, is problematic, making practical use impossible without a great deal of effort. Another possibility is to combine a model-based method with data-based methods. Such an approach is used by Murphey [53] and Masrur [54] to first generate data through modeling, which is then used to train an AI and classify faults in the voltage supply. The basic idea of such a method is advantageous because there is no need for expensive measurements of the motor in the different fault cases. Instead, appropriate modeling is used to generate data on the behavior of the motor in the healthy and fault states. However, a crucial step is missing to ensure practicability. This is because the parameters of the motor are also required to reproduce its behavior. Determining these parameters, whether by different test methods [55,56] or finite element analysis [57], is very time-consuming and thus hinders practical implementation.

In the approach presented here, a method for fault detection of squirrel cage induction motors is demonstrated, which identifies the parameters for modeling in advance from easily measurable quantities. The modeling is based on a multiple coupled circuit model whose parameters are identified using the differential evolution (DE) algorithm by comparing the simulation results with real measured data. Finally, the data set generated by the model is used to train a feedforward neural network for fault detection. The contribution of this paper aims at the practicable application of fault detection in the industry. By combining modeling and machine learning, the monitoring of an induction motor can be conducted with little prior knowledge, low effort, and already existing measurement technology. The structure of this paper leads through the different aspects of the method. In Section 2, the theoretical background of the modeling, the differential evolution algorithm, and neural networks are presented. Then, in Section 3, the interaction of the individual components within the overall framework is described. Section 4 shows the experimental setup in detail, leading to the validation and results in Section 5 and the conclusions in Section 6.

2. Theoretical Background

First, the technical background of the main aspects used in the approach is presented. This includes the modeling of the induction motor with squirrel cage rotor based on the multiple coupled circuit model together with the modified winding function method, which is used to calculate the self and mutual inductances. In addition to the healthy state modeling, the effects of the fault cases on the modeling are described. In addition, the mathematical background of the differential evolution algorithm and neural networks is briefly explained.

2.1. Induction Machine Modeling

A variety of model approaches exist for calculating the behavior of squirrel cage induction motors. Models based on a transformation into an arbitrary reference frame [58] do not allow the precise calculation of faults such as winding short circuits and are therefore not suited for fault detection. The finite element method (FEM), on the other hand, can be used to perform very complex simulations [59,60]. However, this approach is also not optimal for practical fault detection because of the high computational and time requirements. For comprehensive and flexible fault detection, the multiple coupled circuit model [61] is well-suited. This analytical modeling approach is based on the electrical network of the machine. With this type of modeling, the static and dynamic behavior as well as several fault types can be calculated [62]. The inductances for the modeling are usually estimated with the winding function method (WFM). This approach utilizes the distribution of the respective windings and geometrical quantities of the machine [63]. The modified winding function method (MWFM) is an extension of the original approach. Unlike the basic version, it is possible to calculate the inductances with variable air gap thicknesses [64].

2.1.1. Modeling Basic Machine

The theory of the modeling in this chapter originates from Toliyat et al. [61]. The central component of the model is the voltage equation with the corresponding resistances R, leakage inductances L, and inductances M. The voltages and currents of the three stator phases U_S and i_S and the rotor loops U_R and i_R are considered individually. Due to N_R cage bars in the rotor, N_R loop currents and an end ring current i_E exist for the rotor currents i_R (see Figure 1). The squirrel cage rotor causes zero values for the rotor voltages U_R [61]:

$$
\begin{bmatrix} [U_S] \\ [U_R] \end{bmatrix} = \begin{bmatrix} [R_S] & 0 \\ 0 & [R_R] \end{bmatrix} \begin{bmatrix} [i_S] \\ [i_R] \end{bmatrix} + \begin{bmatrix} [L_S] & 0 \\ 0 & [L_R] \end{bmatrix} \frac{d}{dt} \begin{bmatrix} [i_S] \\ [i_R] \end{bmatrix}
$$
$$
+ \frac{d}{dt} \left(\begin{bmatrix} [M_{SS}] & [M_{SR}] \\ [M_{RS}] & [M_{RR}] \end{bmatrix} \begin{bmatrix} [i_S] \\ [i_R] \end{bmatrix} \right)
\tag{1}
$$

$$
[U_S] = \begin{bmatrix} U_{S,a} \\ U_{S,b} \\ U_{S,c} \end{bmatrix} \quad [i_S] = \begin{bmatrix} i_{S,a} \\ i_{S,b} \\ i_{S,c} \end{bmatrix} \quad [U_R] = 0 \quad [i_R] = \begin{bmatrix} i_{R,1} \\ i_{R,2} \\ i_{R,3} \\ \cdots \\ i_{R,N_R} \\ i_E \end{bmatrix}
\tag{2}
$$

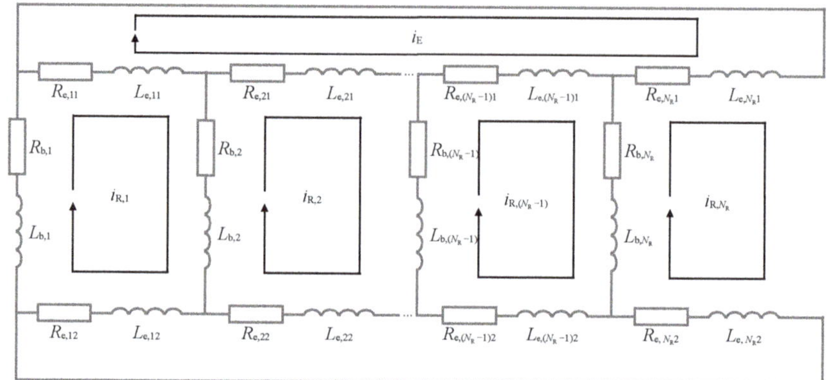

Figure 1. Electrical network of a rotor of a squirrel cage induction motor with the individual loop currents i_R, end ring current i_E, and associated resistances R_e and R_b, as well as leakage inductances L_e and L_b of cage bars and end ring segments.

The individual windings of the stator phases are summarized for the stator resistance R_S and leakage inductance L_S. The resistances R_b and leakage inductances L_b of the cage bars and the resistances R_e and leakage inductances L_e of the end ring segments build the matrices for the resistance R_R and leakage inductance L_R of the rotor (see Figure 1). The leakage inductance matrices L behave analogously to the resistance R and are therefore not shown separately below [61]:

$$[R_S] = \begin{bmatrix} R_S & 0 & 0 \\ 0 & R_S & 0 \\ 0 & 0 & R_S \end{bmatrix} \tag{3}$$

$$[R_R] = \begin{bmatrix} 2(R_b + R_e) & -R_b & 0 & \cdots & -R_b & -R_e \\ -R_b & 2(R_b + R_e) & -R_b & \cdots & 0 & -R_e \\ 0 & -R_b & 2(R_b + R_e) & \cdots & 0 & -R_e \\ \cdots & \cdots & \cdots & \cdots & \cdots & \cdots \\ -R_b & 0 & 0 & \cdots & 2(R_b + R_e) & -R_e \\ -R_e & -R_e & -R_e & \cdots & -R_e & N_R \cdot R_e \end{bmatrix} \tag{4}$$

For the self inductance of stator and rotor, the square matrices M_{SS} and M_{RR} are present. In these matrices the couplings of the stator phases to each other and rotor loops to each other are described. The only difference in the self inductance of the rotor M_{RR} is that no magnetic coupling with the end ring exists. The coupling between stator and rotor results in the mutual inductances M_{SR} and M_{RS}, which specify the relationship between the stator phases and rotor loops. These two matrices are mirror symmetric [61]:

$$[M_{SS}] = \begin{bmatrix} M_{Sa,Sa} & M_{Sa,Sb} & M_{Sa,Sc} \\ M_{Sb,Sa} & M_{Sb,Sb} & M_{Sb,Sc} \\ M_{Sc,Sa} & M_{Sc,Sb} & M_{Sc,Sc} \end{bmatrix} \tag{5}$$

$$[M_{RR}] = \begin{bmatrix} M_{R1,R1} & M_{R1,R2} & \cdots & M_{R1,RN_R} & 0 \\ M_{R2,R1} & M_{R2,R2} & \cdots & M_{R2,RN_R} & 0 \\ \cdots & \cdots & \cdots & \cdots & \cdots \\ M_{RN_R,R1} & M_{RN_R,R2} & \cdots & M_{RN_R,RN_R} & 0 \\ 0 & 0 & \cdots & 0 & 0 \end{bmatrix} \tag{6}$$

$$[M_{SR}] = \begin{bmatrix} M_{Sa,R1} & M_{Sa,R2} & \cdots & M_{Sa,RN_R} & 0 \\ M_{Sb,R1} & M_{Sb,R2} & \cdots & M_{Sb,RN_R} & 0 \\ M_{Sc,R1} & M_{Sc,R2} & \cdots & M_{Sc,RN_R} & 0 \end{bmatrix} \tag{7}$$

$$[M_{RS}] = [M_{SR}]^{\mathrm{T}} \tag{8}$$

In addition to the electrical description, the mechanical equation of motion is also important. Only in combination with the mechanical equation is it possible to calculate the rotor speed ω and the dynamic properties of the electrical machine. For this purpose, the moment of inertia J as well as the torque of the load machine T_L and the electrical machine T_{el} are required. The generated torque T_{el} results from the local derivation of the currents i and the inductances M [61]:

$$\frac{\mathrm{d}}{\mathrm{d}t}\omega = \frac{1}{J}(T_{el} + T_L) \tag{9}$$

$$T_{el} = \frac{1}{2}\begin{bmatrix}[i_S]^T & [i_R]^T\end{bmatrix}\frac{\partial}{\partial\varphi}\begin{bmatrix}[M_{SS}] & [M_{SR}] \\ [M_{RS}] & [M_{RR}]\end{bmatrix}\begin{bmatrix}[i_S] \\ [i_R]\end{bmatrix} \tag{10}$$

2.1.2. Inductance Calculation

The winding function method is used to calculate the self and mutual inductances M. This analytical method assumes an infinite permeability of iron and does not need any symmetry for the winding slots. Consequently, the coupling inductance $M_{A,B}$ between any two windings A and B in an electrical machine can be calculated according to the following equation [63]:

$$M_{A,B}(\varphi) = \mu_0 r l \int_0^{2\pi} n_A(\varphi,\theta) \cdot N_B(\varphi,\theta) \cdot g^{-1}(\varphi,\theta)\mathrm{d}\theta \tag{11}$$

This equation contains the turn function $n_A(\varphi,\theta)$, which describes the local distribution of the windings of A over the circumference θ. In addition, the winding function $N_B(\varphi,\theta)$ appears, which reflects the magnetomotive force of the windings of B. Other parameters used in the calculation are the machine length l, the stator core radius r, the air gap thickness g, and the vacuum permeability μ_0. Thus, the equation allows determining the inductances $M(\varphi)$ for the magnetic coupling between the individual stator phases and rotor loops as a function of the rotation angle φ.

In the presence of a variable air gap, as in the case of an eccentric rotor, the equation must be extended to the modified winding function method because the average of the winding function is no longer zero. Al-Nuaim and Toliyat [64] describe the necessary steps to account for a variable air gap in detail. Finally, only the calculation of the winding function changes, where $\overline{N_B}(\varphi,\theta)$ describes the average value of the winding function [64]:

$$N_B(\varphi,\theta)) = n_B(\varphi,\theta) - \overline{N_B}(\varphi,\theta) \tag{12}$$

$$\overline{N_B}(\varphi,\theta) = \frac{1}{2\pi \cdot \overline{g}^{-1}(\varphi,\theta)} \int_0^{2\pi} n_B(\varphi,\theta) \cdot g^{-1}(\varphi,\theta)\mathrm{d}\theta \tag{13}$$

2.1.3. Fault Implementation

Stator faults

Winding and phase short circuits create a new path in the electrical network. Therefore, the multiple coupled circuit modeling is extended to model these fault cases [62,65]. Figure 2 shows an example of a winding short circuit in the upper phase. In the path shown in red, the short-circuit current i_K flows across the resistor R_K, while the total voltage in the short-circuit path remains at zero.

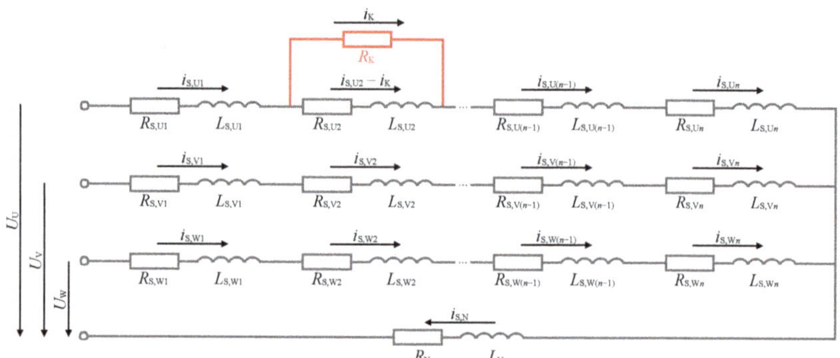

Figure 2. Electrical network of a stator of a squirrel cage induction motor with the phase currents i_S and the winding short-circuit path (red) with short-circuit current i_K and the resistor R_K.

The system of electrical equations must be extended by one line with newly introduced matrices, which describe the short-circuit path accordingly. The matrices of the resistance R and leakage inductance L from Equation (1) no longer resemble a unit matrix since different currents flow in the stator phases and in the short circuit. The short-circuit matrices with the resistances R_K, R_SK, and R_KS and the leakage inductances L_K, L_SK, and L_KS can be derived from the newly created network of the stator. The corresponding magnetic couplings of the short-circuit loop with the stator phases and rotor loops extend the inductance matrix M. In the case of the healthy machine or other fault cases without a short circuit, the newly introduced short-circuit matrices become zero so that the expression finally returns to the form of the initial Equation (1):

$$
\begin{bmatrix} [U_\text{S}] \\ [U_\text{R}] \\ 0 \end{bmatrix} = \begin{bmatrix} [R_\text{S}] & 0 & [R_\text{SK}] \\ 0 & [R_\text{R}] & 0 \\ [R_\text{KS}] & 0 & [R_\text{K}] \end{bmatrix} \begin{bmatrix} [i_\text{S}] \\ [i_\text{R}] \\ [i_\text{K}] \end{bmatrix} + \begin{bmatrix} [L_\text{S}] & 0 & [L_\text{SK}] \\ 0 & [L_\text{R}] & 0 \\ [L_\text{KS}] & 0 & [L_\text{K}] \end{bmatrix} \frac{\text{d}}{\text{d}t} \begin{bmatrix} [i_\text{S}] \\ [i_\text{R}] \\ [i_\text{K}] \end{bmatrix}
$$
$$
+ \frac{\text{d}}{\text{d}t} \left(\begin{bmatrix} [M_\text{SS}] & [M_\text{SR}] & [M_\text{SK}] \\ [M_\text{RS}] & [M_\text{RR}] & [M_\text{RK}] \\ [M_\text{KS}] & [M_\text{KR}] & [M_\text{KK}] \end{bmatrix} \begin{bmatrix} [i_\text{S}] \\ [i_\text{R}] \\ [i_\text{K}] \end{bmatrix} \right)
\tag{14}
$$

To model open phases in the stator, the matrices of the inductance M, leakage inductance L_S, and resistance R_S are adjusted according to the changed winding distribution. Since no current can flow in the affected phase due to the disconnection of the voltage, no windings of the open phase actively contribute to the electrical behavior.

Rotor faults

Broken bars or end rings in the squirrel cage rotor create a new structure of the electrical network in the rotor. For the rotor with a broken bar shown in Figure 3, the first and second rotor loops combine to form a single loop because current flow is no longer possible through the broken bar [62,65].

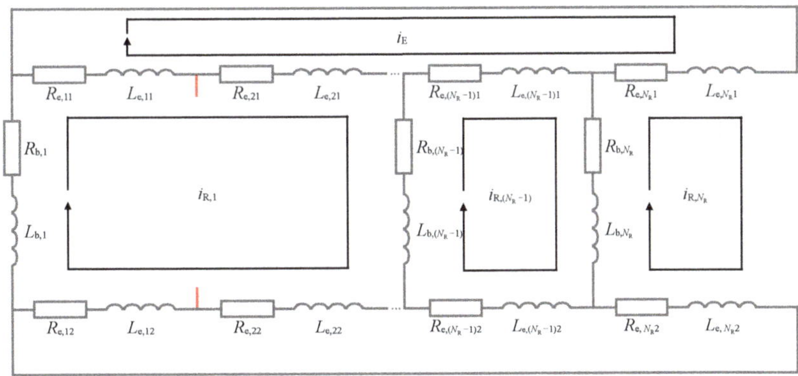

Figure 3. Electrical network of a rotor of a squirrel cage induction motor with the individual loop currents i_R in case of one broken cage bar.

For the multiple coupled circuit model, the elements of the first rotor loop now describe the newly created loop, while the elements of the second rotor loop become zero. This changes the matrix for the resistance R_R, while the adjustment of the matrix for the leakage inductance L_R behaves identically to the resistance matrix. Furthermore, the changed composition of the loops results in modified winding distributions for the calculation of the self and mutual inductances M. The first rotor loop combines a larger rotor section than before, while the second loop no longer has a winding available. The procedure is analogous for multiple broken cage bars or broken end rings:

$$[R_R] = \begin{bmatrix} 2(R_b + 2R_e) & 0 & -R_b & \cdots & -R_b & -2R_e \\ 0 & 0 & 0 & \cdots & 0 & 0 \\ -R_b & 0 & 2(R_b + R_e) & \cdots & 0 & -R_e \\ \cdots & \cdots & \cdots & \cdots & \cdots & \cdots \\ -R_b & 0 & 0 & \cdots & 2(R_b + R_e) & -R_e \\ -2R_e & 0 & -R_e & \cdots & -R_e & N_R \cdot R_e \end{bmatrix} \quad (15)$$

Mechanical faults

For the implementation of static, dynamic, and mixed eccentricity in the modeling, it is necessary to consider the air gap changes over the circumference of the machine. The variable air gap $g(\varphi, \theta)$ has a direct influence on the calculation of the inductances M via the modified winding function method. As an approximate description of the air gap $g(\varphi, \theta)$, the relationship from the following Equation (16) is used, which combines all three forms of eccentricity in one equation. For this purpose, the parameters δ_s and δ_d are introduced, which define the severity of the static and dynamic eccentricity [17]:

$$g(\varphi, \theta) = g \cdot [1 - \delta_s \cdot cos(\theta) - \delta_d \cdot cos(\omega t - \theta)] \quad (16)$$

Eccentricity is also used to model localized bearing faults. The difference is that the eccentricity occurs only when the air gap changes due to the passage of a defect in the bearing. To account for this moment, the model uses the typical frequencies of bearing faults (see Equation (17) as an example for outer ring fault), which is determined as a function of the bearing geometry (N_{bear}, b_d, d_p, β) and the rotor frequency f_R [29,30]:

$$f_o = \left(\frac{N_{bear}}{2} \right) \cdot f_R \cdot \left[1 - \frac{b_d}{d_p} cos(\beta) \right] \quad (17)$$

The modeling of global bearing faults with general roughness must be implemented differently due to its lack of predictability. The distributed roughness results in a slight increase of the load torque, which is modeled by additional added noise.

Voltage supply faults

Failures due to a faulty voltage supply directly affect the input to the model. Depending on the type of fault, the voltages are increased or decreased individually or collectively.

2.2. Differential Evolution Algorithm

The differential evolution algorithm offers an approach for coping with optimization problems. The algorithm uses a population of possible solutions (individuals) that are varied over several iterations with the goal of minimizing a defined fitness function [66]. The hyperparameters are the differential weight F, the crossover probability CR, and the population size NP. An important property of the algorithm is its capability to work with nonlinear and non-derivative problems. The advantage of a population-based approach over methods based on a single individual, such as cyclic coordinate search, lies in the low risk of getting stuck in local minima [67].

The individual steps and operations of the differential evolution algorithm can be seen in Algorithm 1. The mutation for designing new individuals is performed in step two, and the binary crossover between existing individuals and new designed individuals is performed in step four. The steps of the algorithm are executed for every individual in the population NP in every iteration of the optimization process. It is common to select a special mutation scheme such as DE/rand-to-best/1 [68], which has an effect on the design of the new individual z. In this case, the calculation utilizes the currently considered individual x, the best individual $best$, and two randomly chosen individuals a and b by the following equation [68]:

$$z = x + F \cdot (best - x) + F \cdot (a - b) \tag{18}$$

Algorithm 1 Procedure of the individual steps for one iteration of the differential evolution algorithm.

Find the best individual $best$ of the population.
For each individual x:

1. Choose the two random distinct individuals
   a and b.
2. Construct an interim design:
   $z = x + F \cdot (best - x) + F \cdot (a - b)$
3. Choose a random dimension $j \in [1, \dots, n]$
   for optimization in n dimensions.
4. Construct the candidate individual x' using
   binary crossover.

$$x'_i = \begin{cases} z_i, & \text{if } i = j \text{ or with probability } CR \\ x_i, & \text{otherwise} \end{cases}$$

5. Evaluate x' with fitness function.
6. Insert the better design between x and x'
   into the next generation.

2.3. Artificial Neural Network

An artificial neural network (ANN) is a computational model inspired by the structure and function of biological neural networks in the brain [69]. It consists of interconnected neurons that process and transmit information in the form of numerical values. The connections between neurons are modeled by weights learned during training that adjust the strength of the signal transmitted from one neuron to another [70]. In general, a feedforward neural network consists of three types of layers: an input layer, one or more hidden layers, and an output layer. The input layer receives the raw input data and passes it to the first hidden layer. The output y of a single neuron in the hidden layers of an ANN is calculated as the sum of its inputs x multiplied with a weight w plus a bias term b, which is then passed through an activation function $f()$ (e.g., sigmoid or ReLU function) to produce the neuron's output value [71]:

$$y = f\left(\sum w_i \cdot x_i + b\right) \tag{19}$$

The number of neurons in the input and output layers depends on the task at hand, while the number of hidden layers and the number of neurons in each hidden layer can vary greatly depending on the complexity of the problem and the amount of data available [71].

Artificial neural networks are often trained using a supervised learning process in which the network is presented with a set of input data with corresponding output values. The network adjusts its weights to minimize the difference between the predicted output and the actual output. In this backpropagation process, the gradient of the error function with respect to the weights is calculated and updated using gradient descent [69].

During training, an ANN can be prone to overfitting. This means that the network adapts too much to the training data instead of learning generalizable patterns [72]. To prevent such overfitting and to improve the generalization of the network, regularization techniques, such as L1 and L2 regularization, are used. L1 regularization adds a penalty value to the loss function that is proportional to the absolute value of the weights, while L2 regularization adds a penalty value that is proportional to the square of the weights. These penalty terms encourage the network to use smaller weights and reduce the complexity of the model, thus preventing overfitting [73]. Another commonly used regularization technique is the dropout procedure, which randomly drops a portion of the neurons in the network during each training epoch. This forces the network to learn robust features and prevents overfitting by reducing the co-adaptation of neurons [74].

3. Fault Detection Framework

The key component of the fault detection framework is the multiple coupled circuit model in combination with the modified winding function method, which is used to calculate the inductances. The modeling is primarily used to simulate the stator currents that are later used by the neural network as the basis for fault detection. The inputs to the model are the voltages U_S applied to the motor and the torque of the load machine T_L. The outputs are the rotor speed ω and the aforementioned stator currents i_S. An overview of the modeling inputs and outputs is shown in Figure 4.

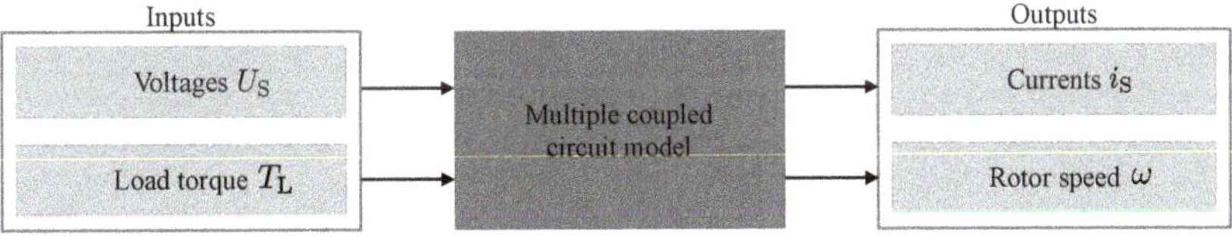

Figure 4. Structure of multiple coupled circuit modeling with inputs (voltages U_S and load torque T_L) and outputs (stator currents i_S and rotor speed ω).

The fault detection framework consists of several steps (see Figure 5), which are presented in the following. First, the model parameters for the multiple coupled circuit model are identified by the differential evolution algorithm. The model with the identified parameters is then used to create a data set of healthy and faulty states. This data set, in turn, allows the learning of fault characteristics by a neural network, which ultimately enables fault detection based on real stator currents.

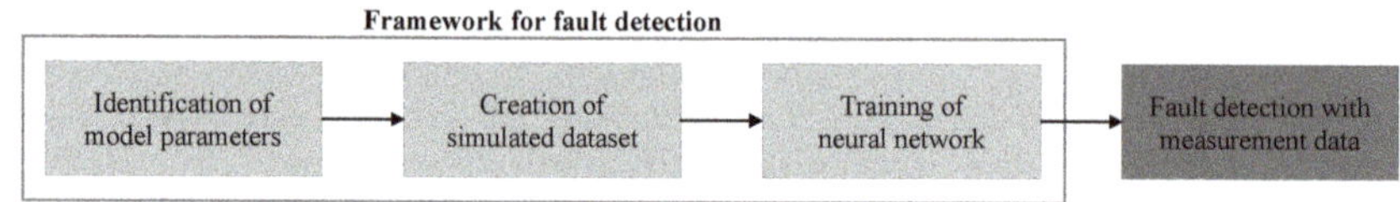

Figure 5. Sequence of the individual steps of the fault detection framework up to the application with real measurement data.

3.1. Parameter Identification

An important condition for the identification of model parameters for use in fault detection is high practicability. Therefore, the basic idea of parameter identification is to use only easily obtainable measurement data and information from the nameplate of the motor (see Figure 6).

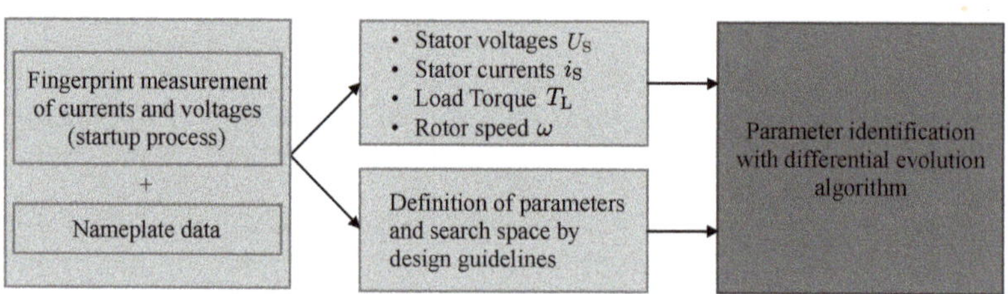

Figure 6. Sequence of the individual steps of the parameter identification from the fingerprint measurement (startup process from standstill to rated operation) and nameplate data to the identification process.

The parameters of the modeling are adjusted over a large number of iterations until the simulated results of the modeling match the real measured data as closely as possible. Consequently, measured data for the inputs and outputs of the modeling are needed to compare the model to reality. For this purpose, a time-based measurement of the stator currents i_S and the applied voltages U_S for the startup process of the motor from standstill to rated load in the healthy state is performed, which serves as a fingerprint for the motor under investigation. From the stator currents i_S, the rotor speed ω is determined by frequency analysis. This is performed indirectly by calculating the slip s from the frequency of the principal slot harmonics (PSH) with the number of rotor bars N_R, the number of pole pairs p, and the supply frequency f_S [28]:

$$f_{PSH} = \left(1 \pm k \cdot N_R \cdot \frac{1-s}{p} \right) \cdot f_s$$

$$\text{with } k = 0, 1, 2, \ldots$$

(20)

Using the rotor speed ω, the currents i_S, and voltages U_S, the load torque T_L is estimated via the efficiency η from the nameplate using the following equation (valid for rated load):

$$T_L = \frac{3 \cdot U_S \cdot i_S}{\eta \cdot \omega}$$

(21)

Thus, all modeling inputs and outputs are known. The parameters required for the modeling are the resistances R_S, R_b, and R_e and the leakage inductances L_S, L_b, and L_e of the stator phases, cage bars, and end ring segments. In addition, the length of the motor l, the radius of the stator core r, the air gap thickness g, the number of bars in the rotor N_R, and the number of windings in the stator w_S with the corresponding winding distribution are required for the calculation of the self and mutual inductances. Upper and lower limits for the search space are defined for these parameters. For this purpose, a rough estimation of the parameters is made based on design guidelines for the respective power class of the motor.

The differential evolution algorithm with the mutation scheme DE/rand-to-best/1 from Algorithm 1 is used to iteratively determine the parameters. The algorithm is not based on the derivative of a function, but instead uses a fitness function. This is an important aspect because the modeling is not differentiable due to the discrete calculation of the inductances. A two-part approach is used for the fitness function, which utilizes the data after the machine achieved continuous operation. First, whether the deviation between the simulated rotor speed ω_{sim} and measured rotor speed ω_{meas} is less than 1% is checked. If this condition is met, only the mean squared error (MSE) between the fast Fourier transforms (FFTs) of the simulated stator currents i_{simFFT} and measured stator currents i_{measFFT} is used for the fitness function; otherwise the mean squared error is multiplied by a penalty term:

$$
\text{fitness} = \begin{cases} MSE(i_{\text{measFFT}}, i_{\text{simFFT}}) & \text{if } \left| \frac{\omega_{\text{meas}} - \omega_{\text{sim}}}{\omega_{\text{meas}}} \right| \leq 0.01 \\ MSE(i_{\text{measFFT}}, i_{\text{simFFT}}) \cdot 10^{10} & \text{if } \left| \frac{\omega_{\text{meas}} - \omega_{\text{sim}}}{\omega_{\text{meas}}} \right| > 0.01 \end{cases} \tag{22}
$$

Comparing the currents at the level of the frequency spectra has the advantage of reflecting specific characteristics of the motor that can be used later for fault detection. Although the rotor speed information is also included in the frequency spectrum, an additional check is made because this component has a comparatively small value.

3.2. Creation of the Data Set

The identified parameters allow the simulation of the motor behavior in a healthy state. To generate data for fault conditions, the modeling is adapted according to the explanations in Section 2.1.3. Depending on the fault case, it is possible to simulate different fault severities. For example, the number of short-circuited windings or the number of broken bars can be varied, and the magnitude of the deviations from the normal value can be varied for the eccentricity and the faulty voltage supplies. With the voltages and the load torque, changing the input variables to the model is also a way to generate more various states and a larger data set.

3.3. Training of the Neural Network

A data set of healthy and faulty states enables the training of a neural network for fault classification. The fault detection is based on the fast Fourier transform of all three stator currents. This has the advantage that the characteristics of the faults are clearly visible in the frequency domain and that a time offset between individual samples is not significant. The frequency spectra are limited to a range of 0 to 1000 Hz, which contain the critical information. In addition, a subtraction with the average of the simulated healthy stator currents is formed for the entire data set (see Figure 7). With this type of normalization, the deviations from the healthy state are learned more sensitively.

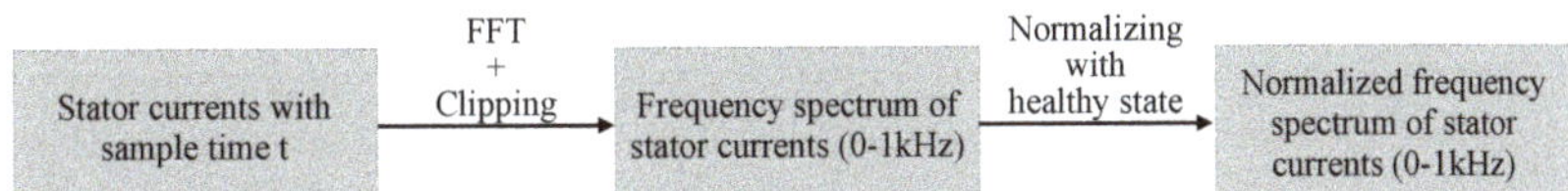

Figure 7. Sequence of the individual steps of the preprocessing with fast Fourier transformation, clipping, and normalization via the subtraction with the healthy state.

An important aspect of the training process is to ensure the best possible generalization. Despite the individually identified parameters, an ideal representation of the motor, especially in the fault cases, cannot be guaranteed. The neural network is considered to be well generalized if the loss for the test data is lower than for the training data. Thus, the goal in training the neural network is to minimize the loss function while maintaining a lower loss for the test data than for the training data. As an additional aspect of generalization, the data set is split into 50% training data and 50% test data, which allows the accuracy and loss results to be compared using the same sample size.

The structure of the feedforward neural network for fault detection consists of an input layer, several hidden layers, and an output layer. The number of neurons of the input layer corresponds to the number of data points of the frequency spectra of the three stator currents (depending on the sampling rate), and the number of neurons of the output layer corresponds to the number of healthy and faulty states. The ReLU activation function is used for the hidden layers, while the softmax function is used for the output layer to generate probabilities. The Adam algorithm is used as an optimizer. In addition, for the purpose of generalization, different regularization techniques are implemented with the L1 and L2 regularization and the dropout technique.

4. Experimental Setup

To prove the functionality, the method is run once completely for an exemplary motor. The motor under investigation is a squirrel cage induction motor with two pole pairs and 1.1 kW of power. The motor is coupled to a controllable load machine and is directly connected to the mains (230 V) in a delta connection. The nameplate data are given in Table 1. Current transformers are used to measure the stator currents. The outputs of the current transformers and the voltages are connected to an analog-to-digital converter, which acquires the analog signals at a sampling rate of 10 kHz.

Table 1. Nameplate data of the examined 1.1 kW squirrel cage induction motor for a delta connection.

Parameter	Value
Rated power P_N	1.1 kW
Rated voltage U_N	400 V
Rated current I_N	2.5 A
Rated rotor speed n_N	1445 1/min
Frequency f	50 Hz
Power factor $cos\varphi$	0.75
Efficiency η	84.4%

To identify the model parameters, the stator currents and applied voltages are measured for 10 s for the startup process from standstill to rated operation in the healthy state. This measurement serves as fingerprint and is used to compare the model with the real motor.

To verify the accuracy of the neural network, additional measurements of the motor in different fault conditions are required. These measurements are for validation purposes only and are used to verify the detection capability of the neural network. The nine different fault conditions from Table 2 are applied to the motor. A total of 3 measurements of the stator currents, each lasting 10 s, are taken in rated operation for the healthy state and each fault state. Splitting the measurements into 0.2 s intervals results in 150 samples per fault case for validation. The measured data in this step are not used for training.

The insertion of each fault into the motor is very different: For the undervoltage and unbalance faults, a variable transformer is used to regulate the input voltages to the motor. For the open phase, the motor is disconnected from one of the three phases during operation. The stator winding is short-circuited by stripping the insulation from two adjacent windings and pressing the resulting contacts directly against each other. For the broken rotor bar, a hole is drilled in the cage bar. Bearing faults in the outer and inner ring are caused by laser cutting. In the case of a global bearing fault, grit is inserted into the bearing (grit size: 0.05 mm, amount: 0.25 g) , which is equivalent to heavy contamination or poor lubrication.

Table 2. List of inserted fault cases on the examined induction motor with 150 samples each with a length of 0.2 s for the validation of the neural network for fault detection.

Motor State	Number of Samples
Healthy state	150
Undervoltage	150
Unsymmetrical Voltage	150
Open phase	150
Broken bar	150
Winding short circuit	150
Mixed eccentricity	150
Bearing—Outer ring fault	150
Bearing—Innen ring fault	150
Bearing—Global fault	150

5. Experimental Results

5.1. Parameter Identification

The fingerprint measurement of the voltages and stator currents is used as the basis for the parameter identification. The speed of the rotor ($\omega = 153.12$ rad/s) is determined from the frequency spectrum of the stator current, and the load torque ($T_L = 7.13$ Nm) is estimated using the stator currents i_S and the voltages U_S together with the efficiency η from the nameplate. The measured values for the inputs and outputs of the model are thus fully available. Of the required parameters, the length of the motor ($l = 0.11$ m) and the radius of the stator core ($r = 0.04$ m) are already known. The number of cage bars ($N_R = 28$) is also determined from the frequency spectrum of the stator currents. For the winding distribution in the stator, a single-layer winding with a total of 36 slots (3 slots per phase and pole pair) is assumed. If this assumption is wrong, it is compensated by adjusting the remaining parameters accordingly. The required parameters are shown in Table 3 together with the estimated search space.

Table 3. Required parameters for the modeling of the examined induction motor with lower and upper limit of the search space (source of limits: power of considered machine) as well as the value identified via the differential evolution algorithm.

Parameter	Lower Limit	Value	Upper Limit	Unit
Air gap thickness g	10^{-4}	1.68×10^{-4}	10^{-2}	m
Number of stator windings per slot w_S	10^1	48	10^3	-
Moment of inertia J	10^{-3}	6.27×10^{-3}	10^{-1}	kg m^2
Stator Resistance R_S	10^0	5.25	10^2	Ω
Stator leakage inductance L_S	10^{-2}	1.29×10^{-1}	10^0	H
Cage bar resistance R_b	10^{-5}	3.27×10^{-5}	10^{-3}	Ω
Cage bar leakage inductance L_b	10^{-8}	9.44×10^{-7}	10^{-6}	H
End ring segment resistance R_e	10^{-5}	2.90×10^{-5}	10^{-3}	Ω
End ring segment leakage inductance L_e	10^{-9}	3.45×10^{-9}	10^{-7}	H

For the hyperparameters of the differential evolution algorithm, default values are chosen so that the difference weight is $F = 0.95$, and the crossover probability is $CR = 0.7$. The population size NP and the number of iterations are both set to 50. The parameter values identified with the algorithm are listed in Table 3. The simulated stator currents with the parameterized model are compared to the real stator currents at the frequency level in Figure 8. The agreement in the basic appearance is very high; yet certain deviations can be seen. These primarily affect the harmonics of the fundamental frequency. Especially the third harmonic, which is influenced by the saturation behavior of the motor, cannot be reproduced with the linear multiple coupled circuit modeling.

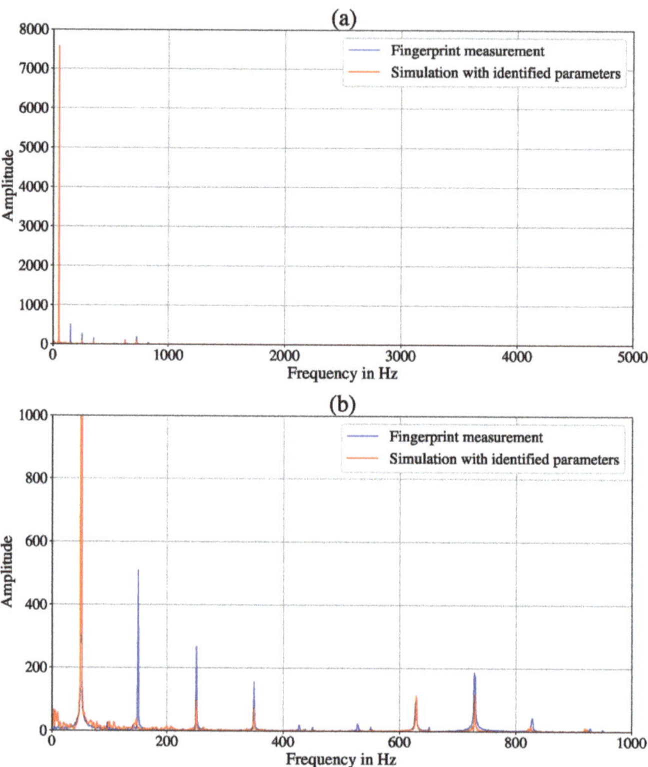

Figure 8. Comparison of the frequency spectra for one of the three stator currents from the fingerprint measurement and the simulation with the parameterized model: (**a**) complete frequency spectrum; (**b**) zoom of the frequency spectrum into the significant range (0 to 1000 Hz).

5.2. Fault Detection

Using the parameterized modeling, 300 samples with a length of 0.2 s are generated for each of the 10 motor states shown in Table 2. Different fault severities are simulated depending on the fault case. In addition, the input voltage to the modeling is varied randomly in the range of ±4 V of the measured voltage. In the next step, data preprocessing is performed. The stator currents are transformed into the frequency domain by a fast Fourier transform and then clipped to the range between 0 and 1000 Hz. In the final step, normalization is performed by subtracting the average of the simulated healthy states from the entire data set.

The preprocessed, simulated data set with 300 samples per state is then used to train a neural network with the highest possible generalization. The inputs to the neural network are the preprocessed frequency spectra of the three stator currents (see Figure 7), and the outputs are the probabilities for each of the ten fault types. The goal of the training process is to keep the loss of the test data lower than the loss of the training data. This is accomplished by manually tuning the hyperparameters of the neural network. The final hyperparameters are summarized in Table 4. For the training data (loss = 1.1458, accuracy = 92.7%) and the test data (loss = 1.1455, accuracy = 93.3%), the confusion matrices are obtained from Figure 9.

Table 4. Tuned hyperparameters of the neural network with high generalization.

Hyperparameter	Values
Hidden layers	3
Number of neurons	[100, 50, 25]
Learning rate	0.001
Dropout	0.25
L1 regularization	0
L2 regularization	0.2
Batch size	32
Epochs	150

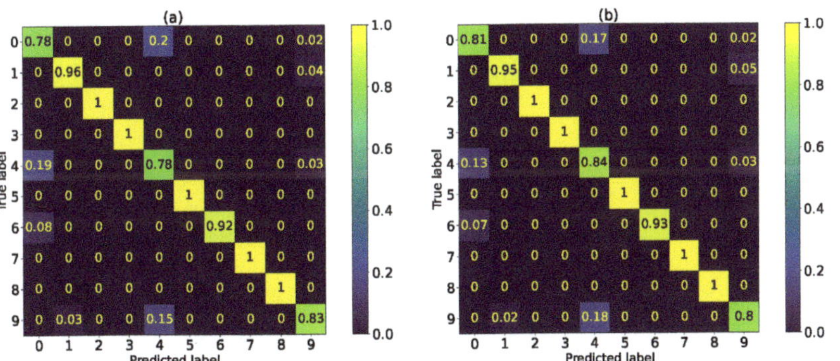

Figure 9. Confusion matrix with results of accuracy for each fault case for the training data (**a**) and test data (**b**). 0: healthy state; 1: undervoltage; 2: unsymmetrical voltage; 3: open phase; 4: broken bar; 5: winding short circuit; 6: mixed eccentricity; 7: bearing—outer ring fault; 8: bearing—inner ring fault; 9: bearing—global fault.

It can be seen that most faults can be classified with an accuracy of almost 100%. However, the neural network has difficulty in distinguishing between the cases of healthy state, broken bar, mixed eccentricity, and global bearing fault. This is due to their high similarity and the small deviation of these faults from the healthy state. In addition, the characteristics of these fault cases are highly dependent on the values of the identified parameters, such as the air gap thickness, the moment of inertia and the electrical and magnetic quantities. To overcome this problem, the parameter identification and data set creation are performed several times. This has the advantage that different sets of parameters are identified, preventing the neural network from focusing on the characteristics of a single set of parameter values. In addition, the repeated parameter identification multiplies the overall size of the data set. The neural network is trained again with a data set simulated on the basis of 10 different parameter sets (3000 samples per state in total), overcoming the previous problems. The corresponding confusion matrices in Figure 10 show that all fault cases can now be classified with an accuracy of over 90% (overall accuracy for training data: 96.9%, for test data: 96.6%).

To verify the trained neural network, the same preprocessing is applied to the measured validation data. The only difference is that the normalization of the data set is now accomplished by a subtraction based on the fingerprint measurement. When the neural network is applied to the validation data, the results turn out differently depending on the fault case as can be seen on the left confusion matrix in Figure 11. It is not possible to detect the three types of bearing faults. They are classified as healthy state. This is due to the fact that the measured deviations of the bearing faults from the healthy state are marginal, and furthermore, the fault features cannot be suitably represented by the modeling. The classification of the remaining fault cases works very well with an overall detection rate of 97.14%. Training the neural network without the bearing faults in the data set also produces a very high accuracy with 94.81% on the validation data (see right confusion matrix in Figure 11).

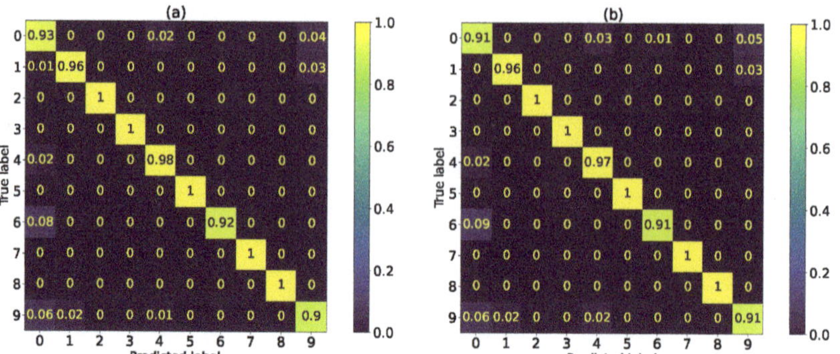

Figure 10. Confusion matrix with results of accuracy for each fault case for the training data (**a**) and test data (**b**) with 10 different parameter sets. 0: healthy state; 1: undervoltage; 2: unsymmetrical voltage; 3: open phase; 4: broken bar; 5: winding short circuit; 6: mixed eccentricity; 7: bearing—outer ring fault; 8: bearing—inner ring fault; 9: bearing—global fault.

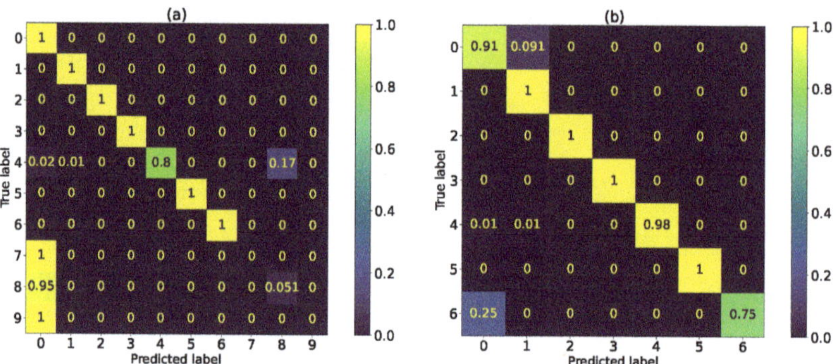

Figure 11. Confusion matrix with results of accuracy for each fault case for the validation data (**a**) and validation data without bearing faults (training data also without bearing faults) (**b**). 0: Healthy state; 1: undervoltage; 2: unsymmetrical voltage; 3: open phase; 4: broken bar; 5: winding short circuit; 6: mixed eccentricity; 7: bearing—outer ring fault; 8: bearing—inner ring fault; 9: bearing—global fault.

6. Conclusions

The presented framework enables early detection of fault conditions in squirrel cage induction motors while providing a high degree of practicability. The contribution of this paper is a fault detection method for industrial applications with little prior knowledge of the motor and low measurement effort. By combining analytical modeling with parameter identification based on easily obtained data, the behavior of the monitored motor can be well reproduced. The data set simulated by the modeling enables a neural network to learn the characteristics of stator, rotor, mechanical, and voltage supply faults and to detect them in real measured data. This demonstrates that the transfer of the simulated fault characteristics to real fault cases is possible with the help of machine learning. A drawback is that bearing faults are not detected. Furthermore, the severity of the faults cannot be determined since only the major qualitative deviations have been examined so far.

The presented method combines the strengths of different approaches and mitigates their disadvantages. The prior knowledge about the effects of the fault cases is already included in the modeling and can be applied to the respective motor by means of the parameter identification. Thus, no costly measurements are required to train a neural network, but only the simulation of different fault cases to generate a sufficiently large data set. Possible inaccuracies of the modeling are concealed by the neural network by learning the qualitative characteristics of the fault cases. Therefore, an exact quantitative accuracy of the model is not necessary. These aspects clearly distinguish the presented method from pure model-, signal- or data-based approaches. Another unique point is the high practicability of the framework since the parameter identification with the differential evolution algorithm can be performed based on easily obtained measurement data and information from the nameplate.

Furthermore, the method offers a high degree of flexibility. On the one hand, this applies to parameter identification, where the desired parameters can be selected depending on the application. On the other hand, the modeling itself is also flexible so that for other machine types, such as doubly-fed induction generators or synchronous motors with permanent magnets, the model can be adapted accordingly, and the method can still be carried out. Thus, the presented approach offers high transferability to different motor types and applications.

Further work will apply more sophisticated machine learning methods to improve the detection accuracy. This should also strengthen the robustness and generalization for the transferability of the fault characteristics from the simulated data to real data. Additionally, a method for the independent detection of bearing faults based on acoustic or vibration data will be developed. In combination with the presented framework, this should cover the detection of several possible fault cases for squirrel cage induction motors.

Author Contributions: Conceptualization, M.B., C.K. and M.L.; methodology, M.B. and C.K.; software, M.B.; validation, M.B.; formal analysis, M.B.; investigation, M.B.; resources, C.K. and M.L.; data curation, M.B.; writing—original draft preparation, M.B.; writing—review and editing, C.K. and M.L.; visualization, M.B.; supervision, C.K. and M.L.; project administration, C.K. and M.L.; funding acquisition, M.L. All authors have read and agreed to the published version of the manuscript.

Funding: This research was funded by the Federal Ministry for Economic Affairs and Climate Action, grant number 16KN088835.

Data Availability Statement: Not applicable.

Conflicts of Interest: The authors declare no conflict of interest.

Abbreviations

The following abbreviations are used in this manuscript:

AE	Autoencoder
AI	Artificial Intelligence
ANN	Artificial neural network
CNN	Convolutional neural network
DBN	Deep belief network
DE	Differential evolution algorithm
DWT	Discrete wavelet transform
FEM	Finite element method
FFNN	Feedforward neural network
FFT	Fast Fourier transform
HHT	Hilbert–Huang transform
kNN	k-Nearest Neighbors
MCSA	Motor current signature analysis
ML	Machine learning
MSE	Mean squared error
MWFM	Modified winding function method

PSH	Principal slot harmonics
RNN	Recurrent Neural Network
SVM	Support vector machine
WFM	Winding function method
WVD	Wigner-Ville distribution

References

1. Benbouzid, M.E.H. A review of induction motors signature analysis as a medium for faults detection. *IEEE Trans. Ind. Electron.* **2000**, *47*, 984–993. [CrossRef]
2. Bellini, A.; Filippetti, F.; Tassoni, C.; Capolino, G.-A. Advances in Diagnostic Techniques for Induction Machines. *IEEE Trans. Ind. Electron.* **2008**, *55*, 142–149. [CrossRef]
3. Zhang, P.; Du, Y.; Habetler, T.G.; Lu, B. A Survey of Condition Monitoring and Protection Methods for Medium-Voltage Induction Motors. *IEEE Trans. Ind. Appl.* **2011**, *55*, 34–46. [CrossRef]
4. Gao, Z.; Cecati, C.; Ding, S.X. A Survey of Fault Diagnosis and Fault-Tolerant Techniques—Part I: Fault Diagnosis With Model-Based and Signal-Based Approaches. *IEEE Trans. Ind. Electron.* **2015**, *62*, 3757–3767. [CrossRef]
5. Benbouzid, M.E.H.; Vieira, M.; Theys, C. Induction motors' faults detection and localization using stator current advanced signal processing techniques. *IEEE Trans. Power Electron.* **1999**, *14*, 14–22. [CrossRef]
6. Thomson, W.T.; Fenger, M. Current signature analysis to detect induction motor faults. *IEEE Ind. Appl. Mag.* **2001**, *7*, 26–34. [CrossRef]
7. Jung, J.-H.; Lee, J.-J.; Kwon, B.-H. Online Diagnosis of Induction Motors Using MCSA. *IEEE Trans. Ind. Electron.* **2006**, *53*, 1842–1852. [CrossRef]
8. Cruz, S.M.A.; Cardoso, A.J.M. Stator winding fault diagnosis in three-phase synchronous and asynchronous motors, by the extended Park's vector approach. *IEEE Trans. Ind. Appl.* **2001**, *37*, 1227–1233. [CrossRef]
9. Cardoso, M.A.; Cruz, S.M.A.; Fonseca, D.S.B. Inter-turn stator winding fault diagnosis in three-phase induction motors, by Park's vector approach. *IEEE Trans. Energy Convers.* **1999**, *14*, 595–598. [CrossRef]
10. CusidÓCusido, J.; Romeral, L.; Ortega, J.A.; Rosero, J.A.; Espinosa, A.G. Fault Detection in Induction Machines Using Power Spectral Density in Wavelet Decomposition. *IEEE Trans. Ind. Electron.* **2008**, *55*, 633–643. [CrossRef]
11. Bouzida, A.; Touhami, O.; Ibtiouen, R.; Belouchrani, A.; Fadel, M.; Rezzoug, A. Fault Diagnosis in Industrial Induction Machines Through Discrete Wavelet Transform. *IEEE Trans. Ind. Electron.* **2001**, *37*, 1227–1233. [CrossRef]
12. Zamudio-Ramirez, I.; Antonino-Daviu, J.A.; Osornio-Rios, R.A.; de Jesus Romero-Troncoso, R.; Razik, H. Detection of Winding Asymmetries in Wound-Rotor Induction Motors via Transient Analysis of the External Magnetic Field. *IEEE Trans. Ind. Electron.* **2020**, *67*, 5050–5059. [CrossRef]
13. Climente-Alarcon, V.; Antonino-Daviu, J.A.; Riera-Guasp, M.; Puche-Panadero, R.; Escobar, L. Application of the Wigner—Ville distribution for the detection of rotor asymmetries and eccentricity through high-order harmonics. *Electr. Power Syst. Res.* **2012**, *91*, 28–36. [CrossRef]
14. Soualhi, A.; Medjaher, K.; Zerhouni, N. Bearing health monitoring based on hilbert huang transform, support vector machine, and regression. *IEEE Trans. Instrum. Meas.* **2015**, *64*, 52–62. [CrossRef]
15. Benbouzid, M.E.H.; Kliman, G.B. What stator current processing-based technique to use for induction motor rotor faults diagnosis? *IEEE Trans. Energy Convers.* **2003**, *18*, 238–244. [CrossRef]
16. Henao, H.; Capolino, G.-A.; Fernandez-Cabanas, M.; Filippetti, F.; Bruzzese, C.; Strangas, E.; Pusca, R.; Estima, J.; Riera-Guasp, M.; Hedayati-Kia, S. Trends in Fault Diagnosis for Electrical Machines: A Review of Diagnostic Techniques. *IEEE Ind. Electron. Mag.* **2014**, *8*, 31–42. [CrossRef]
17. Dorrell, D.G.; Thomson, W.T.; Roach, S. Analysis of airgap flux, current, and vibration signals as a function of the combination of static and dynamic airgap eccentricity in 3-phase induction motors. *IEEE Trans. Ind. Appl.* **1997**, *33*, 24–34. [CrossRef]
18. Immovilli, F.; Bellini, A.; Rubini, R.; Tassoni, C. Diagnosis of Bearing Faults in Induction Machines by Vibration or Current Signals: A Critical Comparison. *IEEE Trans. Ind. Appl.* **2010**, *46*, 1350–1359. [CrossRef]
19. Henao, H.; Demian, C.; Capolino, G.-A. A frequency-domain detection of stator winding faults in induction machines using an external flux sensor. *IEEE Trans. Ind. Appl.* **2003**, *39*, 1272–1279. [CrossRef]
20. Park, Y.; Choi, H.; Shin, J.; Park, J.; Lee, S.B.; Jo, H. Airgap Flux Based Detection and Classification of Induction Motor Rotor and Load Defects During the Starting Transient. *IEEE Trans. Ind. Electron.* **2020**, *67*, 10075–10084. [CrossRef]
21. Zamudio-Ramírez, I.; Osornio-Rios, R.A.; Antonino-Daviu, J.A. Smart Sensor for Fault Detection in Induction Motors Based on the Combined Analysis of Stray-Flux and Current Signals: A Flexible, Robust Approach. *IEEE Ind. Appl. Mag.* **2022**, *28*, 56–66. [CrossRef]
22. Choudhary, A.; Goyal, D.; Letha, S.S. Infrared Thermography-Based Fault Diagnosis of Induction Motor Bearings Using Machine Learning. *IEEE Sensors J.* **2021**, *21*, 1727–1734. [CrossRef]
23. Choudhary, A.; Mian, T.; Fatima, S.; Panigrahi, B.K. Passive Thermography Based Bearing Fault Diagnosis Using Transfer Learning With Varying Working Conditions. *IEEE Sensors J.* **2023**, *23*, 4628–4637. [CrossRef]
24. Joksimovic, G.M.; Penman, J. The detection of inter-turn short circuits in the stator windings of operating motors. *IEEE Trans. Ind. Electron.* **2000**, *47*, 1078–1084. [CrossRef]

25. Tallam, R.M.; Habetler, T.G.; Harley, R.G. Transient model for induction machines with stator winding turn faults. *IEEE Trans. Ind. Appl.* **2002**, *38*, 632–637. [CrossRef]

26. Kliman, G.B.; Koegl, R.A.; Stein, J.; Endicott, R.D.; Madden, M.W. Noninvasive detection of broken rotor bars in operating induction motors. *IEEE Trans. Energy Convers.* **1988**, *3*, 873–879. [CrossRef]

27. Bellini, A.; Filippetti, F.; Franceschini, G.; Tassoni, C.; Kliman, G.B. Quantitative evaluation of induction motor broken bars by means of electrical signature analysis. *IEEE Trans. Ind. Appl.* **2001**, *37*, 1248–1255. [CrossRef]

28. Nandi, S.; Ahmed, S.; Toliyat, H.A. Detection of rotor slot and other eccentricity related harmonics in a three phase induction motor with different rotor cages. *IEEE Trans. Energy Convers.* **2001**, *16*, 253–260. [CrossRef]

29. Schoen, R.R.; Habetler, T.G.; Kamran, F.; Bartfield, R.G. Motor bearing damage detection using stator current monitoring. *IEEE Trans. Ind. Appl.* **1995**, *31*, 1274–1279. [CrossRef]

30. Blodt, M.; Granjon, P.; Raison, B.; Rostaing, G. Models for Bearing Damage Detection in Induction Motors Using Stator Current Monitoring. *IEEE Trans. Ind. Electron.* **2008**, *55*, 1813–1822. [CrossRef]

31. Filippetti, F.; Franceschini, G.; Tassoni, C.; Vas, P. AI techniques in induction machines diagnosis including the speed ripple effect. *IEEE Trans. Ind. Appl.* **1998**, *34*, 98–108. [CrossRef]

32. Riera-Guasp, M.; Antonino-Daviu, J.A.; Capolino, G.-A. Advances in Electrical Machine, Power Electronic, and Drive Condition Monitoring and Fault Detection: State of the Art. *IEEE Trans. Ind. Electron.* **2015**, *62*, 1746–1759. [CrossRef]

33. Filippetti, F.; Franceschini, G.; Tassoni, C.; Vas, P. Recent developments of induction motor drives fault diagnosis using AI techniques. *IEEE Trans. Ind. Electron.* **2000**, *47*, 994–1004. [CrossRef]

34. Singh, M.; Shaik, A.G. Incipient Fault Detection in Stator Windings of an Induction Motor Using Stockwell Transform and SVM. *IEEE Trans. Instrum. Meas.* **2020**, *69*, 9496–9504. [CrossRef]

35. Ali, M.Z.; Shabbir, M.N.S.K.; Liang, X.; Zhang, Y.; Hu, T. Machine learningbased fault diagnosis for single- and multi-faults in induction motors using measured stator currents and vibration signals. *IEEE Trans. Ind. Appl.* **2019**, *55*, 2378–2391. [CrossRef]

36. Ibrahim, A.; Anayi, F.; Packianather, M.; Alomari, O.A. New Hybrid Invasive Weed Optimization and Machine Learning Approach for Fault Detection. *Energies* **2022**, *15*, 1488. [CrossRef]

37. Su, H.; Chong, K.T. Induction machine condition monitoring using neural network modeling. *IEEE Trans. Ind. Electron.* **2007**, *54*, 241–249. [CrossRef]

38. Schoen, R.R.; Lin, B.K.; Habetler, T.G.; Schlag, J.H.; Farag, S. An unsupervised, on-line system for induction motor fault detection using stator current monitoring. *IEEE Trans. Ind. Appl.* **1995**, *31*, 1280–1286. [CrossRef]

39. Martins, J.F.; Pires, V.F.; Pires, A.J. Unsupervised neural-network-based algorithm for an on-line diagnosis of three-phase induction motor stator fault. *IEEE Trans. Ind. Electron.* **2007**, *54*, 259–264. [CrossRef]

40. Ince, T.; Kiranyaz, S.; Eren, L.; Askar, M.; Gabbouj, M. Real-Time Motor Fault Detection by 1-D Convolutional Neural Networks. *IEEE Trans. Ind. Electron.* **2016**, *63*, 7067–7075. [CrossRef]

41. Sun, W.; Zhao, R.; Yan, R.; Shao, S.; Chen, X. Convolutional discriminative feature learning for induction motor fault diagnosis. *IEEE Trans. Ind. Inform.* **2017**, *13*, 1350–1359. [CrossRef]

42. Zhao, R.; Wang, D.; Yan, R.; Mao, K.; Shen, F.; Wang, J. Machine Health Monitoring Using Local Feature-Based Gated Recurrent Unit Networks. *IEEE Trans. Ind. Electron.* **2018**, *65*, 1539–1548. [CrossRef]

43. Zhao, R.; Wang, J.; Yan, R.; Mao, K. Machine health monitoring with lstm networks. In Proceedings of the 2016 10th International Conference on Sensing Technology (ICST), Nanjing, China, 11–13 November 2016; pp. 1–6.

44. Sun, J.; Yan, C.; Wen, J. Intelligent bearing fault diagnosis method combining compressed data acquisition and deep learning. *IEEE Trans. Instrum. Meas.* **2018**, *67*, 185–195. [CrossRef]

45. Chen, Z.; Li, W. Multisensor Feature Fusion for Bearing Fault Diagnosis Using Sparse Autoencoder and Deep Belief Network. *IEEE Trans. Instrum. Meas.* **2017**, *66*, 1693–1702. [CrossRef]

46. Seshadrinath, J.; Singh, B.; Panigrahi, B.K. Vibration analysis based interturn fault diagnosis in induction machines. *IEEE Trans. Ind. Inform.* **2014**, *10*, 340–350. [CrossRef]

47. Husari, F.; Seshadrinath, J. Incipient Interturn Fault Detection and Severity Evaluation in Electric Drive System Using Hybrid HCNN-SVM Based Model. *IEEE Trans. Ind. Inform.* **2022**, *18*, 1823–1832. [CrossRef]

48. Nejjari, H.; Benbouzid, M.E.H. Monitoring and diagnosis of induction motors electrical faults using a current park's vector pattern learning approach. *IEEE Trans. Ind. Appl.* **2000**, *36*, 730–735. [CrossRef]

49. Martinez-Herrera, A.L.; Ferrucho-Alvarez, E.R.; Ledesma-Carrillo, L.M.; Mata-Chavez, R.I.; Lopez-Ramirez, M.; Cabal-Yepez, E. Multiple Fault Detection in Induction Motors through Homogeneity and Kurtosis Computation. *Energies* **2022**, *15*, 1541. [CrossRef]

50. Tran, M.-Q.; Liu, M.-K.; Tran, Q.-V.; Nguyen, T.-K. Effective Fault Diagnosis Based on Wavelet and Convolutional Attention Neural Network for Induction Motors. *IEEE Trans. Instrum. Meas.* **2022**, *71*, 1–13. [CrossRef]

51. Guo, X.; Chen, L.; Shen, C. Hierarchical adaptive deep convolution neural network and its application to bearing fault diagnosis. *Measurement* **2016**, *93*, 490–502. [CrossRef]

52. Zhang, S.; Zhang, S.; Wang, B.; Habetler, T.G. Deep Learning Algorithms for Bearing Fault Diagnostics—A Comprehensive Review. *IEEE Access* **2020**, *8*, 29857–29881. [CrossRef]

53. Murphey, Y.L.; Masrur, M.A.; Chen, Z.; Zhang, B. Model-based fault diagnosis in electric drives using machine learning. *IEEE/ASME Trans. Mechatr.* **2006**, *11*, 290–303. [CrossRef]

54. Masrur, M.A.; Chen, Z.; Zhang, B.; Murphey, Y.L. Model-based fault diagnosis in electric drive inverters using artificial neural network. In Proceedings of the 2007 IEEE Power Engineering Society General Meeting, Tampa, FL, USA, 24–28 June 2007; pp. 1–7.
55. Nürnberg, W.; Hanitsch, W. *Die Prüfung Elektrischer Maschinen*; Springer: Berlin/Heidelberg, Germany, 1987.
56. International Electrotechnical Commission (IEC). *IEC 60034-28:2012*; Rotating Electrical Machines—Part 28: Test Methods for Determining Quantities of Equivalent Circuit Diagrams for Three-Phase Low-Voltage Cage Induction Motors. IEC: Geneva, Switzerland, 2012.
57. Rosendahl, J. *Ursachen und Auswirkungen von Windungs- und Phasenschlüssen im Stator Großer Synchronmaschinen*; Shaker: Aachen, Germany, 2010.
58. Krause, P.; Wasynczuk, O.; Sudhoff, S.; Pekarek, S. *Analysis of Electric Machinery and Drive Systems*; Wiley-IEEE Press: Hoboken, NJ, USA, 2013.
59. Lubin, T.; Hamiti, T.; Razik, H.; Rezzoug, A. Comparison between finite-element analysis and winding function theory for inductances and torque calculation of a synchronous reluctance machine. *IEEE Trans. Magn.* **2007**, *43*, 3406–3410. [CrossRef]
60. Nandi, S.; Bharadwaj, R.; Toliyat, H. Performance analysis of a three-phase induction motor under mixed eccentricity condition. *IEEE Trans. Energy Convers.* **2002**, *17*, 392–399. [CrossRef]
61. Toliyat, H.; Lipo, T.; White, J. Analysis of a concentrated winding induction machine for adjustable speed drive applications. i. motor analysis. *IEEE Trans. Energy Convers.* **1991**, *6*, 679–683. [CrossRef]
62. Toliyat, H.A.; Lipo, T.A. Transient analysis of cage induction machines under stator, rotor bar and end ring faults. *IEEE Trans. Energy Convers.* **1995**, *10*, 241–247. [CrossRef]
63. Luo, X.; Liao, Y.; Toliyat, H.; El-Antably, A.; Lipo, T. Multiple coupled circuit modeling of induction machines. *IEEE Trans. Ind. Appl.* **1995**, *31*, 311–318.
64. Al-Nuaim, N.; Toliyat, H. A novel method for modeling dynamic air-gap eccentricity in synchronous machines based on modified winding function theory. *IEEE Trans. Energy Convers.* **1998**, *13*, 156–162. [CrossRef]
65. Tang, J.; Chen, J.; Dong, K.; Yang, Y.; Lv, H.; Liu, Z. Modeling and Evaluation of Stator and Rotor Faults for Induction Motors. *Energies* **2020**, *13*, 33. [CrossRef]
66. Storn, R.; Price, K. Differential Evolution—A Simple and Efficient Heuristic for global optimization over Continuous Spaces. *J. Glob. Optim.* **1997**, *11*, 341–359. [CrossRef]
67. Kochenderfer, M.J.; Wheeler, T.A. *Algorithms for Optimization*; MIT Press: Cambridge, MA, USA, 2019.
68. Storn, R. On the usage of differential evolution for function optimization. In Proceedings of the North American Fuzzy Information Processing, Berkeley, CA, USA, 19–22 June 1996; pp. 519–523.
69. Rumelhart, D.E.; Hinton, G.E.; Williams, R.J. Learning representations by back-propagating errors. *Nature* **1986**, *323*, 533–536. [CrossRef]
70. Bishop, C.M. *Neural Networks for Pattern Recognition*; Oxford University Press: Oxford, UK, 1995.
71. Goodfellow, I.; Bengio, Y.; Courville, A. *Deep Learning*; MIT Press: Cambridge, MA, USA, 2016.
72. LeCun, Y.; Bengio, Y.; Hinton, G. Deep learning. *Nature* **2015**, *521*, 436–444. [CrossRef]
73. Ng, A. Feature selection, L1 vs. L2 regularization, and rotational invariance. In Proceedings of the Twenty-First International Conference on Machine Learning, Banff, AB, Canada, 4–8 July 2004.
74. Srivastava, N.; Hinton, G.; Krizhevsky, A.; Sutskever, I.; Salakhutdinov, R. Dropout: A simple way to prevent neural networks from overfitting. *J. Mach. Learn. Res.* **2014**, *1*, 1929–1958.

Article

A Generalized Fault Tolerant Control Based on Back EMF Feedforward Compensation: Derivation and Application on Induction Motors Drives

Mahdi Tousizadeh [1,*], **Amirmehdi Yazdani** [2,*], **Hang Seng Che** [1], **Hai Wang** [2], **Amin Mahmoudi** [3] **and Nasrudin Abd Rahim** [1]

1   Higher Institution Centre of Excellence (HICoE), UM Power Energy Dedicated Advanced Centre (UMPEDAC), Level 4, Wisma R&D, University of Malaya, Jalan Pantai Baharu, Kuala Lumpur 59990, Malaysia

2   College of Science, Health, Engineering and Education, Murdoch University, Perth 6150, Australia

3   College of Science and Engineering, Flinders University, Adelaide 5042, Australia

*   Correspondence: toosizadeh@um.edu.my (M.T.); amirmehdi.yazdani@murdoch.edu.au (A.Y.)

Abstract: In this paper, a fault-tolerant three-phase induction drive based on field-oriented control is studied, and an analytical approach is proposed to elucidate the limitations of FOC in flux-torque regulation from the controller perspective. With an open-phase fault, the disturbance terms appear in the controller reference frame and degrade the controller performance when operating in a d-q plane with DC quantities. In addition, the hardware reconfiguration, which is essential to operate faulted three-phase drives, causes substantial change in the way the control parameters v_d, v_q are reflected onto the machine terminals. An accurate understanding of the feedforward term, by considering the open-phase fault and the hardware modifications, is provided to re-enable the FOC in presence of an open-phase fault. Furthermore, the concept of feedforward term derivation is generically extended to cover multiphase induction drives encountering an open-phase fault whereby no hardware reconfiguration is intended. The proposed method is explained based on a symmetrical six-phase induction and can be extended to drives with a higher number of phases. The effectiveness of the proposed derivation method, which is required to form a feedforward fault-tolerant controller, is verified and compared through the simulation and experiment, ensuring smooth operation in postfault mode.

Keywords: induction motors; fault-tolerant control; AC machines; back EMF; feedforward compensation

Citation: Tousizadeh, M.; Yazdani, A.; Che, H.S.; Wang, H.; Mahmoudi, A.; Rahim, N.A. A Generalized Fault Tolerant Control Based on Back EMF Feedforward Compensation: Derivation and Application on Induction Motors Drives. *Energies* **2023**, *16*, 51. https://doi.org/10.3390/en16010051

Academic Editors: Daniel Morinigo-Sotelo, Joan Pons-Llinares and Rene Romero-Troncoso

Received: 21 November 2022
Revised: 12 December 2022
Accepted: 13 December 2022
Published: 21 December 2022

1. Introduction

Adjustable speed AC motor drives are, in general, susceptible to failure, especially in the power section where the stress is on the power switches and/or motor windings [1,2]. Since the failure might cause the whole drive to shut down, reliability is a key feature in applications in which a failure can cause safety issues. For example, fault-tolerant control of three-phase adjustable speed drives in automotive applications has recently attracted significant attention [3–6]. Thus, the need for an effective fault-tolerant method that can be embedded into the existing motor drives is practically favorable.

Since a three-phase machine with wye-connected stator winding will be effectively reduced to a "single-phase machine" under an OPF, topological reconfiguration is necessary for three-phase fault-tolerant drives to retain two degrees of freedom. In the literature, there are several feasible topologies used to reconfigure the three-phase AC drives [7–10]. Among these topologies, the majority of fault-tolerant three-phase drives utilize an additional inverter leg connected to the neutral point of the three-phase machine [11–14] as shown in Figure 1. Unless otherwise stated, this is considered the standard topology for fault-tolerant three-phase drive hereafter [1,11,15]. This is to allow neutral current to return back to dc-link.

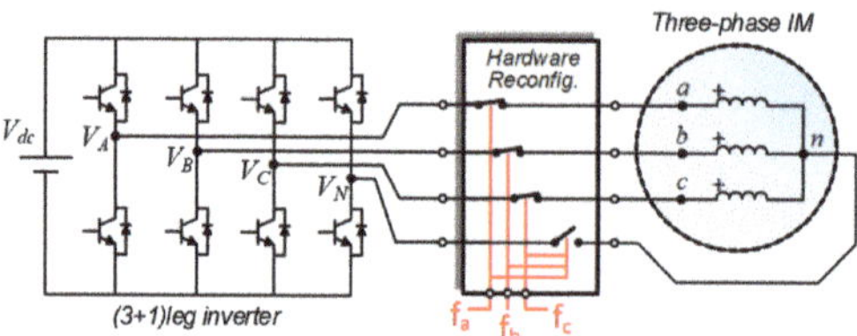

Figure 1. Fault-tolerant AC drive with (3 + 1) leg inverter and switches to emulate the open phase fault and reconfiguration. The signals fa, fb, and fc are designed to emulate OPF on phase a, b, or c.

For multiphase (more than three phases) machines, however, the higher DOF allows the drive system to be inherently fault-tolerant towards OPF without the need for any hardware reconfiguration, as long as there are still three or more phases intact. Due to this higher fault tolerance, multiphase machines are often favored over their three-phase counterparts as fault-tolerant drives [16–19]. Regardless of the number of phases, OPFs always result in the loss of DOF in a drive and degrade the control performance if no mitigation measures are taken. In light of this, various fault-tolerant control techniques have been proposed in the past.

The mathematical model of the machine under OPF with a reduced order transformation matrix is attempted in [14,20–23]. It takes into account the reduced DOF and the unbalanced condition of the machine. While these methods were shown to be robust to machine parameter detuning, the re-derivation of machine models and reduced order transformation matrices are mathematically complex and unique for specific faults.

Alternatively, the original machine model and transformation can be maintained, and therefore, the effect of fault on the machine must be compensated by modifying the controller to handle double frequency in a d-q plane. For fault-tolerant three-phase drive, [13,24] demonstrated that the OPF gives rise to a negative sequence component and a negative sequence controller is needed to retain the current control performance in a positive sequence. For multiphase machines, analysis using the vector space decomposition method shows that OPFs create coupling between the torque-producing current components that are otherwise decoupled under healthy conditions [25,26]. The non-flux-and-torque-producing current components, also known as the x-y currents, are proportionally regulated to achieve fault tolerance [25,27]. These studies, for both three-phase and multiphase machines, utilize closed-loop feedback control methods where the unwanted current components are controlled using designated feedback current loop. However, as with any feedback control method, the dynamic performance during the transient will depend on the tuning of the controllers' parameters.

More recently, several research studies have highlighted the superiority of the feedforward compensation method being combined with a resonant controller for three-phase and six-phase drives [24,28]. The sensitivity of stator resistance to temperate, hence the inaccuracy of the feedforward term, is stated to be the main motivation for introducing an additional current controller to control the neutral current. However, this is not the only solution as the feedforward terms derived in the [11,12,29] are irrespective of stator resistance, which makes it robust against temperature variations. An accurate compensation term injected in a feedforward manner is shown to be effective for PMSM drives [11,12]. Similarly, for three-phase induction motors, feedforward compensation methods were introduced in [30,31] using the zero-sequence component.

Despite the documented research in the area of feedforward compensation methods for fault-tolerant induction motor drives, the following research questions still remain unaddressed:

1. How the concept of feedforward compensation can be realized from a control perspective in the context of FOC-driven AC drives?
2. How the feedforward compensation technique developed for three-phase machines can be extended for multiphase machines?

This paper is an extended version of the primary work presented in [29] to address the highlighted research questions. It expands the preliminary investigation on feedforward derivation and compensation techniques and the main contributions of this study are as follows:

1. An accurate feedforward compensation method based on FOC control and the open-phase fault is systematically derived that can be readily integrated into any three-phase AC drive with minimal modifications to the FOC controller. Furthermore, the stationary reference frame is used to apply the feedforward compensation as it simplifies the derivation complexities without any compromise in effectiveness.
2. The concept of feedforward term derivation is generically extended to a six-phase drive where the back EMF term is still the dominant part of the feedforward term but injected into a different plane to retain the control of the machine.

The organization of this paper is as follows. Section 2 discusses the fault tolerant control of a three-phase induction machine in both stationary and rotation reference frames, where the impacts of OPF on the mapping of the controlled variables to the machine variables are elucidated. In Section 3, the discussion is extended to a multiphase machine using a symmetrical six-phase machine as an example. Section 4 shows the experimental results, where the performances of the proposed feedforward compensation methods are verified using lab-scale three-phase and symmetrical six-phase induction machines. Finally, conclusions are given in Section 5.

2. Fault-Tolerant Control of Three-Phase Induction Machines

2.1. Mathematical Model of Three-Phase IM under RFOC

The dynamic model of the induction machine is usually given in the SRF d-q variables (and zero-sequence variable), which can be obtained from the phase variables using magnitude-invariant Clarke–Park transformation [32] as shown in (1):

$$T_3 = \frac{2}{3} \begin{bmatrix} \cos(\omega t) & \cos(\omega t - \delta) & \cos(\omega t + \delta) \\ -\sin(\omega t) & -\sin(\omega t - \delta) & -\sin(\omega t + \delta) \\ 0.5 & 0.5 & 0.5 \end{bmatrix} \tag{1}$$

where ω is the synchronous frequency of the machine and δ is the displacement factor of $2\pi/3$ a three-phase machine.

Based on the RFOC approach, the dynamic behavior of the induction machine can be expressed in terms of the stator voltage equations as follows:

$$\begin{bmatrix} v_{ds} \\ v_{qs} \\ v_{0s} \end{bmatrix} = \begin{bmatrix} R_s + \sigma L_s \rho & -\omega \sigma L_s & 0 \\ \omega \sigma L_s & R_s + \sigma L_s \rho & 0 \\ 0 & 0 & R_{s0} + L_0 \rho \end{bmatrix} \cdot \begin{bmatrix} i_{ds} \\ i_{qs} \\ i_{0s} \end{bmatrix} + \begin{bmatrix} \omega \frac{L_m}{L_r} \rho \\ \omega \frac{L_m}{L_r} \\ 0 \end{bmatrix} \cdot \psi_{dr} \tag{2}$$

where R_s, R_{s0}, L_m, L_r, L_0, and σ are the stator resistance in the SRF plane, zero sequence resistance, magnetizing inductance, rotor self-inductance, zero sequence inductance and leakage factor of the induction machine, respectively. The leakage factor is defined to be $\sigma = 1 - (L_m^2 / L_s L_r)$. The synchronous angular speed of the machine in the electrical domain and rotor flux is denoted by ω and ψ_{dr}, respectively, where the symbol ρ represents the time-derivative of the variable. The rotor flux under RFOC is controlled directly by i_{ds}, known as flux current, to form a first-order system as

$$\psi_{dr} = \frac{L_m i_{ds}}{1 + \tau_r \rho} \tag{3}$$

with the rotor time constant τ_r being the ration of rotor self-inductance over the rotor resistance R_r.

2.2. Relation between Control Variables and Machine Variables

2.2.1. Healthy Operation

A typical three-phase induction motor drive system connected in a wye configuration and controlled under RFOC is depicted in Figure 2a. The main components of the drive are given in separate modules to elucidate how the control variables would be eventually mapped onto the machine under the healthy and postfault configuration.

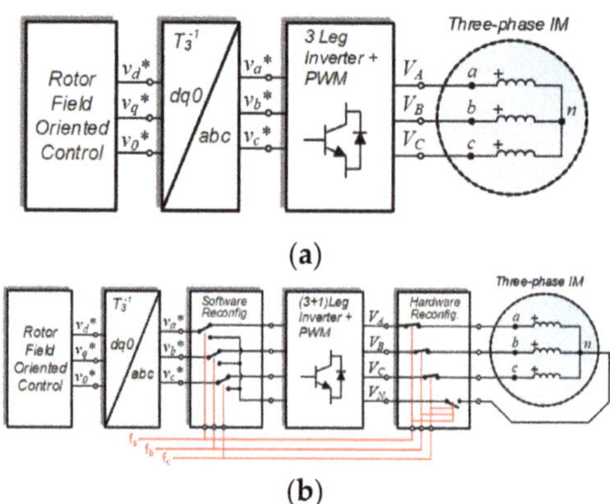

Figure 2. Three-phase induction motor drive with RFOC controller: (**a**) Typical topology for healthy operation, (**b**) Reconfigurable fault-tolerant topology with fault emulation signals f_n (n: a, b, c).

Starting with the machine on the rightmost part of Figure 2a, the phase windings are supplied through the leg voltage of the inverter V_A, V_B, and V_C. This configuration explicitly implies that the motor phase voltages v_{an}, v_{bn}, and v_{cn} are indirectly defined by the leg voltages of the inverter, as detailed in (4).

$$
\begin{bmatrix} v_{an} \\ v_{bn} \\ v_{cn} \end{bmatrix} = \frac{2}{3} \begin{bmatrix} 1 & -0.5 & -0.5 \\ -0.5 & 1 & -0.5 \\ -0.5 & -0.5 & 1 \end{bmatrix} \begin{bmatrix} V_A \\ V_B \\ V_C \end{bmatrix}
\tag{4}
$$

The relation in (4) is valid for a balanced motor with all phases having equal impedance, which is the case for a healthy drive.

On the inverter block, the carrier-based PWM helps to form a voltage amplifier with a fixed gain of $K = V_{dc}/2$ that converts the modulating signals $v_a{}^*$, $v_b{}^*$, and $v_c{}^*$ to leg voltages. The lumped transfer function of the inverter together with Sine PWM is given in a matrix in (5), assuming that inverter non-idealities are negligible, and no homopolar voltage is being injected.

$$
\begin{bmatrix} V_A \\ V_B \\ V_C \end{bmatrix} = \begin{bmatrix} K & 0 & 0 \\ 0 & K & 0 \\ 0 & 0 & K \end{bmatrix} \begin{bmatrix} v_a^* \\ v_b^* \\ v_c^* \end{bmatrix}
\tag{5}
$$

One step before PWM, the control variables $v_d{}^*$, $v_q{}^*$, and $v_0{}^*$ are basically transformed into modulating signals by applying the inverse Clarke–Park transformation given in (1).

$$
\begin{bmatrix} v_a^* \\ v_b^* \\ v_c^* \end{bmatrix} = T_3^{-1} \begin{bmatrix} v_d^* \\ v_q^* \\ v_0^* \end{bmatrix}
\tag{6}
$$

By substituting (6) into (5) and (5) into (4), the relation between the control variable $v_d{}^*$ and $v_q{}^*$ and phase voltages received by the machine in matrix form is obtained to be as follow.

$$\begin{bmatrix} v_{an} \\ v_{bn} \\ v_{cn} \end{bmatrix} = K \begin{bmatrix} \cos(\omega t) & -\sin(\omega t) & 0 \\ \cos(\omega t - \delta) & -\sin(\omega t - \delta) & 0 \\ \cos(\omega t + \delta) & -\sin(\omega t + \delta) & 0 \end{bmatrix} \begin{bmatrix} v_d^* \\ v_q^* \\ v_0^* \end{bmatrix} \tag{7}$$

By applying (1) to both side of (7), the machine phase voltages would also be transformed into *d-q-0* space, bringing everything to the same page (*d-q-0*), as stated in (8).

$$\begin{bmatrix} v_{ds} \\ v_{qs} \\ v_{0s} \end{bmatrix} = K \begin{bmatrix} 1 & 0 & 0 \\ 0 & 1 & 0 \\ 0 & 0 & 0 \end{bmatrix} \begin{bmatrix} v_d^* \\ v_q^* \\ v_0^* \end{bmatrix} \tag{8}$$

The terms v_{ds}, v_{qs}, and v_{0s} in (8) are in fact the transformed version of the stator phase voltages that correspond to the machine model in (2).

Accordingly, a couple of notable conclusions can be made from (8) which are valid for any wye connected three-phase AC drive, under healthy operation as:

C1: The voltage received by the motor in the synchronous frame (v_{ds} and v_{qs}) are directly proportional to the control variables (v_d^* and v_q^*) by a fixed gain of K. This unique property of healthy drive allows current controllers (such as PI controllers) in the SRF to control the flux and torque currents effectively;

C2: The zero-sequence voltage v_{0s} reached to the machine winding is not linked to v_0^* of the controller, making it decoupled from control variables (v_d^* and v_q^*).

2.2.2. Postfault Operation for Three-Phase Induction Motor Drive

Upon generation of fault flag f_n (n: a, b, c) in Figure 2b for the postfault topology, the following modifications are applied [11,13] to provide a path for neutral current to flow back to the dc-link:

*m*1: the fourth leg is clamped to the motor neutral point (according to hardware reconfiguration block in Figure 2b);

*m*2: the modulating signal of the faulted leg is switched over to the fourth leg (according to software reconfiguration block in Figure 2b).

To facilitate a quick transition to postfault operation for three-phase wye-connected drives, the fault must be detected in the first place, followed by hardware and software reconfigurations. However, fault detection is beyond the scope of this study, and the fault flag is created manually.

For the sake of simplicity, the fault and reconfigurations are simultaneously emulated by activating the corresponding fault signal, based on what has been shown in Figure 2b with red lines. Therefore, the relation stated in (8) needs to be re-examined, as the topology of the drive has been modified.

Assuming an OPF in phase *a*, i.e., by activation of f_a, the motor phase voltages in terms of leg voltage in postfault mode should be redefined as follow.

$$\begin{bmatrix} v_{an} \\ v_{bn} \\ v_{cn} \end{bmatrix} = \begin{bmatrix} 0 & 0 & 0 \\ -1 & 1 & 0 \\ -1 & 0 & 1 \end{bmatrix} \begin{bmatrix} V_N \\ V_B \\ V_C \end{bmatrix} + \begin{bmatrix} E_a \\ 0 \\ 0 \end{bmatrix} \tag{9}$$

where E_a is the back EMF voltage on the faulted phase and V_N is the leg voltage of the fourth leg. This induced voltage is basically due to the existence of rotating MMF in the machine. Since the motor phase *a* is disconnected from an inverter, the back EMF voltage E_a is no longer directly controllable by any inverter leg voltage, and therefore, is represented in a separate matrix in (9).

Once the modification *m2* is executed, the active inverter leg voltages as a function of modulating signals would be the same as (5), with V_A being substituted with V_N.

$$\begin{bmatrix} V_N \\ V_B \\ V_C \end{bmatrix} = \begin{bmatrix} K & 0 & 0 \\ 0 & K & 0 \\ 0 & 0 & K \end{bmatrix} \begin{bmatrix} v_a^* \\ v_b^* \\ v_c^* \end{bmatrix} \tag{10}$$

Taking the same step as in (6) the relation between the control variables and phase voltage on the motor is obtained and stated in (11).

$$\begin{bmatrix} v_{an} \\ v_{bn} \\ v_{cn} \end{bmatrix} = \sqrt{3}K \begin{bmatrix} 0 & 0 & 0 \\ \cos(\theta_1) & -\sin(\theta_1) & 0 \\ \cos(\theta_2) & -\sin(\theta_2) & 0 \end{bmatrix} \begin{bmatrix} v_d^* \\ v_q^* \\ v_0^* \end{bmatrix} + \begin{bmatrix} E_a \\ 0 \\ 0 \end{bmatrix} \tag{11}$$

$$\theta_1 = \omega t - \delta - \tfrac{\pi}{6} \quad , \quad \theta_2 = \omega t + \delta + \tfrac{\pi}{6}$$

One step further, by applying the Clarke–Park transformation in (1) onto both sides of (11), the voltage relation in *d-q*-0 space after reconfiguration is found to be as follows:

$$\begin{bmatrix} v_{ds} \\ v_{qs} \\ v_{0s} \end{bmatrix} = K \begin{bmatrix} 1 & 0 & 0 \\ 0 & 1 & 0 \\ -\cos(\omega t) & \sin(\omega t) & 0 \end{bmatrix} \begin{bmatrix} v_d^* \\ v_q^* \\ v_0^* \end{bmatrix} + \frac{1}{3} \begin{bmatrix} 2E_a \cos(\omega t) \\ -2E_a \sin(\omega t) \\ E_a \end{bmatrix} \tag{12}$$

which is no longer the same as the healthy case stated in (8).

Equation (12) reveals that there exists a substantial double frequency AC disturbance on the machine winding in the *d-q* plane, whereas it is supposed to have all quantities in DC. The reconfigured drive, along with modifications *m1* and *m2*, leads to the following observations:

Observation 1. *The OPF together with reconfiguration introduces double-frequency AC disturbance terms appearing on the d-q plane with the magnitude being proportional to the back EMF of the faulted phase.*

Observation 2. *The zero-sequence circuit of the motor is excited through the d-q voltages of the controller (v_d^* and v_q^*), and therefore, the machine voltages on the d-q-0 plane are now coupled under OPF.*

Observation 3. *The zero-sequence reference voltage on the controller side, v_0^*, still has no impact on either the zero sequence or the d-q components of the machine. Therefore, it can be ignored.*

As with the case of faulted PMSM discussed in [11], issue X1 is the main cause of control performance degradation in faulted three-phase induction drive even after converter reconfiguration. When PI controllers are used, the AC disturbance voltages disrupt the regulation of the *d-q* currents, due to the inability of PI controllers to completely suppress AC signals.

2.3. Feedforward Compensation for Fault-Tolerant Three-Phase Induction Motor Drive

From (12), it can be observed that by compensating the disturbance terms appearing on the right-hand side of the equation, one-to-one mapping of *d-q* voltages between the controller and the machine voltages will be restored. This can be done by calculating the terms and subtracting them from the control variables in the *d-q* plane in a feedforward manner, in a similar way as in [11].

$$v_{d\_ff} = \frac{-2}{3K} E_a \cos(\omega t) \quad , \quad v_{q\_ff} = \frac{2}{3K} E_a \sin(\omega t) \tag{13}$$

Alternatively, if the inverse Park transformation is applied to (12), the relation between controller and machine variables in the stationary reference frame (α-β-0) can be simplified as follows.

$$\begin{bmatrix} v_{\alpha s} \\ v_{\beta s} \\ v_{0s} \end{bmatrix} = K \begin{bmatrix} 1 & 0 & 0 \\ 0 & 1 & 0 \\ -1 & 0 & 0 \end{bmatrix} \begin{bmatrix} v_{\alpha}^{*} \\ v_{\beta}^{*} \\ v_{0}^{*} \end{bmatrix} + \frac{1}{3} \begin{bmatrix} 2E_a \\ 0 \\ E_a \end{bmatrix} \tag{14}$$

In this form, the disturbance terms need to be compensated in the α-β plane, as stated in (15).

$$v_{\alpha\_ff} = \frac{-2}{3K} E_a \quad , \quad v_{\beta\_ff} = 0 \tag{15}$$

Once the feedforward terms are added according to Figure 3, the AC disturbance terms in (12) or (14) would disappear such that;

i.   The C1 would hold true, and therefore, the RFOC would take over the control of the machine, just like the healthy drive, and

ii.   Unlike the healthy drive, there would be a non-zero voltage v_{0s} appearing on the zero-sequence circuit of the machine that is a function of control variables.

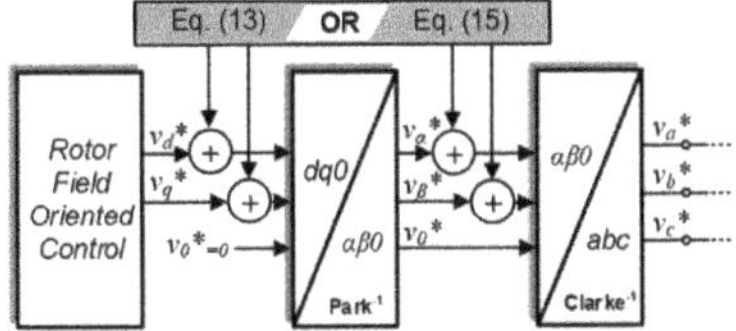

Figure 3. The two alternative planes for the injection of feedforward term(s) for three-phase IM drive.

After adding the feedforward term stated in (13) or (15), the zero-sequence voltage in postfault mode according to (12) or (14) would become

$$v_{0s} = -Kv_d^* \cos(\omega t) + Kv_q^* \sin(\omega t) + E_a = -Kv_{\alpha}^* + E_a \tag{16}$$

This should be in agreement with the machine equations given in (2) at steady state. Since C1 is valid after feedforward injection (i.e., $Kv_d{}^* = v_{ds}$, and $Kv_q{}^* = v_{qs}$), the back EMF voltage E_a can be calculated by revisiting machine equations so that (16) would become

$$R_{s0}i_{0s} + L_0 \rho i_{0s} = -\left(R_s i_{ds} - \omega\sigma L_s i_{qs}\right)\cos(\omega t) + \\ \left(R_s i_{qs} + \omega\sigma L_s i_{ds} + \omega L_m \psi_{dr}/L_r\right)\sin(\omega t) + E_a \tag{17}$$

where the current relation in postfault mode is

$$i_{0s} = -i_d \cos(\omega t) + i_q \sin(\omega t) = -i_{\alpha} \tag{18}$$

Finally, by substituting (18) into (17) and neglecting the resistive terms, the steady state representation of back EMF voltage E_a is found to be as follows.

$$E_a = -\omega\left[((\sigma L_s - L_0)i_{ds} + \psi_{dr}L_m/L_r)\sin(\omega t) + (\sigma L_s - L_0)i_{qs}\cos(\omega t)\right] \tag{19}$$

Considering the feedforward terms obtained from (13) and (19), it is made clear that the back EMF voltage in (19), hence the feedforward term (13), is irrespective of stator resistance value. This important relation explicitly rules out the dependency of this feedforward compensation method on the machine temperature. Furthermore, a general analogy between the two types of machines, i.e., induction machine and PMSM, can be established. Due to the absence of saliency in the rotor structure of the IM, the σL_s in IM are found to be equivalent to L_d and L_q of the PMSM. Likewise, the equivalent term corresponding to the permanent magnet flux (λ_{pm}) of PMSM is found to be $\psi_{dr}{}^*L_m/L_r$. This agreement suggests a further extension of the discussion to be detailed on the multiphase drives.

3. Fault-Tolerant Control of Six-Phase Induction Machines

3.1. Six-Phase Induction Machine Model under Rotor Field Oriented Control (RFOC)

For multiphase machines, analysis is usually performed based on the Vector Space Decomposition (VSD) model [33], where the machine variables can be decoupled into flux-and-torque producing α-β components, non-flux-and-torque producing x-y components and zero sequence 0_1-0_2 components. The concept of VSD transformation for multiphase machines with different phase numbers has been well-addressed in the literature on multiphase machines [26,34,35] and hence not dealt with further in this paper for brevity. To facilitate the subsequent discussion, a symmetrical six-phase induction machine with is used as a case study to represent a multiphase induction machine. The VSD model for a symmetrical six-phase induction machine controlled using RFOC is given in (20).

$$
\begin{bmatrix} v_{ds} \\ v_{qs} \\ v_{xs} \\ v_{ys} \\ v_{01} \\ v_{02} \end{bmatrix} = \begin{bmatrix} R_s + \sigma L_s \rho & -\omega \sigma L_s & 0 & 0 & 0 & 0 \\ \omega \sigma L_s & R_s + \sigma L_s \rho & 0 & 0 & 0 & 0 \\ 0 & 0 & R_{sxy} + L_{xy}\rho & 0 & 0 & 0 \\ 0 & 0 & 0 & R_{sxy} + L_{xy}\rho & 0 & 0 \\ 0 & 0 & 0 & 0 & R_{s0} + L_0\rho & 0 \\ 0 & 0 & 0 & 0 & 0 & R_{s0} + L_0\rho \end{bmatrix} \cdot \begin{bmatrix} i_{ds} \\ i_{qs} \\ i_{xs} \\ i_{ys} \\ i_{01} \\ i_{02} \end{bmatrix} + \begin{bmatrix} \omega \frac{L_m}{L_r}\rho \\ \omega \frac{L_m}{L_r} \\ 0 \\ 0 \\ 0 \\ 0 \end{bmatrix} \cdot \psi_{dr} \quad (20)
$$

By using rotational transformation, the α-β subspace can be rotated to form the synchronous d-q subspace, where the control of the machine in RFOC will be identical to that of a three-phase machine. Therefore, the rotor flux of a six-phase machine under RFOC is obtained in the same way as (3).

However, the x-y and 0_1-0_2 planes remain in a stationary reference frame as they are represented by a simple R-L circuit with no coupling to the rotor flux and do not contribute to flux-and-torque production.

3.2. Relation between Controller and Machine Variables

3.2.1. Healthy Operation

For a six-leg inverter driving a six-phase machine with star-connected stator winding and two isolated neutral, as shown in Figure 4, the motor phase voltages are a function of the inverter leg voltages as follows:

$$
\begin{bmatrix} v_{a1n1} \\ v_{b1n1} \\ v_{c1n1} \\ v_{a2n2} \\ v_{b2n2} \\ v_{c2n2} \end{bmatrix} = \frac{1}{3} \begin{bmatrix} 2 & -1 & -1 & 0 & 0 & 0 \\ -1 & 2 & -1 & 0 & 0 & 0 \\ -1 & -1 & 2 & 0 & 0 & 0 \\ 0 & 0 & 0 & 2 & -1 & -1 \\ 0 & 0 & 0 & -1 & 2 & -1 \\ 0 & 0 & 0 & -1 & -1 & 2 \end{bmatrix} \begin{bmatrix} V_{A1} \\ V_{B1} \\ V_{C1} \\ V_{A2} \\ V_{B2} \\ V_{C2} \end{bmatrix} \quad (21)
$$

where the leg voltages of (21) are determined through six modulating signals, as per Figure 4, in a similar condition as the three-phase case explained in (5).

$$
[V_{A1}\ V_{B1}\ V_{C1}\ V_{A2}\ V_{B2}\ V_{C2}]^T = K[v_{a1}^*\ v_{b1}^*\ v_{c1}^*\ v_{a2}^*\ v_{b2}^*\ v_{c2}^*]^T \quad (22)
$$

The modulating signals for the six-phase machine can be obtained by transforming the control variables using the extended inverse Clarke transformation (for symmetrical six-phase) as follow.

$$\begin{bmatrix} v_{a1}^* \\ v_{b1}^* \\ v_{c1}^* \\ v_{a2}^* \\ v_{b2}^* \\ v_{c2}^* \end{bmatrix} = \begin{bmatrix} 1 & 0 & 1 & 0 & 1.4142 & 0 \\ -0.5 & 0.866 & -0.5 & -0.866 & 1.4142 & 0 \\ -0.5 & -0.866 & -0.5 & 0.866 & 1.4142 & 0 \\ 0.5 & 0.866 & -0.5 & 0.866 & 0 & 1.4142 \\ -1 & 0 & 1 & 0 & 0 & 1.4142 \\ 0.5 & -0.866 & -0.5 & -0.866 & 0 & 1.4142 \end{bmatrix} \begin{bmatrix} v_\alpha^* \\ v_\beta^* \\ v_x^* \\ v_y^* \\ v_{01}^* \\ v_{02}^* \end{bmatrix} \tag{23}$$

Finally, by substituting (23) into (21)–(22) and applying extended Clarke transformation, the voltage relation in healthy operation arrives at (24) as follows.

$$\begin{bmatrix} v_{\alpha s} \\ v_{\beta s} \\ v_{x s} \\ v_{y s} \\ v_{01 s} \\ v_{02 s} \end{bmatrix} = K \begin{bmatrix} 1 & 0 & 0 & 0 & 0 & 0 \\ 0 & 1 & 0 & 0 & 0 & 0 \\ 0 & 0 & 1 & 0 & 0 & 0 \\ 0 & 0 & 0 & 1 & 0 & 0 \\ 0 & 0 & 0 & 0 & 0 & 0 \\ 0 & 0 & 0 & 0 & 0 & 0 \end{bmatrix} \begin{bmatrix} v_\alpha^* \\ v_\beta^* \\ v_x^* \\ v_y^* \\ v_{01}^* \\ v_{02}^* \end{bmatrix} \tag{24}$$

The relation between control variables and machine voltages for a symmetrical six-phase machine in (24) is harmonious to (8) whereby the α-β components, corresponding to a d-q plane, are directly controllable, as well as x-y components. Moreover, the v_{01s} and v_{02s} are still uncontrollable and isolated from one another since the machine is configured with two isolated neutrals. Till this point, the conclusions C1 and C2 are also valid for (24), however, the postfault relation is yet to be derived.

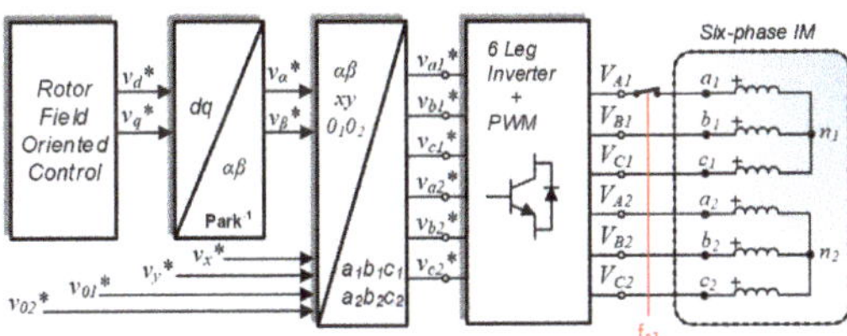

Figure 4. Six-phase induction motor drive (two isolated neutral) with RFOC and one reconfigurable OPF on phase a_1. The signal f_{a1} is to emulate the OPF.

3.2.2. Postfault Operation

By introducing an OPF, emulated by a switch being triggered via f_{a1} in Figure 4, phase a_1 gets disconnected. As stated earlier, the six-phase drives do not require any hardware reconfiguration, as there exists a minimum DOF to control the machine in presence of an OPF. Nevertheless, the OPF alters the relation between the leg voltage and phase voltage of the machine, represented by the following matrix.

$$\begin{bmatrix} v_{a1n1} \\ v_{b1n1} \\ v_{c1n1} \\ v_{a2n2} \\ v_{b2n2} \\ v_{c2n2} \end{bmatrix} = \frac{1}{3} \begin{bmatrix} 0 & 0 & 0 & 0 & 0 & 0 \\ 0 & 1.5 & -1.5 & 0 & 0 & 0 \\ 0 & -1.5 & 1.5 & 0 & 0 & 0 \\ 0 & 0 & 0 & 2 & -1 & -1 \\ 0 & 0 & 0 & -1 & 2 & -1 \\ 0 & 0 & 0 & -1 & -1 & 2 \end{bmatrix} \begin{bmatrix} V_{A1} \\ V_{B1} \\ V_{C1} \\ V_{A2} \\ V_{B2} \\ V_{C2} \end{bmatrix} + \begin{bmatrix} E_{a1} \\ -0.5E_{a1} \\ -0.5E_{a1} \\ 0 \\ 0 \\ 0 \end{bmatrix} \tag{25}$$

With the connection between phase a_1 and leg V_{A1} being open-circuited in postfault, the remaining phases in the $a_1b_1c_1$ winding set will receive an AC term proportional to the back EMF of this faulted phase, E_{a1}, as given in Equation (25). By replacing (25) with (21) and repeating the steps in section III-B-1, the voltage relation in postfault for a symmetrical six-phase machine would be given as (26).

$$
\begin{bmatrix} v_{\alpha s} \\ v_{\beta s} \\ v_{x s} \\ v_{y s} \\ v_{01s} \\ v_{02s} \end{bmatrix} = K \begin{bmatrix} 0.5 & 0 & -0.5 & 0 & 0 & 0 \\ 0 & 1 & 0 & 0 & 0 & 0 \\ -0.5 & 0 & 0.5 & 0 & 0 & 0 \\ 0 & 0 & 0 & 1 & 0 & 0 \\ 0 & 0 & 0 & 0 & 0 & 0 \\ 0 & 0 & 0 & 0 & 0 & 0 \end{bmatrix} \begin{bmatrix} v_{\alpha}^{*} \\ v_{\beta}^{*} \\ v_{x}^{*} \\ v_{y}^{*} \\ v_{01}^{*} \\ v_{02}^{*} \end{bmatrix} + \begin{bmatrix} 0.5E_{a1} \\ 0 \\ 0.5E_{a1} \\ 0 \\ 0 \\ 0 \end{bmatrix} \tag{26}
$$

Similar to (14), OPF in a six-phase machine introduces AC disturbance in the α-axis with the fundamental frequency. It is not hard to derive that this disturbance will appear as double frequency AC disturbances in the d-q frame (0.5 E_{a1} cos(ωt) on the d-axis, $-0.5\ E_{a1}$ sin(ωt) on the q-axis), analogous to what appeared in (12), which will cause the degradation of RFOC performance. Furthermore, (26) shows that there is a coupling between α- and x-axes, which is similar to the coupling between α- and 0-axes for the case of the three-phase machine in (14).

3.3. Feedforward Compensation for Fault Tolerant Six-Phase Induction Motor Drive

To cancel the disturbance due to OPF in the α-β frame, the following steps are taken: Firstly, since the reference voltages in the α- and x-axes have opposite and equal coefficients in (26), the reference for the x-axis is derived directly from the α-axis and set to be exactly opposite as:

$$
v_{x}^{*} = -v_{\alpha}^{*} \quad , \quad v_{y}^{*} = v_{\beta}^{*} \tag{27}
$$

By substituting (27) into (26), the disturbance in the α-axis can be eliminated by adding a feedforward voltage to the $v_{\alpha}{}^{*}$ as follows.

$$
v_{\alpha\_ff} = \frac{-E_{a1}}{K} \quad , \quad v_{\beta\_ff} = 0 \tag{28}
$$

The implementation of (27) and (28) is illustrated in Figure 5.

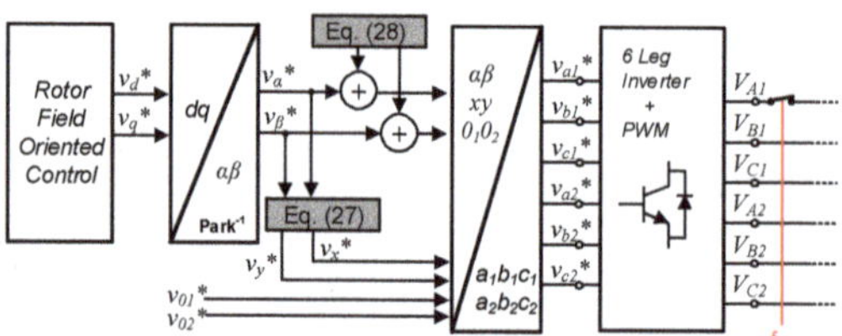

Figure 5. The injection of feedforward term(s) in symmetrical six-phase IM drive. The signal f_{a1} is to emulate the OPF.

As with a three-phase machine, the feedforward term requires the knowledge of the back EMF of the faulted phase. Through the application of (27) and (28), the voltage in the x-axis applied to the machine after injection of the feedforward term in (26) would be obtained as (29).

$$
v_{xs} = -Kv_{d}^{*}\cos(\omega t) + Kv_{q}^{*}\sin(\omega t) + E_{a1} = -Kv_{\alpha}^{*} + E_{a1} \tag{29}
$$

Using a similar approach as in the three-phase case, the back EMF voltage E_{a1} needs to be derived. The voltage in the x-axis from the machine equation in (20) needs to be substituted into (29) to get E_{a1} as follows.

$$
E_{a1} = -\omega\left[\left((\sigma L_s - L_{xy})i_{ds} + \psi_{dr}L_m/L_r\right)\sin(\omega t) + (\sigma L_s - L_{xy})i_{qs}\cos(\omega t)\right] \tag{30}
$$

Equation (30) formulates the back EMF voltage of the lost phase on S6 in terms of the α-β parameters and operating. However, compared to three-phase IM, the zero sequence parameters are replaced with x-y, yet independent of stator resistance value. The effectiveness of the derived feedforward terms for symmetrical three- and six-phase machines is verified in the following section.

4. Results and Discussion

This section demonstrates the results of an experimental test conducted for the effectiveness and robustness validation of the proposed feedforward fault-tolerant control methodology. The experiment is performed using symmetrical three- and six-phase IM drives with the details given in Table 1. As shown in Figure 6, the motors are mechanically coupled with a passive load (1.8 kW PMSM feeding an adjustable resistor bank) and an incremental encoder (resolution 5000 pulse/rev) that is used to feedback on the speed. The phase current of the motor is measured through a six-channel current sensor (based on LEM current transducer). The motors are powered by a six-leg custom-made 12 kW inverter being supplied from a DC power supply (TDK Lambda GEN600-8.5). The RFOC control is implemented on the dSPACE DS1103 digital controller with a 5 kHz switching frequency.

Table 1. kW three-phase and 0.55 kW symmetrical six-phase induction motors.

	3-Phase IM	6-Phase IM
Power	1000 W	550 W
Phase Voltage	220 V	240 V
Phase Current	2.7 A	1.45 A
Speed	2800 RPM	1390 RPM
Frequency	50 Hz	50 Hz
Magnetizing Inductance Lm	490 mH	420 mH
Stator Leakage Inductance Lls	13 mH	6 mH
Stator Leakage Inductance Lxy	-	3.6 mH
Rotor Leakage Inductance Llr	13 mH	78 mH
Rotor Resistance	5.9 Ω	5.77 Ω
Flux Current, i_d	1.4 A	0.75 A

Figure 6. The experimental test rig for three-phase and six-phase induction motor drives under OPF.

The experiment is started by running the motors in healthy operation and after a while, a trigger signal is manually generated by the user to emulate the OPF using relay contacts. For the case of three-phase the modifications $m1$ and $m2$, as described in Figure 2b, are executed instantaneously together with OPF.

Figure 7 illustrates the transient section before and after OPF happening to the three-phase IM drive described in Figure 2b whereby both $m1$ and $m2$ are executed. Being in the postfault mode, the feedforward term shown in Figure 7b is calculated according to (15) and injected at t = 0.2 s. The irregularity of the phase currents before injection of the feedforward term, shown in Figure 7c, is highlighting the inability of the conventional PI current controller even if the DOF is more than 2. This is because the PI controller in the context of FOC is designed to handle DC quantities only. However, after t = 0.2 s the disturbances originating from OPF are canceled out by injecting the feedforward voltage. It eventually allows disturbance-free operation of the RFOC re-enabling the two PI current controllers to track the set point. From 0.2 s onward in Figure 7c, the waveform of phases b and c start to become equal in magnitude and 60 degrees apart to generate circular rotating MMF with two windings only.

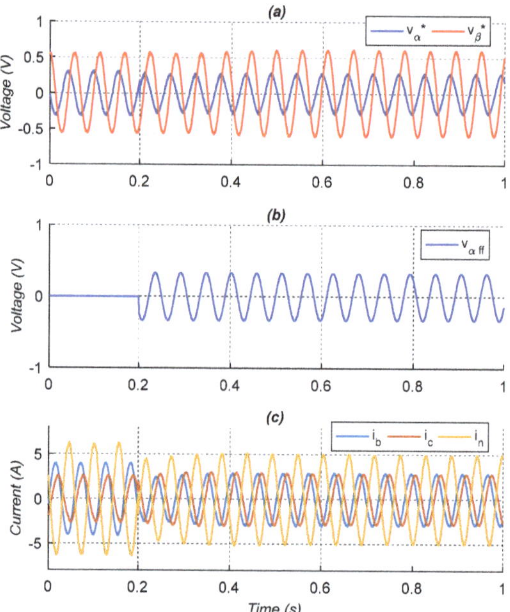

Figure 7. Postfault experimental result for the three-phase IM drive with feedforward injected at 0.2 s: (**a**) α-β voltage, (**b**) feedforward voltage, (**c**) phase and neutral current.

The experimental results show that feedforward injection successfully cancels out the disturbance, and hence, enables the current controller to regulate the circular trajectory of the α-β current, as illustrated in Figure 8a (highlighted by a circle). Unlike the healthy operation, the Figure 8b shows the elliptical shape of reference α-β voltage that has been supplied to the motor to have a circular trajectory of the α-β current in postfault mode. From the perspective of a rotating reference frame, the feedforward injection eventually blocks the severe double frequency oscillations in d-q current as well as mechanical speed, as depicted in Figure 9. Using the same approach as the three-phase IM, the symmetrical six-phase IM is driven in healthy mode first and one OPF is created on phase a_1 by means of relay contact. The waveform in Figure 10b shows the feedforward voltage that is obtained using (30) and has been injected according to (28) at t = 0.2 s. Due to OPF being phase a_1

and the neutral configuration of the drive, the phase current of b_1 and c_1 are forced to have equal and opposite magnitude, however, by implementing (27) together with feedforward injection from (28) at t = 0.2 s, the phase current of the set 2 is regulated to be unequal to restore circular current trajectory in the α-β plane, as illustrated in Figure 11a.

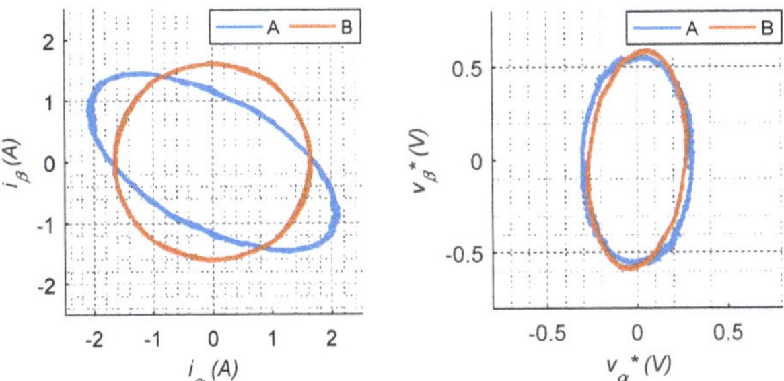

Figure 8. Trajectory of the current and voltage in α-β frame. A: before injection of feedforward, B: after injection of feedforward for three-phase IM.

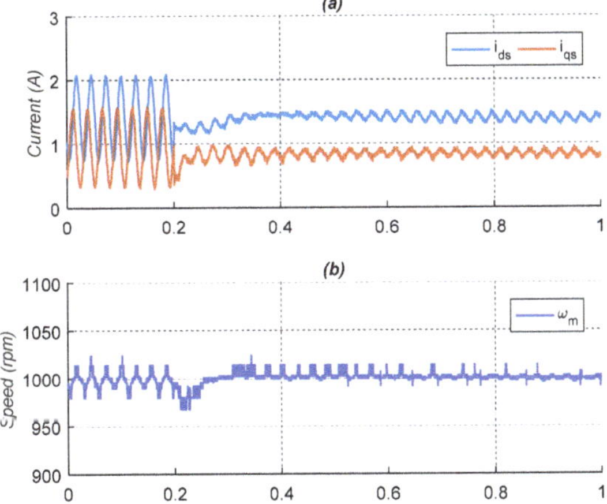

Figure 9. Postfault waveform of (**a**) the d-q current and (**b**) mechanical speed from the three-phase IM drive before and after feedforward injection at 0.2 s.

The suppression of double frequency AC oscillations in d-q current as well as mechanical speed in Figure 12 confirms the effectiveness of feedforward compensation on the S6 machine. It should be highlighted that feedforward compensation in the case of multiphase machines helps to remove AC disturbance terms caused by OPF. However, the speed oscillations of the symmetrical six-phase machine in Figure 12b due to one OPF are comparably lower than the three-phase counterpart in Figure 9b. This is one of the claimed advantages of multiphase drives in terms of fault tolerance which has been well-addressed in the literature. On top of tolerating the OPF fault in a feedforward manner, additional current control methods might be applied for multiphase drives to run the motor in maximum torque or minimum loss mode. However, this is not the case for the three-phase drives with an OPF.

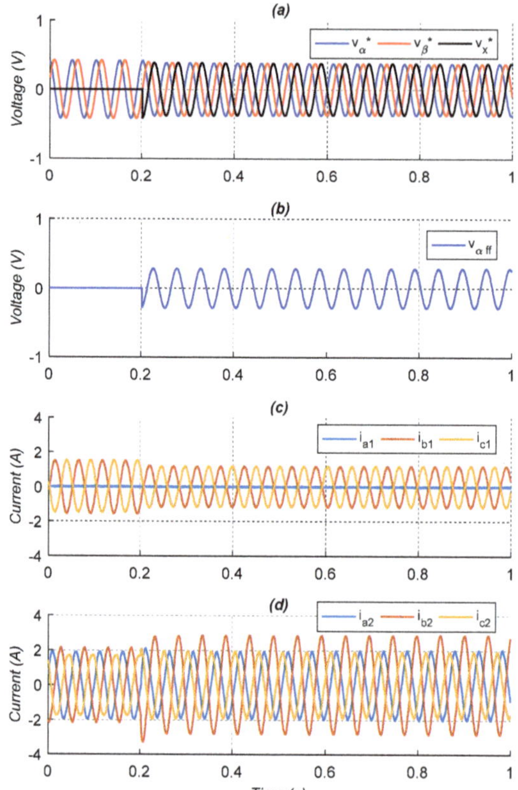

Figure 10. Postfault experimental result for the symmetrical six-phase IM drive with feedforward injected at 0.2 s: (**a**) α-β and x voltage, (**b**) feedforward voltage, (**c**) phase current of set1, (**d**) phase current of set 2.

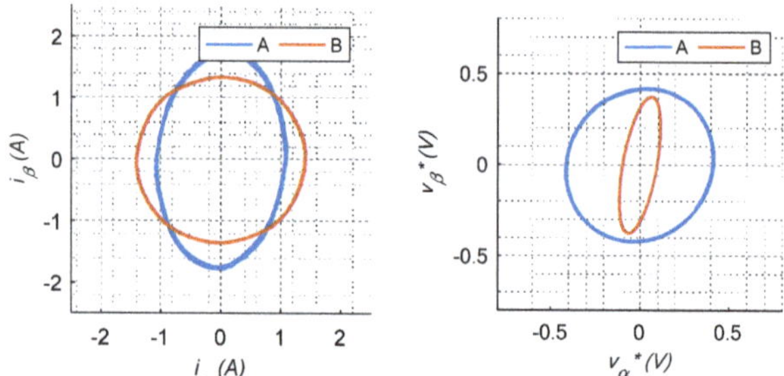

Figure 11. Trajectory of the current and voltage in α-β frame. A: before injection of feedforward, B: after injection of feedforward for symmetrical six-phase IM.

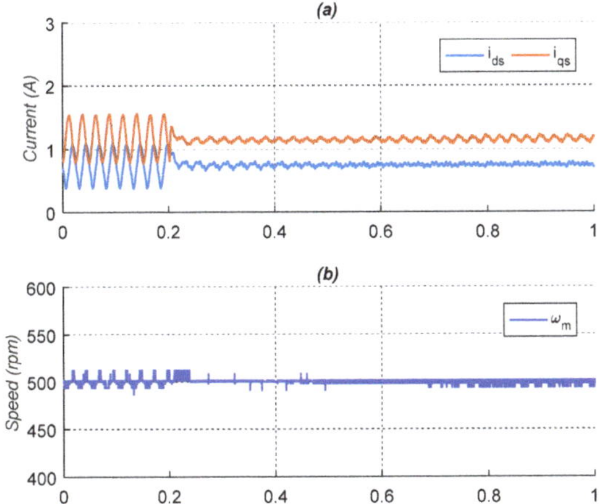

Figure 12. Postfault waveform of (**a**) the *d-q* current and (**b**) mechanical speed from the symmetrical six-phase IM drive before and after feedforward injection at 0.2 s.

5. Conclusions

In this paper, a generic analytical method is proposed to formulate the governing variables of FOC-driven AC drives under healthy and faulted conditions. This method considers the imposed and mandatory changes to the drive after OPF, if any, to specify how the control variables would be reflected in the machine terminals. The feedforward terms are subsequently derived based on a comparison of postfault relation to the healthy mode. The proposed method explicitly and generically formulates the feedforward terms to cancel out the undesired AC oscillatory terms expressed in both rotating and stationary reference frames. The experimental results of the symmetrical three- and six-phase machines verify the effectiveness of the proposed analytical method. Besides, the following salient findings can be noted:

- The feedforward compensation method, previously introduced for three-phase PMSM, has been re-derived in a generic way and applied to three-phase induction machines.
- The feedforward compensation on the stationary α-β reference frame is introduced instead of the *d-q* frame to make it immune to any error due to rotational transformations.
- The concept of feedforward compensation is further extended to multiphase induction machines, using a symmetrical six-induction machine as an example.
- It was shown that the feedforward term for a multiphase machine, like its three-phase counterpart, is a function of the back EMF voltage of the faulted phase, and independent of stator resistance value.

The future line of this study includes an investigation of the rapid fault detection schemes [36] incorporating the impact of small transient oscillation and DC offset [34] to be embedded into industrial drives, as well as commercial EVs [35,37].

Author Contributions: Conceptualization, M.T., A.Y. and H.S.C.; Methodology, M.T. and A.Y.; Software, M.T. and H.W.; Validation, M.T., A.Y. and H.S.C.; Formal analysis, M.T., H.S.C. and A.M.; Investigation, M.T.; Resources, H.W. and A.M.; Writing—original draft, M.T.; Writing—review and editing, M.T. and A.Y.; Supervision, H.S.C.; Project administration, M.T. and N.A.R.; Funding acquisition, H.S.C. and N.A.R. All authors have read and agreed to the published version of the manuscript.

Funding: This research was funded by Malaysian Ministry of Science and Technology, grant number FP090-2020.

Acknowledgments: Multiphase Machines Parameter Estimation Based on Drive Integrated Variable Extraction (DrIVE) Approach FP090-2020.

Conflicts of Interest: The authors declare no conflict of interest.

Nomenclature

AC	Alternating current
DC	Direct current
DOF	Degrees of freedom
DSRF	Double synchronous reference frame
FOC	Field-oriented control
FTC	Fault-tolerant control
MMF	Magnetomotive Force
OPF	Open-phase fault
PMSM	Permanent magnet synchronous machine
RFOC	Rotor field-oriented control
S6	Symmetrical six-phase
SRF	Synchronous reference frame

References

1. Mirafzal, B. Survey of Fault-Tolerance Techniques for Three-Phase Voltage Source Inverters. *IEEE Trans. Ind. Electron.* **2014**, *61*, 5192–5202. [CrossRef]
2. Behjati, H.; Davoudi, A. Reliability Analysis Framework for Structural Redundancy in Power Semiconductors. *IEEE Trans. Ind. Electron.* **2012**, *60*, 4376–4386. [CrossRef]
3. Benbouzid, M.E.H.; Diallo, D.; Zeraoulia, M. Advanced Fault-Tolerant Control of Induction-Motor Drives for EV/HEV Traction Applications: From Conventional to Modern and Intelligent Control Techniques. *IEEE Trans. Veh. Technol.* **2007**, *56*, 519–528. [CrossRef]
4. Naidu, M.; Gopalakrishnan, S.; Nehl, T.W. Fault-Tolerant Permanent Magnet Motor Drive Topologies for Automotive X-By-Wire Systems. *IEEE Trans. Ind. Appl.* **2010**, *46*, 841–848. [CrossRef]
5. Wang, R.; Wang, J. Fault-Tolerant Control With Active Fault Diagnosis for Four-Wheel Independently Driven Electric Ground Vehicles. *IEEE Trans. Veh. Technol.* **2011**, *60*, 4276–4287. [CrossRef]
6. Zhang, W.; Xu, D.; Enjeti, P.N.; Li, H.; Hawke, J.T.; Krishnamoorthy, H.S. Survey on Fault-Tolerant Techniques for Power Electronic Converters. *IEEE Trans. Power Electron.* **2014**, *29*, 6319–6331. [CrossRef]
7. Tousizadeh, M.; Che, H.S.; Selvaraj, J.; Rahim, N.A.; Ooi, B.-T. Performance Comparison of Fault-Tolerant Three-Phase Induction Motor Drives Considering Current and Voltage Limits. *IEEE Trans. Ind. Electron.* **2018**, *66*, 2639–2648. [CrossRef]
8. Wang, R.; Zhao, J.; Liu, Y. A Comprehensive Investigation of Four-Switch Three-Phase Voltage Source Inverter Based on Double Fourier Integral Analysis. *IEEE Trans. Power Electron.* **2011**, *26*, 2774–2787. [CrossRef]
9. Van Der Broeck, H.W.; Van Wyk, J.D. A Comparative Investigation of a Three-Phase Induction Machine Drive with a Component Minimized Voltage-Fed Inverter under Different Control Options. *IEEE Trans. Ind. Appl.* **1984**, *20*, 309–320. [CrossRef]
10. Liu, T.-H.; Fu, J.-R.; Lipo, T. A strategy for improving reliability of field oriented controlled induction motor drives. *IEEE Trans. Ind. Appl.* **1993**, *29*, 910–918. [CrossRef]
11. Bolognani, S.; Zordan, M.; Zigliotto, M. Experimental fault-tolerant control of a PMSM drive. *IEEE Trans. Ind. Electron.* **2000**, *47*, 1134–1141. [CrossRef]
12. Bianchi, N.; Bolognani, S.; Zigliotto, M.; Zordan, M. Innovative remedial strategies for inverter faults in IPM synchronous motor drives. *IEEE Trans. Energy Convers.* **2003**, *18*, 306–314. [CrossRef]
13. Correa, M.B.D.R.; Jacobina, C.B.; da Silva, E.C.; Lima, A.N. An induction motor drive system with improved fault tolerance. *IEEE Trans. Ind. Appl.* **2001**, *37*, 873–879. [CrossRef]
14. Gaeta, A.; Scelba, G.; Consoli, A. Modeling and Control of Three-Phase PMSMs Under Open-Phase Fault. *IEEE Trans. Ind. Appl.* **2013**, *49*, 74–83. [CrossRef]
15. Tousizadeh, M.; Che, H.S.; Abdel-Khalik, A.S.; Munim, W.; Selvaraj, J.; Rahim, N.A. Effects of flux derating methods on torque production of fault-tolerant polyphase inductiondrives. *IET Electr. Power Appl.* **2021**, *15*, 616–628. [CrossRef]
16. Lu, H.; Li, J.; Qu, R.; Ye, D.; Lu, Y. Fault-Tolerant Predictive Control of Six-Phase PMSM Drives Based on Pulsewidth Modulation. *IEEE Trans. Ind. Electron.* **2019**, *66*, 4992–5003. [CrossRef]
17. Peng, Z.; Zheng, Z.; Li, Y.; Liu, Z. Machine Drives Based on Virtual Winding Method. In Proceedings of the 2017 IEEE Transportation Electrification Conference and Expo (ITEC), Chicago, IL, USA, 22–24 June 2017; pp. 252–256.
18. Sun, J.; Liu, Z.; Zheng, Z.; Li, Y. An Online Global Fault-Tolerant Control Strategy for Symmetrical Multiphase Machines With Minimum Losses in Full Torque Production Range. *IEEE Trans. Power Electron.* **2020**, *35*, 2819–2830. [CrossRef]
19. Moraes, T.D.S.; Nguyen, N.K.; Semail, E.; Meinguet, F.; Guerin, M. Dual-Multiphase Motor Drives for Fault-Tolerant Applications: Power Electronic Structures and Control Strategies. *IEEE Trans. Power Electron.* **2017**, *33*, 572–580. [CrossRef]
20. Zhao, Y.; Lipo, T. Modeling and control of a multi-phase induction machine with structural unbalance: Part I. Machine modeling and multi-dimensional current regulation. *IEEE Trans. Energy Convers.* **1996**, *11*, 570–577. [CrossRef]

21. Ryu, H.-M.; Kim, J.-W.; Sul, S.-K. Synchronous Frame Current Control of Multi-Phase Synchronous Motor —Part II Asymmetric Fault Condition due to Open Phases. *IEEE Trans. Ind. Appl.* **2006**, *42*, 1062–1070.
22. Zhou, H.; Yang, G.; Wang, J. Modeling, Analysis, and Control for the Rectifier of Hybrid HVdc Systems for DFIG-Based Wind Farms. *IEEE Trans. Energy Convers.* **2011**, *26*, 340–353. [CrossRef]
23. Zhou, X.; Sun, J.; Li, H.; Song, X. High Performance Three-Phase PMSM Open-Phase Fault-Tolerant Method Based on Reference Frame Transformation. *IEEE Trans. Ind. Electron.* **2018**, *66*, 7571–7580. [CrossRef]
24. Xu, J.; Guo, S.; Guo, H.; Tian, X. Fault-Tolerant Current Control of Six-Phase Permanent Magnet Motor With Multifrequency Quasi-Proportional-Resonant Control and Feedforward Compensation for Aerospace Drives. *IEEE Trans. Power Electron.* **2023**, *38*, 283–293. [CrossRef]
25. Tani, A.; Mengoni, M.; Zarri, L.; Serra, G.; Casadei, D. Control of Multiphase Induction Motors With an Odd Number of Phases Under Open-Circuit Phase Faults. *IEEE Trans. Power Electron.* **2011**, *27*, 565–577. [CrossRef]
26. Che, H.S.; Duran, M.J.; Levi, E.; Jones, M.; Hew, W.-P.; Rahim, N.A. Postfault Operation of an Asymmetrical Six-Phase Induction Machine With Single and Two Isolated Neutral Points. *IEEE Trans. Power Electron.* **2013**, *29*, 5406–5416. [CrossRef]
27. Munim, W.N.W.A.; Duran, M.J.; Che, H.S.; Bermudez, M.; Gonzalez-Prieto, I.; Rahim, N.A. A Unified Analysis of the Fault Tolerance Capability in Six-Phase Induction Motor Drives. *IEEE Trans. Power Electron.* **2016**, *32*, 7824–7836. [CrossRef]
28. Guo, Y.; Wu, L.; Huang, X.; Fang, Y.; Liu, J. Adaptive Torque Ripple Suppression Methods of Three-Phase PMSM During Single-Phase Open-Circuit Fault-Tolerant Operation. *IEEE Trans. Ind. Appl.* **2020**, *56*, 4955–4965. [CrossRef]
29. Tousizadeh, M.; Yazdani, A.M.; Che, H.S.; Rahim, N.A. Feedforward Fault-Tolerant Control for Three-Phase Induction Motor Drives with Single Open Circuit Fault. In Proceedings of the 2019 International Conference on Power and Energy Systems (ICPES), Perth, WA, Australia, 10–12 December 2019; pp. 1–6. [CrossRef]
30. Jasim, O.; Sumner, M.; Gerada, C.; Arellano-Padilla, J. Development of a new fault-tolerant induction motor control strategy using an enhanced equivalent circuit model. *IET Electr. Power Appl.* **2011**, *5*, 618–627. [CrossRef]
31. Tousizadeh, M.; Che, H.S.; Selvaraj, J.; Abd Rahim, N.; Ooi, B.T. Fault-Tolerant Field Oriented Control of Three-Phase Induction Motor based on Unified Feed-forward Method. *IEEE Trans. Power Electron* **2019**, *34*, 7172–7183. [CrossRef]
32. Krause, P.C.; Wasynczuk, O.; Sudhoff, S.D. *Analysis of Electric Machinery and Drive Systems*, 3rd ed.; IEEE Press: Piscataway, NJ, USA, 2002.
33. Levi, E. Multiphase Electric Machines for Variable-Speed Applications. *IEEE Trans. Ind. Electron.* **2008**, *55*, 1893–1909. [CrossRef]
34. Mule, G.J.A.; Munim, W.N.W.A.; Tousizadeh, M.; Abidin, A.F. Post-fault tolerant of symmetrical six-phase induction machine under open circuit fault with single and two isolated neutral points using graphical user interface. *AIP Conf. Proc.* **2019**, 2173. [CrossRef]
35. Che, H.S.; Duran, M.; Levi, E.; Jones, M.; Hew, W.P.; Rahim, N.A. Post-fault operation of an asymmetrical six-phase induction machine with single and two isolated neutral points. In Proceedings of the 2013 IEEE Energy Conversion Congress and Exposition, Denver, CO, USA, 15–19 September 2013. [CrossRef]
36. Ghanooni, P.; Habibi, H.; Yazdani, A.; Wang, H.; MahmoudZadeh, S.; Mahmoudi, A. Rapid Detection of Small Faults and Oscillations in Synchronous Generator Systems Using GMDH Neural Networks and High-Gain Observers. *Electronics* **2021**, *10*, 2637. [CrossRef]
37. Roshandel, E.; Mahmoudi, A.; Kahourzade, S.; Yazdani, A.; Shafiullah, G. Losses in Efficiency Maps of Electric Vehicles: An Overview. *Energies* **2021**, *14*, 7805. [CrossRef]

 *energies*

Article

Prony Method Estimation for Motor Current Signal Analysis Diagnostics in Rotor Cage Induction Motors

Luis Alonso Trujillo Guajardo *, Miguel Angel Platas Garza, Johnny Rodríguez Maldonado, Mario Alberto González Vázquez, Luis Humberto Rodríguez Alfaro and Fernando Salinas Salinas

Universidad Autónoma de Nuevo León, UANL, FIME, Av. Universidad S/N Ciudad Universitaria, San Nicolás de los Garza C.P. 66451, NL, Mexico; miguel.platasgrz@uanl.edu.mx (M.A.P.G.); johnny.rodriguezml@uanl.edu.mx (J.R.M.); mario.gonzalezvzq@uanl.edu.mx (M.A.G.V.); luis.rodriguezlf@uanl.edu.mx (L.H.R.A.); fernando.salinassln@uanl.edu.mx (F.S.S.)
* Correspondence: luis.trujillogjr@uanl.edu.mx; Tel.: +52-818-329-4020

Abstract: This article presents an evaluation of Prony method and its implementation considerations for motor current signal analysis diagnostics in rotor cage induction motors. The broken rotor bar fault signature in current signals is evaluated using Prony method, where its advantages in comparison with fast Fourier transform are presented. The broken rotor bar fault signature could occur during the life cycle operation of induction motors, so that is why an effective early detection estimation technique of this fault could prevent an insulation failure or heavy damage, leaving the motor out of service. First, an overview of cage winding defects in rotor cage induction motors is presented. Next, Prony method and its considerations for the implementation in current signature analysis are described. Then, the performance of Prony method using numerical simulations is evaluated. Lastly, an assessment of Prony method as a tool for current signal analysis diagnostics is performed using a laboratory test system where real signals of an induction motor with broken rotor bar operated with/without a variable frequency drive are analyzed. The summary results of the estimation (amplitudes and frequencies) are presented in the results and discussion section.

Keywords: Prony method; broken rotor bar; fast Fourier transform; current signal analysis

Citation: Trujillo Guajardo, L.A.; Platas Garza, M.A.; Rodríguez Maldonado, J.; González Vázquez, M.A.; Rodríguez Alfaro, L.H.; Salinas Salinas, F. Prony Method Estimation for Motor Current Signal Analysis Diagnostics in Rotor Cage Induction Motors. *Energies* **2022**, *15*, 3513. https://doi.org/10.3390/en15103513

Academic Editors: Daniel Morinigo-Sotelo, Rene Romero-Troncoso and Joan Pons-Llinares

Received: 18 March 2022
Accepted: 9 May 2022
Published: 11 May 2022

1. Introduction

Electric motors, particularly rotor cage induction motors (RCIM) are considered for most of industry applications, because of its operation performance and low maintenance cost. According to statistical studies performed by IEEE and Electric Power Research Institute, a percentage of 8–9% of the total RCIM faults occur in the rotor [1], where broken rotor bars (BRB) or cracked/broken end rings (CBER) are the most common issues in RCIM operation. It is well-known that the cage of RCIM is made typically of aluminum and in some cases of copper [2]. BRB or CBER will have a high probability of appearing when, for example a RCIM is operated considering several direct-on-line (DOL) starts in a short period of time or RCIM with high inertial loads, these conditions will put the RCIM through an excessive centrifugal, thermal, and mechanical stress. It is important to mention that a BRB or CBER cause a reduction in RCIM operation performance, for example, having problems to move its load mainly due to an unbalanced flux in the rotor causing a reduction in the output torque [3] and heavy damage to the RCIM. So, to prevent a heavy damage situation, BRB or CBER must be detected early in a noninvasive way, where as a part of condition monitoring of RCIM, motor current signature analysis (MCSA) is used to determine problems such as BRB, CBER, or abnormal air gap eccentricity (AAGE) at the rotor cage when the motor is under normal operation.

It is important that BRB or CBER are accurately detected using MCSA. However, in some cases there exist "false positives", which means an unnecessary shut down and transportation to a qualified repair facility, and if the RCIM is a large machine there will be higher costs from loss in production and maintenance expenses than the cost of a new motor [4]. On the other hand, if there is a "false negative" the motor will be kept in operation and a catastrophic failure will occur if the broken rotor bar reaches the stator winding.

Recently, in the last three years, research has been going on to detect RCIM and BRB faults in a noninvasive way and several methods and techniques have been presented in the literature. For example in [5], an analysis using a new algorithm based on the Park's transformation, where the direct and quadrature current components were analyzed for BRB fault detection and identification; in [6], finite element software is proposed to be used to detect motor broken bar mechanical fault by detecting magnetic flux density fluctuations; in [7], a novel methodology based on motor current signal analysis and contrast estimation is introduced for BRB detection, where a textural feature "Contrast" commonly used for image classification in combination with fuzzy logic classifier is proposed for BRB detection; in [8], an intelligent multi-agentsystem (MAS) is proposed to make decisions on the fault conditioning of a three-phase squirrel cage induction motor where also artificial intelligent methods are used; in [9], a two-stage approach for three-phase induction motors diagnosis based on mutual information measures of the current signals, principal component analysis, and intelligent systems is proposed; in [10], an approach based on the analysis of the startup transient current signal through the current signal homogeneity and the fourth central moment (kurtosis) analysis is presented, where these features are used for training a feed-forward, backpropagation artificial neural network used as a classifier; in [11] a scheme to detect broken bar faults and discriminate the severity of faults under starting conditions is presented, where a successive variable mode decomposition (SVMD) is applied to analyze the stator starting current to extract the fault component, and the signal reconstruction is proposed to maximize the energy of the fault component, and so the signature frequency could be detected; in [12], an end-ring wear detection through a multicomponent approach is researched; in [13], an estimate of the fundamental frequency component from an optimization point of view is proposed; in [14], magnetic flux condition monitoring is reviewed in detail, and it is focused on the diagnosis of different types of faults in the most common rotating electric machines used in industry; in [15–19], several methods focus on RCIM using variable frequency drives where the proposed methods have in common start up transient analysis, and others considers bar breakage harmonics evolution for BRB diagnostics; finally, in [20], a zero-setting protection element, which uses the current signature method is proposed to detect broken rotor bars, where the research has been applied in commercial protection relay equipment, and considers the use of an alpha current signal to obtain the frequency spectrum of the signal to detect the BRB fault signature. Nevertheless, MCSA as condition monitoring has been popular to help diagnose RCIM problems since the 1970s and it is commonly used in online test equipment because it only needs voltage and current probes. In online test equipment for RCIM, the tester records voltage and current signals with a defined resolution and observation window which is selected prior to perform a test. Data acquisition is enabled so that the tester software can record the signals and spectral analysis using fast Fourier transform (FFT) [21] can be performed. It should be mentioned that FFT is the widely used digital signal processing technique for this application, where fault signature frequencies can be detected, and a diagnostic can be issued [22]. However, FFT has some limitations that could lead to misdiagnosis of BRB: If the tester user performs the measurement without knowing in detail the different conditions or situations in which a misdiagnosis of BRB can occur, for these scenarios if a load variation during the acquisition of signals occur, if the machine is under low load conditions, or if a short sampling time or length of the recorded signal is not adequate (too short) the spectral analysis of the measured signal will estimate frequencies that are not really there, or simply the BRB fault signature frequencies will not be detected [23]. This is why it is important to mention that knowing how the data acquisition and recording is performed in order to find the limitations is highly recommended, for the main purpose of being able to have an accurate diagnosis, whether there is a BRB fault or not. The reader should know that rotor cage faults such as BRB or

CBER produce a magnetic asymmetry in the air gap, and the asymmetry will be adding specific frequencies which are close to the fundamental frequency that will be appearing at motor line current measurement.

Different techniques for BRB detection have been studied in the last 10 years, for example [5–19,23–28], where some of these techniques require fixed observation windows and others considers recursive algorithms, artificially intelligent methods, startup transient analysis, and new developed transforms such as the dragon transform, these techniques that were studied and proposed for BRB fault signature detection have one point in common, which is the complexity of the practical implementation due to its considerations and calculations, nevertheless, none of these techniques are used nowadays in commercial online test diagnostics (OTD) equipment, the estimation technique used is a FFT-based tool which has its estimation limitations. To overcome the limitations of the FFT-based tool, in this work, the parametric estimation technique Prony method has been proposed as a BRB detection technique to increase the accuracy of the detection of rotor cage fault signature frequencies, particularly BRB sideband frequencies. In the recent years, only five articles have considered the Prony method for BRB fault detection [29–33]. For example, [29] considered using the Prony method to estimate frequency and amplitudes of a signal under analysis using different windows length of recorded data samples. [30] proposed the use of Prony method with other technique, such as Hilbert transform and discrete wavelet transform. In [31], the Prony method was used to evaluate different load conditions at a defined sampling frequency. [32] proposed the use of Prony method in combination with singular value decomposition (SVD) filtering technique and Multiple Signal Classification (MUSIC). [33] proposed the use of a modified Prony method in combination with MUSIC algorithm. In [29,31], the Prony method is used for BRB diagnosis and an analysis of different lengths of data at different load conditions is presented, however, no analysis and validation considering VFD is reported which is a common industry application. In this work, in order to fulfill the application considerations and validation that where not considered in the mentioned references, relevant and application details described in Sections 3 and 4 are presented, where also VFD operation at different speeds for full load condition is evaluated to guarantee an accurate detection of BRB and a diagnosis of the cage of the rotor can be accurately determined.

In this investigation, numerical simulated signal and a laboratory test system with a faulted RCIM being operated and controlled by a VFD under different speed conditions is considered for the validation of the method for its practical application for detection of cage rotor faults, and a comparison with FFT analysis is also considered to show the limitations of using FFT for this particular application. In this approach, the considerations required for the application of Prony method for an accurate detection of cage rotor fault signature frequencies and its magnitudes are described. First, a fixed window of data samples is defined. Then, the signal under analysis is digitally processed and downsampled. Moreover, a low pass and DC filter is applied to the downsampled signal, where the processed output signal is the input signal for Prony method estimation algorithm. The purpose of subsampling is to reduce the computational calculation of the algorithm, because commercial equipment for OTD considers a defined high sampling frequency, which is not needed to estimate the low frequencies related with the rotor fault; also a low pass and DC filter stage is needed to eliminate undesired frequencies in the signal, and to reduce the noise of the signal to the minimum, so an accurate estimate of the amplitudes and frequencies of the BRB signature components can be achieved.

The paper is organized as follows. First, an overview of diagnostics on cage winding defects is described in Section 2. The Prony method and its application considerations are discussed in Section 3. Then, a study case considering numerical simulated signals and a laboratory test system for MCSA where numerical and experimental results comparison between FFT analysis and Prony method estimation are presented in Section 4. Finally, a summary of the results achieved is presented in Section 5, where the performance of Prony method for this application is also discussed.

The main contribution of this article is that Prony method can be considered as a diagnostic tool in BRB OTD, where its application could be feasible, and with lesser recorded data signal values, in comparison with FFT analysis, a good estimate of sideband frequencies and its amplitudes can be obtained, the method can be used as a new tool to improve the accuracy and sensitivity of OTD in RCIM, and it could be implemented in the existing or new software in online test equipment's for electric motors with no need of hardware updates. The methodology was validated by using simulation signals and real data signals, and its effectiveness is presented.

2. Overview of Diagnostics on Cage Winding Defects

In this section, fundamental aspects of broken rotor bar side band frequencies (BRB_{sbf}) required for the detection of these components during the estimation of a current signal under analysis are presented. When a RCIM has a fracture or break in its cage (BRB or CBER), see Figure 1, the effects within the motor result in voltages at specific frequencies as presented in Equation (1), mainly due to the alteration of the magnetic field at the cage of the rotor.

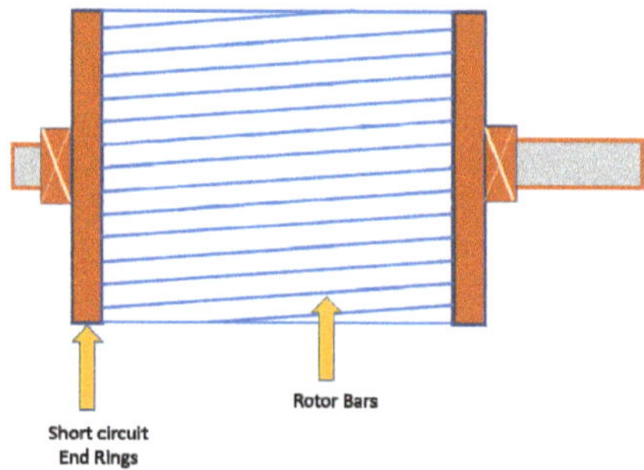

Figure 1. Illustration of a caged rotor of a RCIM.

BRB and CBER in a RCIM are mainly caused due to a maloperation, such as, too many sequential direct on-line starts where time delay between starts is not adequate according to manufacturer specification, which causes high starting currents, also, an incorrect match of torque-speed curve of RCIM and torque-speed curve of the load. These operation conditions add excessive centrifugal, mechanical, and thermal stress at the slip ring which can result in a BRB or CBER fault. Due to a BRB or CBER, an unbalanced rotor flux occurs and an effect in motor operation performance appears, for example, line currents oscillation, torque pulsation, decreased average torque, and excessive vibration [3,4]. If the BRB or CBER fault are not detected, an insulation failure or heavy damage could occur, and a replacement of the RCIM will be required.

The BRB_{sbf} in a RCIM considers the supply fundamental frequency (f_0), the slip of the motor (s), and its harmonic value (h), which represent the number of broken rotor bars that could appear in the RCIM [3,4]. The equation for BRB_{sbf} in a RCIM is presented in Equation (1):

$$BRB_{sbf} = f_0(1 \pm 2 * h * s) \tag{1}$$

The BRB_{sbf} affects the RCIM operation if the energy of the frequencies is within the fault indication value of a break in the cage rotor circuit; typical failure signature values considered are presented in Table 1.

Some physical conditions affect the magnitude of BRB_{sbf}, which can affect the resulting diagnostic during a MCSA. Some of these conditions are, change in load and slip with a fixed rotor cage defect, faulty bar to end ring joints creating an asymmetrical cage, porosity and consequential arcing in aluminum die-cast cage rotors, partially broken rotor bars, actual broken bars still making contact with an end ring and bars which are cracked from the top of the bar, but just a percentage of the total depth of the bar [4]. It is important to mention that it is not possible to predict exact severity of cage defects/number of BRB or CBER from the magnitudes in dB of the BRB_{sbf} with respect to f_0, only an estimate of the condition of the cage winding can be defined as presented in Table 1.

Table 1. Rotor condition sideband frequency failure signature.

Energy, dB	Rotor Condition
>60	Excellent
54–60	Good
48–54	Moderate
42–48	High resistance connection or cracked bars
36–42	Broken rotor bars will show in vibration analysis
30–36	Multiple $\frac{cracked}{broken}$ bars, possible ring problems
<30	Severe rotor faults

The accuracy of the estimation of magnitudes in dB of BRB_{sbf} of a motor current signal under analysis depends on the digital signal processing (DSP) technique, see Figure 2. Nowadays, modern on-line test equipment use the fast Fourier transform (FFT) spectral analysis to detect harmonics and other frequencies such as BRB_{sbf}; this is a commonly used technique but it has its limitations, for example, (a) the motor current signals measurements should be performed during steady state operation of the motor, so the estimation of frequencies and its magnitudes detected could be accurate, (b) the recorded current signal should have between 10 and 120 s of acquire data without any load variations during acquisition, in terms of signals cycles for a 60 Hz or 50 Hz, a minimum of 600 or 500 cycles of recorded signal should be required at least to be able to detect BRB_{sbf}, because these frequencies are too close to the fundamental frequency. In the case of motor operation with VFD, the number of signal cycles will be proportional to the operation frequency considering the same length between 10 and 120 s of acquired data. It should be mentioned that in case of a motor under low load or no-load operation, the cage winding circulating currents will be minimum or null, so at this condition the BRB_{sbf} cannot be detected with the estimation technique being used.

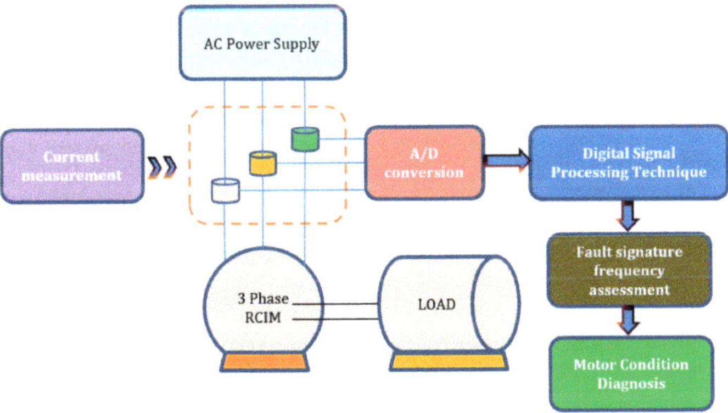

Figure 2. General scheme of MCSA for RCIM.

3. Prony Method Estimation for Motor Current Signal Analysis

Prony method is a signal processing technique based on signal estimation, which extracts desired information from an equally spaced sampled signal and builds a series of damped complex exponentials to approximate the sampled signal by solving a set of linear equations. The Prony algorithm and its practical implementation are presented

where the motor current signals are recorded at full load condition at nominal speed (60 Hz) and at different operating speeds. This will allow to perform a complete evaluation and validation of the method for OTD of RCIM diagnostics application.

4.1. Assessment of Numerical Simulation of Broken Rotor Bar Current Signal

For this analysis, the simulated current signal considering two side band frequencies is used to estimate its signal parameters using Prony method, as presented in (6), (7), (10), and (11). In Figure 3, 2 s simulated signal is used for the analysis, where the full signal considers two sideband frequencies and its individual frequency components are shown for a fundamental frequency of 60 Hz and a slip value of 0.0089 describing the behavior of a loaded motor, where (1) is used to obtain the two sideband frequencies in the signal, so the signal in Figure 3a can be used for the Prony analysis. In Table 2, the harmonic order, frequency, amplitude of fundamental frequency, and the calculated BRB_{sbf} of the signal in Figure 3 that will be used for the analysis are presented.

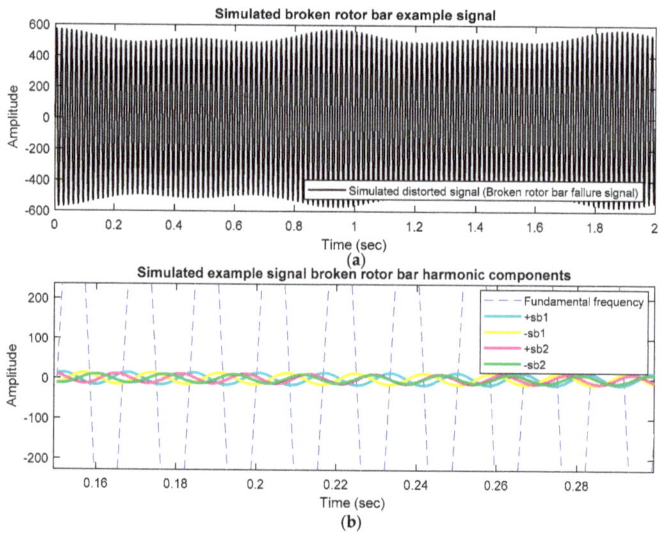

Figure 3. Simulated signal example. (**a**) Distorted signal. (**b**) Broken rotor bar harmonic components.

Table 2. Simulated example signal with broken rotor bar harmonic components.

Harmonic Order	Frequency, (Hz)	Amplitude
1	60	516
1.0178	61.06 (+sb1)	15
0.9822	58.93 (−sb1)	14
1.0356	62.13 (+sb2)	12
0.9644	57.86 (−sb2)	11

Prony Estimation Results

The signal in Figure 3a is used for analysis with a sampling frequency of 64 samples per cycle and a window of data of 8 cycles. The sampling frequency of 64 samples/cycle and 8 cycles of window data length are considered because with a fixed data window with lesser data a good and accurate estimation could be achieved. It is important to mention that with more data considered for the estimation process (more than 8 cycles for simulated signal in Figure 3a), an increase in computational burden will occur and more time will be required to obtain accurate estimated frequencies and amplitudes. In Figure 4a, the Prony estimated and original signal are compared, and it can be observed that there is no considerable error between them, and the estimated spectrum in Figure 4b of signal in Figure 4a is obtained from the Prony estimation results calculated from (7) and (10), the estimation results are presented in Table 3.

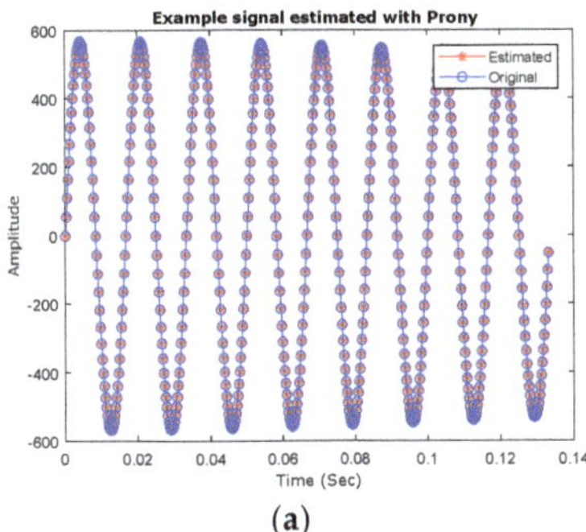
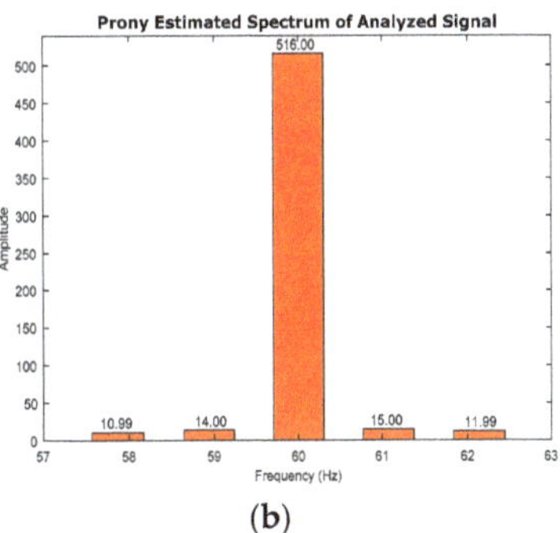

Figure 4. Simulated example signal validation. (a) Estimated and original signal comparison. (b) Estimated Prony spectrum.

Table 3. Simulated example, Prony estimated signal parameters.

Estimated Signal Parameters	Frequency, (Hz)	Amplitude
	60	516.00
	61.06	15.00
Example signal	58.93	14.00
	62.13	11.99
	57.86	10.99

The Prony estimation results in Table 3 are obtained using a defined window of data of the simulated example signal in Figure 3a; 8 cycles and a sampling frequency of 64 samples per cycle are considered. Only the estimated signal amplitude and frequency are presented in Table 3, because these parameters are the ones that can be used to detect BRB_{sbf} of the signal under analysis. It should be mentioned that the damping results and phase angle of the signal parameter estimation correspond to the model of the Prony signal presented in Equation (2) and are required only to form the estimated Prony signal and then the MSE is calculated so the optimum parameter estimates could be obtained. Good estimation results can be achieved considering a window data length of 8 cycles and a sampling frequency of 64 samples per cycle for the detection of two sideband frequencies for broken rotor bars.

4.2. Assessment of Real Broken Rotor Bar Current Signal from a Laboratory Test System

For this analysis, a laboratory test system considering a $\frac{1}{2}$ HP RCIM with a crack in one ring of the rotor, a data acquisition system, and a VFD are used for the current signal analysis using the Prony method; it should be mentioned that different operation conditions are considered for the analysis. Moreover, a comparison of FFT results and Prony estimation results are presented so that the advantages of using Prony method can be highlighted for OTD application.

4.2.1. Laboratory Test System

In Figure 5, the laboratory test system used for the data acquisition of current signals of the RCIM under analysis is presented. For the analysis, the current signals are required, so the signals are measured at the motor terminals; in Figure 5a, a VFD is used to control the speed of the RCIM which will be operating at full load condition. The nameplate data of the RCIM are as follows 208 V, 1.98 A, 60 Hz, 1730 rpm and are presented in Appendix A in Table A1. In Figure 5b, the load is set at 2.05 N-m, which is the condition for full load operation for the RCIM. In Figure 5c, the data acquisition system used is a National Instruments cRIO-9045 with LabVIEW software and Tektronix current sensors A622 with the setting of 100 mv/A. Figure 6 shows the RCIM rotor of RCIM in Figure 5b which has a cracked ring and a broken rotor bar.

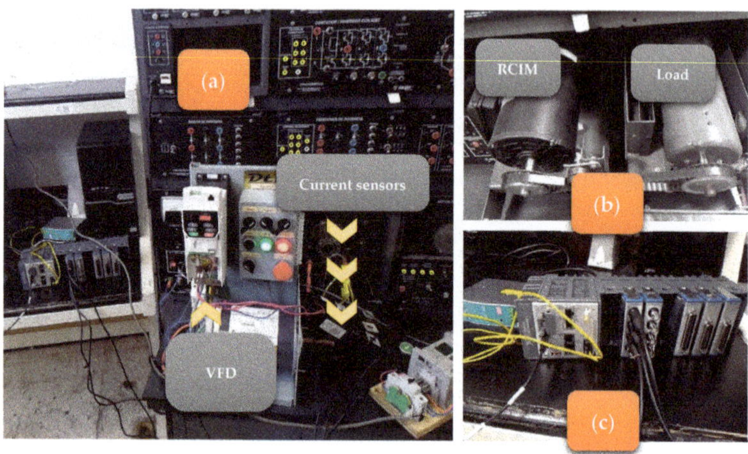

Figure 5. Laboratory test system. (**a**) Full system. (**b**) RCIM under analysis and its load. (**c**) National Instruments data acquisition system compact RIO.

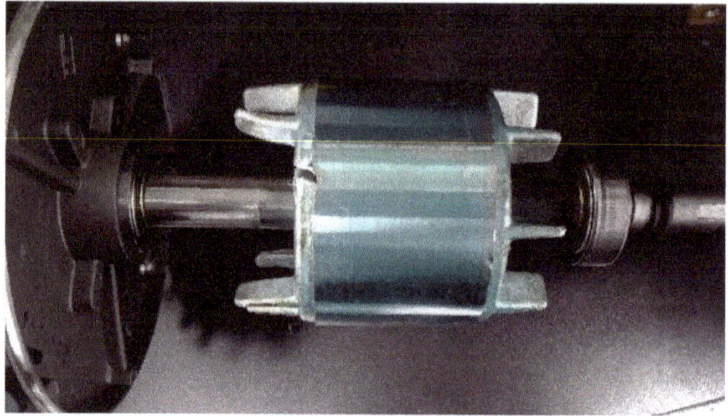

Figure 6. RCIM cracked ring rotor and broken rotor bar for analysis.

Figure 7 presents the current signal measurement of motor line current of phase A, only 1 s of the recorded signal of the 5 s total length is presented for visualization purposes, current signal in Figure 7 was obtained at 60 Hz operation of the motor (full speed), and full load condition without the operation of the VFD. It should be mentioned that the sideband frequencies considered for the detection of the cracked ring or broken rotor bar were calculated with (1) considering a full load slip of 0.0388, 2 sideband frequency components were calculated for 60 Hz as fundamental component: 64.66 Hz, 55.33 Hz, 69.33 Hz, and 50.66 Hz. Hence, the signal in Figure 7 will be used for the Prony analysis and FFT analysis for 60 Hz. In the following sections, current signal measurements at different VFD speed operation conditions (50 Hz, 40 Hz, 30 Hz, 20 Hz, and 10 Hz) are analyzed to validate the proposed Prony method for OTD application in RCIM, and FFT will also be analyzed to compare and highlight the advantages of using Prony method for OTD in RCIM.

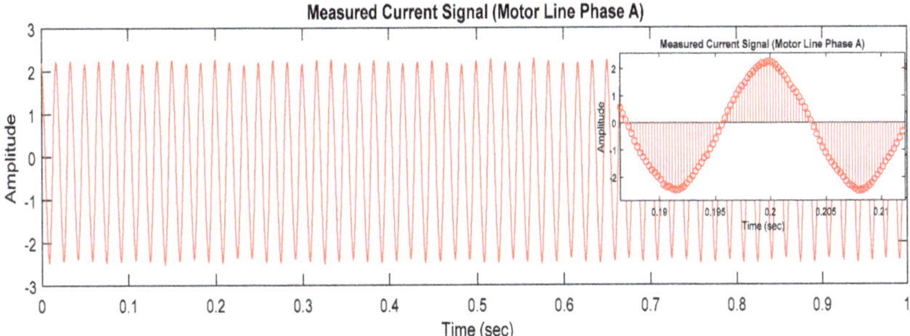

Figure 7. Current signal measurement of motor line current of phase A with a sampling frequency of 64 samples per cycle at 60 Hz with no VFD operation.

4.2.2. Fast Fourier Transform Estimation Results

In this section a FFT analysis of the measured current signals from the test system in Figure 5 is performed using the VFD. For the purpose of analysis, only motor line current of Phase A is considered and the length of the recorded signal is 5 s. The sampling frequency of the signal is 64 samples per cycle, and the FFT estimation results are presented for different speed operations at full load of the RCIM; the output frequencies for the VFD, which are related to rotor speeds, are 60 Hz, 50 Hz, 40 Hz, 30 Hz, 20 Hz, and 10 Hz.

It could be observed that for a signal record of 5 s, for every speed variation 60 Hz, 50 Hz, 40 Hz, 30 Hz, 20 Hz, and 10 Hz in Figures 8–13, the sideband frequencies considered for each speed operation condition are not detected. It should be mentioned that at least 10 s of the signal must be recorded to determine if a problem with the rotor is present, as described in Section 2. Table 4 presents two sideband frequencies of the RCIM calculated from (1) for each speed operation condition frequency. BRB_{sbf} will be used to compare the results obtained from the FFT estimations and also the information defined in Table 1.

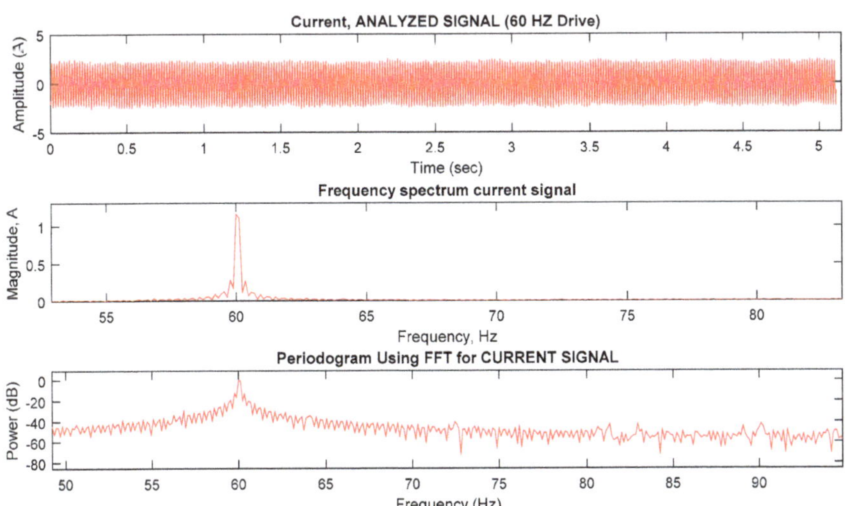

Figure 8. FFT analysis of current signal measurement of motor line current of phase A, frequency spectrum, and periodogram at 60 Hz speed operation.

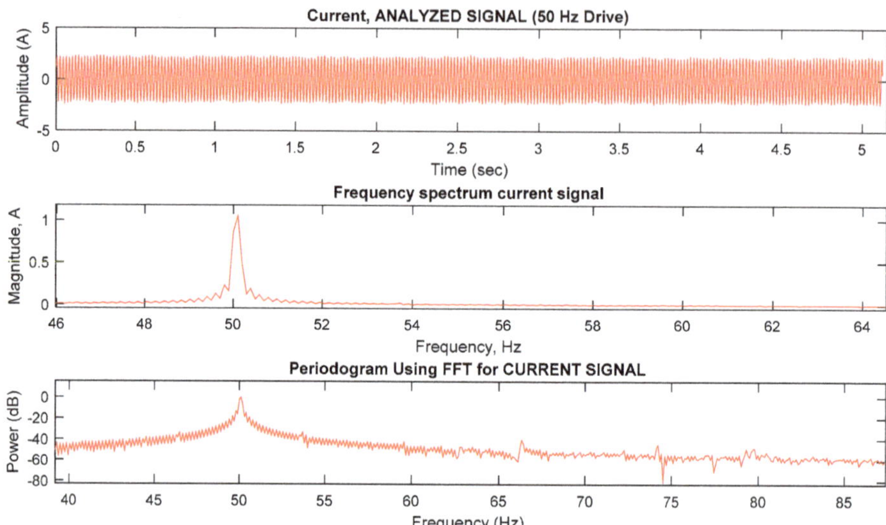

Figure 9. FFT analysis of current signal measurement of motor line current of phase A, frequency spectrum, and periodogram at 50 Hz speed operation.

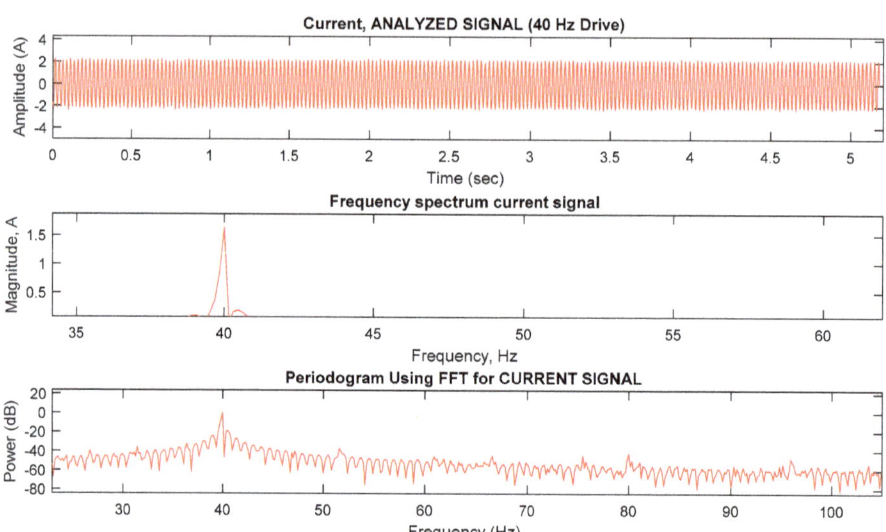

Figure 10. FFT analysis of current signal measurement of motor line current of phase A, frequency spectrum, and periodogram at 40 Hz speed operation.

As shown in FFT spectrum and periodogram from Figures 8–13 and Table 5, frequencies presented in Table 4 are not detected for each VFD speed operation condition, only the fundamental frequency is detected. This occurs mainly due to the closeness of the frequencies to the fundamental frequency, where more signal data are required (at least 10 s) so the FFT could detect the sideband frequencies. The periodogram is included because it is used to highlight the sideband frequencies in dB, which is the unit where a severity of a faulted rotor of RCIM can be measured, as it is presented in Table 1. Hence, one of the disadvantages of the FFT analysis for OTD for RCIM, is that it is not possible for the detection of the BRB_{sbf} with less than 10 s of recorded current signal, where it is evident in the results from Figures 8–13 and Table 5 that the sideband frequencies are not detected, so larger data windows and no variations in load of the motor under analysis are required for FFT analysis to detect BRB_{sbf}.

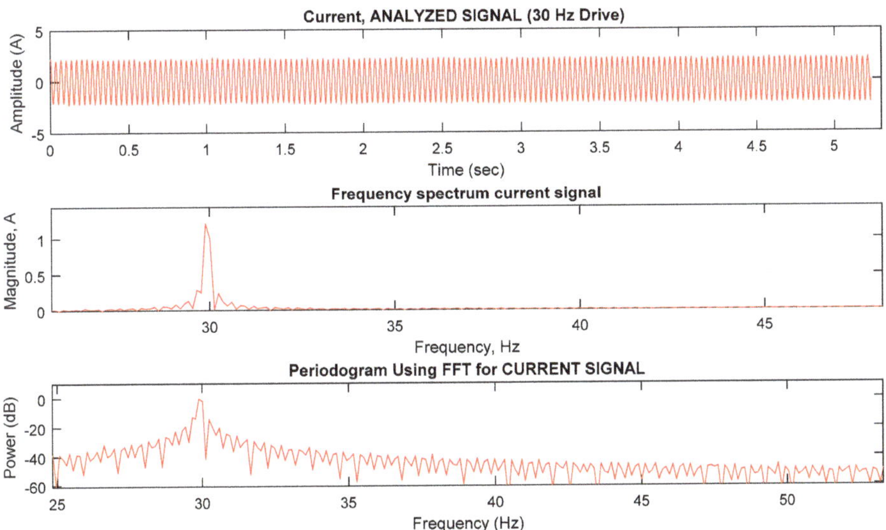

Figure 11. FFT analysis of current signal measurement of motor line current of phase A, frequency spectrum, and periodogram at 30 Hz speed operation.

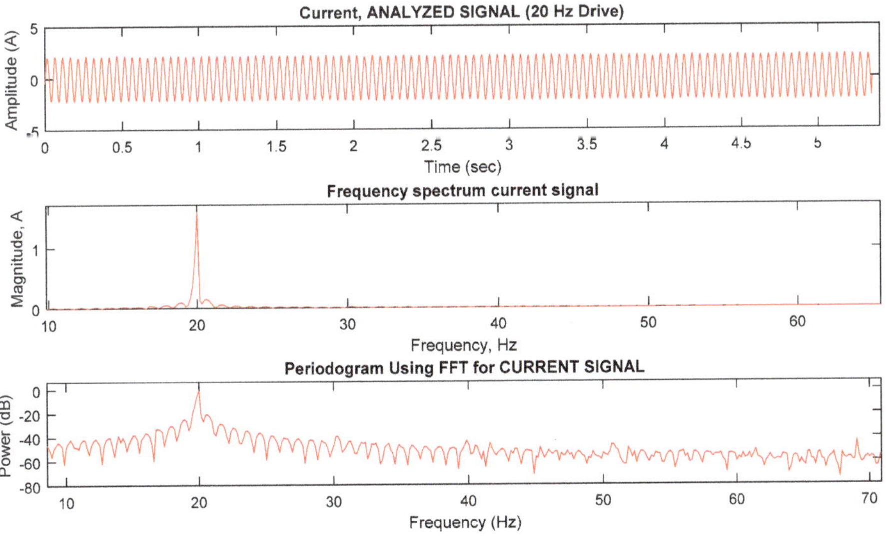

Figure 12. FFT analysis of current signal measurement of motor line current of phase A, frequency spectrum, and periodogram at 20 Hz speed operation.

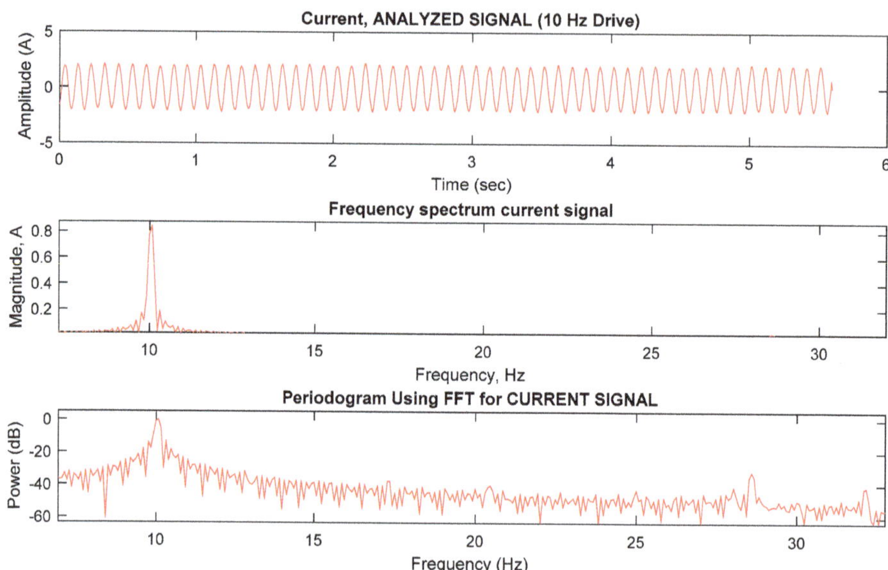

Figure 13. FFT analysis of current signal measurement of motor line current of phase A, frequency spectrum, and periodogram at 10 Hz speed operation.

Table 4. BRB_{sbf} for each speed operation condition at VFD.

Full Load Slip, (s)	Sideband Frequencies	Frequency, (Hz)					
	Speed operation Condition, Fundamental frequency	60	50	40	30	20	10
0.0388	Sb1+	64.66	53.88	43.11	32.33	21.55	10.77
	Sb1−	55.33	46.11	36.88	27.66	18.44	9.22
	Sb2+	69.33	57.77	46.22	34.66	23.11	11.55
	Sb2−	50.66	42.22	33.77	25.33	16.88	8.44

4.2.3. Prony Estimation Results

In this section, the proposed Prony method estimation for OTD application for RCIM diagnostics is evaluated for different speed operation conditions of a RCIM at full load using the test system in Figure 5. The recorded signals used for the analysis are the current signals from Figures 8–13 for each speed operation condition, where the difference between the analysis from FFT and Prony, is that the estimation of signal frequency and amplitude parameters will be obtained only by considering 25 cycles of each signal instead of 5 s. The sampling frequency of the signals is the same for each one of the 64 samples per cycle. The following steps should be followed to obtain the optimum signal parameters of frequency and amplitude, so we can search the BRB_{sbf} presented in Table 4 in the Prony estimation results. First, a downsampling of the signal under analysis is required. Then, the downsampled signal needs to be filtered (low pass filter and a DC filter), so that Prony can estimate the frequencies with a minimum error, because the method is sensitive to noise. Next, the downsampled and filtered signal is used for the Prony estimation calculation as presented in Section 3.

Table 5. FFT results for each speed operation condition at VFD.

Estimated Signal Parameters	True Frequency Values (Hz)		FFT Estimation Results		
			Frequency (Hz)	Amplitude (A)	Power (dB)
	$(1 + 2\,s)\,f$ [Sb1−]	55.33	59.77	0.285	−12.20
	Fundamental (f)	60.00	60.00	1.161	0
	$(1 - 2\,s)\,f$ [Sb1+]	64.66	60.35	0.277	−12.42
	$(1 + 2\,s)\,f$ [Sb1−]	46.11	49.8	0.232	−13.16
	Fundamental (f)	50.00	50.1	1.055	0
	$(1 - 2\,s)\,f$ [Sb1+]	53.88	50.39	0.192	−14.76
	$(1 + 2\,s)\,f$ [Sb1−]	36.88	39.06	0.104	−23.9
Full Load	Fundamental (f)	40.00	40.00	1.641	0
Slip s = 0.0388	$(1 - 2\,s)\,f$ [Sb1+]	43.11	40.47	0.196	−18.44
	$(1 + 2\,s)\,f$ [Sb1−]	27.66	29.65	0.285	−12.58
	Fundamental (f)	30.00	29.88	1.216	0
	$(1 - 2\,s)\,f$ [Sb1+]	32.33	30.23	0.241	−14.03
	$(1 + 2\,s)\,f$ [Sb1−]	18.44	18.91	0.096	−24.41
	Fundamental (f)	20.00	20.00	1.61	0
	$(1 - 2\,s)\,f$ [Sb1+]	21.55	20.63	0.151	−20.52
	$(1 + 2\,s)\,f$ [Sb1−]	9.22	9.21	0.108	−23.1
	Fundamental (f)	10.00	10.00	1.543	0
	$(1 - 2\,s)\,f$ [Sb1+]	10.77	11.41	0.071	−26.68

Downsampling

The recorded current signal is considered at 64 samples per cycle; nevertheless, a downsampling to 16 samples per cycle, as shown in Figure 14, is recommended to reduce the computational effort during the estimation process with Prony method (if a real time application is considered), and it is justified that due to the low frequencies (BRB_{sbf}) that are required to estimate there is no need for a high resolution recorded signal. This condition could be modified according to the needs of specific frequency detection for diagnostics application in MCSA.

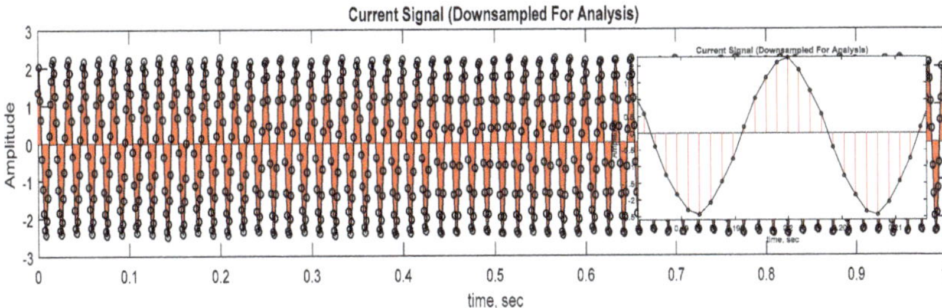

Figure 14. Downsampled current signal of measurement of motor line current of phase A at 60 Hz operation.

Low Pass and DC Filter

As mentioned in Section 3, if the recorded signal under analysis includes noise or other harmonics of no interest, the signal must be filtered to eliminate unwanted frequencies, so that the Prony estimation could be accurate. This step is carried out prior to using the signal as input signal for the estimation of signal parameters. For this particular application, a low pass filter of order 4 with a cut off frequency of harmonic order of 5 (300 Hz for 60 Hz) is used, see Figure 15; it is important to consider that the cutoff frequency will change if a VFD with a specific speed frequency operation (change in fundamental frequency) is considered. Moreover, a DC filter should be considered after the low pass filter, mainly because in a recorded signal, a DC offset will appear due to the measurement equipment (current sensors), see Figure 16, some equipment have zero adjustment to prevent a DC offset to appear in a recorded signal. If low pass filter and DC filter are not considered in signal processing prior to using the signal for Prony estimation it can cause a significant error in the estimated signal parameters, or other non-real frequencies will appear in the estimation results which can lead to a misinterpretation or a misdiagnosis.

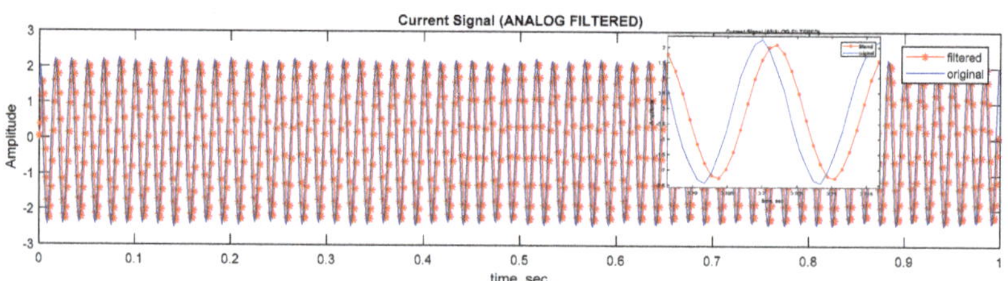

Figure 15. Downsampled low pass filtered current signal of measurement of motor line current of phase A at 60 Hz operation.

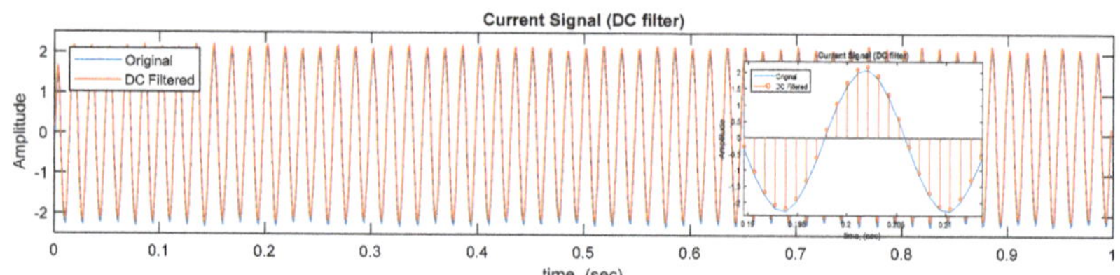

Figure 16. Downsampled low pass filtered and DC filtered current signal of measurement of motor line current of phase A at 60 Hz operation.

Signal Parameter Estimation

In order to obtain a good estimate of signal parameters using Prony method for OTD in RCIM, the considerations mentioned in Section 3 and Sections Downsampling and Low Pass and DC Filter must be applied, then the current signal under analysis, see Figure 17, will be used as the input signal to obtain the parameter estimation. It should be mentioned that a sliding window of data is used in this section to obtain the estimates, in the first analysis are considered only 5 sliding data windows for 60 Hz operation speed, where at each window of data a set of estimated parameters will be obtained with the main purpose to validate Prony method estimation application to detect BRB_{sbf}. Then, an analysis of one data window for each speed frequency operation with VFD (50 Hz, 40 Hz, 30 Hz, 20 Hz, and 10 Hz) is considered, and the BRB_{sbf} to be detected will be the ones indicated in Table 4. Hence, it is important to mention that the window length of data to be analyzed will be of 25 cycles for each speed operation condition.

Once the input signal to the Prony method has been digitally processed under the considerations mentioned in Sections Down Sampling and Low Pass and DC Filter, it now can be used in Prony method to determine an accurate estimate of signal parameters and BRB_{sbf}. Table 6 shows the estimation results using the methodology described in Section 3, where in order to validate the Prony estimation results five sliding data window are considered, so five calculations of Prony method are made, so the reader can observe that the estimated frequencies and its amplitudes correspond to the BRB_{sbf} defined in Table 4; also, it could be observed that a third pair of sideband frequencies also is detected.

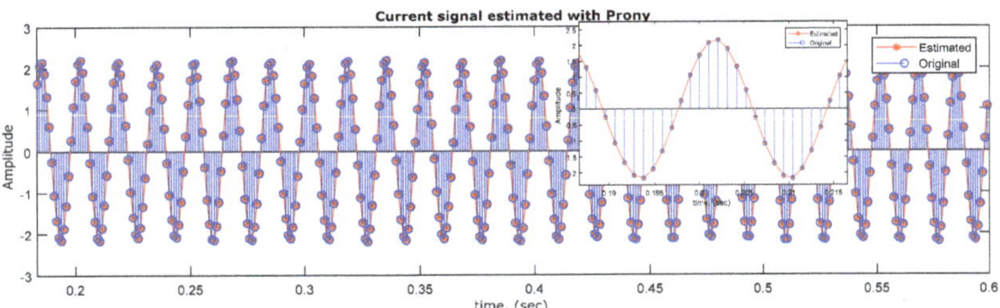

Figure 17. Sliding data window (Prony estimated and original signal) of 25 cycles of current signal measurement of motor line current of phase A at 60 Hz operation.

Table 6. Measured current signal estimated parameters.

Estimated Signal Parameters	Data Window 1		Data Window 2		Data Window 3		Data Window 4		Data Window 5	
	Frequency (Hz)	Amplitude	Frequency (Hz)	Amplitude	Frequency (Hz)	Amplitude	Frequency (Hz)	Amplitude	Frequency (Hz)	Amplitude
Sb3−	44.03	0.0536	44.03	0.0540	43.97	0.0551	43.90	0.0566	43.91	0.0552
Sb2−	51.28	0.0207	51.28	0.0208	51.30	0.0206	51.26	0.0149	51.25	0.0162
Sb1−	55.44	0.0426	55.44	0.0427	55.46	0.0423	55.42	0.0385	55.44	0.0391
Fundamental	60.02	2.1440	60.02	2.1441	60.02	2.1444	60.02	2.1418	60.02	2.1425
Sb1+	64.71	0.0279	64.71	0.0278	64.69	0.0273	64.68	0.0291	64.66	0.0284
Sb2+	69.66	0.0197	69.68	0.0195	69.72	0.0182	69.55	0.0204	69.58	0.0182
Sb3+	75.96	0.0248	75.97	0.0248	75.97	0.0248	75.97	0.0251	75.98	0.0250

To determine the severity condition of the rotor, the energy in dB needs to be obtained from a periodogram spectrum and to be compared with rotor fault signature in Table 1. First, the frequency spectrum of the results obtained for each sliding data window in Table 6 is presented in Figure 18, then, with the information obtained in Table 6, energy is plotted in a periodogram as shown in Figure 19. It could be observed that with frequency spectrum low magnitudes are observed, this is the main reason of the periodogram importance to plot these magnitudes in energy values in dB, so the BRB_{sbf} of the RCIM under analysis could be more evident.

As observed in Figure 19, the energy levels are between 31.56 and 41.42 dB of the BRB_{sbf} for the sliding data windows evaluated, the energy estimated values correspond to multiple cracked or broken bars or in its case a ring problem as described in Table 1. In Figure 6, a cracked ring rotor and broken rotor bar is used in the test system to perform the current signal analysis, so it is evident that an accurate diagnostic could be achieved with Prony method parameter estimation.

To complete the most common operative scenarios of RCIM for the experimental validation of Prony method, as described in Section 4.2, current signals are measured at motor line terminal phase A in the laboratory test system from Figure 5 using a VFD at different speed operation conditions defined in Table 4, at full load condition, where also its BRB_{sbf} are determined so a diagnostic of the RCIM can be obtained from the Prony estimation results.

In Figure 20, the digitally processed current signals (original) for each frequency speed operation are compared with the Prony estimated signal obtained from (2) using the estimated parameter results in Table 7, it can be observed in Figure 20 that for each VFD operation frequency the difference between the signals is null, this can be quantified by calculating MSE curve fitting in (12), the MSE curve fitting results has a value of 1.7889×10^{-4} for 50 Hz, 5.2202×10^{-4} for 40 Hz, 6.4835×10^{-4} for 30 Hz, 4.6944×10^{-4} for 20 Hz, and 2.4833×10^{-4} for 10 Hz; these results confirm that an accurate estimate of the BRB_{sbf} and diagnostic is achieved.

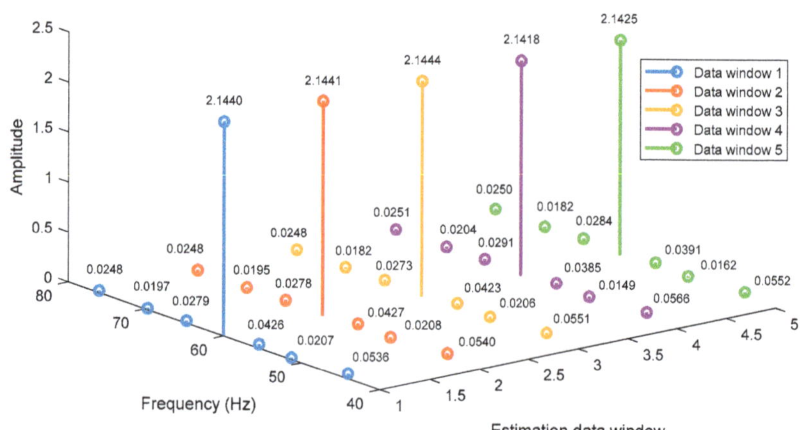

Figure 18. Prony estimated frequency spectrum at 60 Hz frequency speed operation scenario.

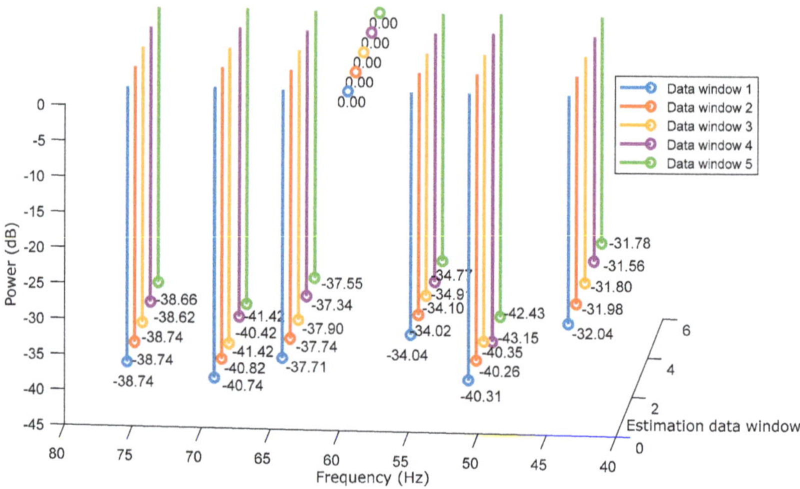

Figure 19. Prony estimated periodogram at 60 Hz frequency speed operation scenario.

In Table 7, the Prony estimation results of signal parameters are presented, where it can be observed that for each VFD speed operation, its sideband frequencies (True frequency values), amplitudes, and energy in dB are accurately estimated, where the obtained energy in dB corresponds to the real damage severity condition presented in Table 1 of the rotor of the RCIM under analysis. First, the frequency spectrum of the results obtained for each VFD speed operation frequency in Table 7 is presented in Figure 21, then, for visualization purpose of the results 7, the Periodogram for each Prony estimated signal parameters from Figure 20 is presented in Figure 22, where its fundamental frequency is normalized to 0 dB.

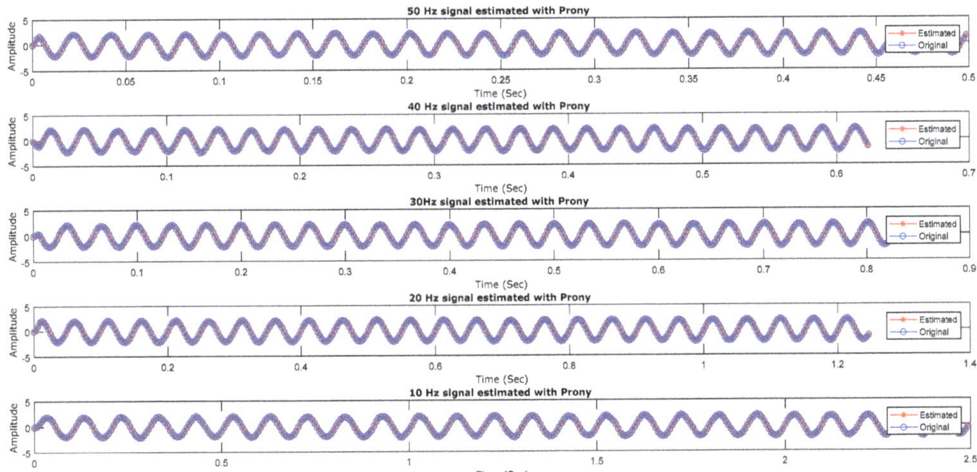

Figure 20. Data window selected for analysis of 25 cycles of current signal measurement of motor line current of phase A at VFD operation frequencies of 50 Hz, 40 Hz, 30 Hz, 20 Hz, and 10 Hz (Prony estimated and original signal).

Table 7. Measured current signal estimated parameters (VFD, motor at full load).

Estimated Signal Parameters	Measured Signal (Current Phase A)					
	True Frequency Values (Hz)		**Prony Method Estimation Results**			
			Frequency (Hz)	**Amplitude (A)**	**Power (dB)**	**MSE Curve Fitting**
	$(1 + 2\,s)\,f\,[Sb1-]$	46.11	46.89	0.0222	-39.76	
	Fundamental (f)	50.00	50.07	2.1599	0	1.7889×10^{-4}
	$(1 - 2\,s)\,f\,[Sb1+]$	53.88	54.16	0.0192	-41.02	
	$(1 + 2\,s)\,f\,[Sb1-]$	36.88	36.50	0.0327	-36.41	
	Fundamental (f)	40.00	39.96	2.1628	0	5.2202×10^{-4}
	$(1 - 2\,s)\,f\,[Sb1+]$	43.11	43.64	0.0140	-43.78	
	$(1 + 2\,s)\,f\,[Sb1-]$	27.66	26.78	0.0231	-39.37	
Full Load Slip s = 0.0388	Fundamental (f)	30.00	29.92	2.1475	0	6.4835×10^{-4}
	$(1 - 2\,s)\,f\,[Sb1+]$	32.33	33.23	0.0424	-34.09	
	$(1 + 2\,s)\,f\,[Sb1-]$	18.44	18.01	0.0400	-34.41	
	Fundamental (f)	20.00	19.95	2.1011	0	4.6944×10^{-4}
	$(1 - 2\,s)\,f\,[Sb1+]$	21.55	21.59	0.0277	-37.60	
	$(1 + 2\,s)\,f\,[Sb1-]$	9.22	9.68	0.0305	-36.39	
	Fundamental (f)	10.00	10.05	2.0117	0	2.4833×10^{-4}
	$(1 - 2\,s)\,f\,[Sb1+]$	10.77	10.60	0.0420	-33.61	

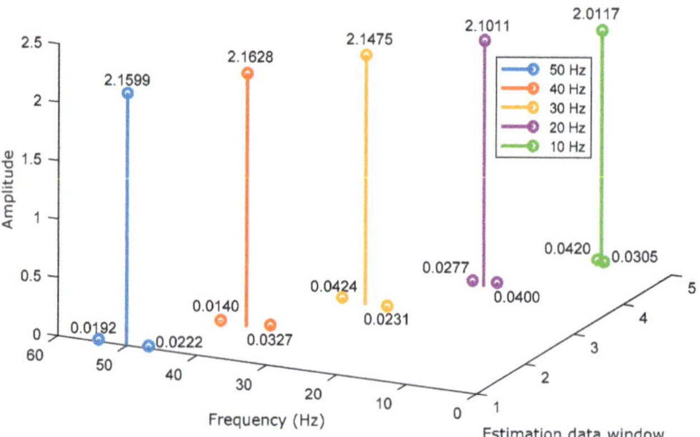

Figure 21. Prony estimated spectrum of VFD frequency speed operation scenarios.

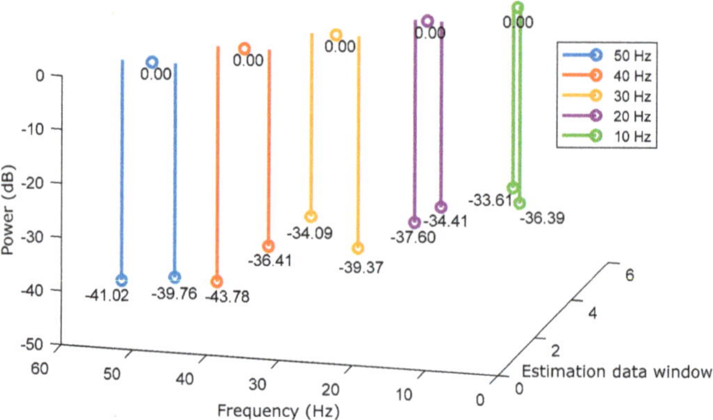

Figure 22. Prony estimated periodogram of VFD frequency speed operation scenarios.

5. Results and Discussion

A results comparison is performed between FFT analysis (using a 5 s recorded current signal) and Prony method (using a 25 cycles of recorded current signal) in this work. It is important to mention that 25 cycles have been selected because that is the minimum data window length with good and accurate results, other tests from 1–24 cycles were performed but no good estimation is achieved, so the best possible application option for online test diagnostics equipment with the minimum possible estimation time is 25 cycles. Moreover, a very important detail to mention is that the computational estimation time increase exponentially for each increase in the sampling frequency, so this is the main reason why a subsampling to 16 samples per cycle is considered. In Sections 4.2.2 and 4.2.3, the estimation results at different VFD speed operation conditions are presented, and Table 8 shows a summary of the results for both methods. Note from Table 8 that for each VFD speed operation condition, the FFT results are not accurate and BRB_{sbf} cannot be detected having a 5 s recorded signal, but if Prony estimation results are observed and compared with the True frequency values, then use Table 1 to determine the severity condition of the cage of the rotor, it can be validated that the estimated energy in dB corresponds to a cracked ring and/or BRB, which is the real condition of the rotor shown in Figure 6.

Table 8. Summary of current signal estimated parameters using FFT and Prony method (VFD, motor at full load).

Estimated Signal Parameters	True Frequency Values (Hz)		Measured Signal (Current Phase A)						
			FFT Estimation Results			Prony Method Estimation Results			
			Frequency (Hz)	Amplitude (A)	Power (dB)	Frequency (Hz)	Amplitude (A)	Power (dB)	MSE Curve Fitting
	$(1 + 2\,s)\,f$ [Sb1−]	55.33	59.77	0.285	−12.20	55.44	0.0426	−34.04	
	Fundamental (f)	60.00	60.00	1.161	0	60.02	2.1440	0	6.4593×10^{-4}
	$(1 − 2\,s)\,f$ [Sb1+]	64.66	60.35	0.277	−12.42	64.71	0.0279	−37.71	
	$(1 + 2\,s)\,f$ [Sb1−]	46.11	49.8	0.232	−13.16	46.89	0.0222	−39.76	
	Fundamental (f)	50.00	50.1	1.055	0	50.07	2.1599	0	1.7889×10^{-4}
	$(1 − 2\,s)\,f$ [Sb1+]	53.88	50.39	0.192	−14.76	54.16	0.0192	−41.02	
	$(1 + 2\,s)\,f$ [Sb1−]	36.88	39.06	0.104	−23.9	36.50	0.0327	−36.41	
Full Load Slip s = 0.0388	Fundamental (f)	40.00	40.00	1.641	0	39.96	2.1628	0	5.2202×10^{-4}
	$(1 − s)\,f$ [Sb1+]	43.11	40.47	0.196	−18.44	43.64	0.0140	−43.78	
	$(1 + 2\,s)\,f$ [Sb1−]	27.66	29.65	0.285	−12.58	26.78	0.0231	−39.37	
	Fundamental (f)	30.00	29.88	1.216	0	29.92	2.1475	0	6.4835×10^{-4}
	$(1 − 2\,s)\,f$ [Sb1+]	32.33	30.23	0.241	−14.03	33.23	0.0424	−34.09	
	$(1 + 2\,s)\,f$ [Sb1−]	18.44	18.91	0.096	−24.41	18.01	0.0400	−34.41	
	Fundamental (f)	20.00	20.00	1.61	0	19.95	2.1011	0	4.6944×10^{-4}
	$(1 − 2\,s)\,f$ [Sb1+]	21.55	20.63	0.151	−20.52	21.59	0.0277	−37.60	
	$(1 + 2\,s)\,f$ [Sb1−]	9.22	9.21	0.108	−23.1	9.68	0.0305	−36.39	
	Fundamental (f)	10.00	10.00	1.543	0	10.05	2.0117	0	2.4833×10^{-4}
	$(1 − 2\,s)\,f$ [Sb1+]	10.77	11.41	0.071	−26.68	10.60	0.0420	−33.61	

6. Conclusions

It is important to mention that a RCIM should be operating at full load in order to detect BRB_{sbf}, because at full load condition, induced currents circulate at the cage of the rotor and when a measurement of the motor line current is performed, these frequencies will appear; the main difficulty is to extract or detect these BRB_{sbf} because they are frequency components that are too close to the fundamental frequency and some considerations have to be made so that a digital signal processing technique can be used to estimate these frequencies accurately and a diagnostic of the cage of the rotor can be defined. As it is commonly known, FFT analysis is used for RCIM OTD, where this signal processing technique has been used widely in most of online test equipment for RCIM OTD but one of the main disadvantages of the technique for this application is that it requires at least 10 s of recorded signal to give a good estimate of the BRB_{sbf}. Hence, Prony method is proposed to be used as a BRB_{sbf} detection technique to be applied in OTD equipment for RCIM, where at least 25 cycles of a current signal are needed to obtain an accurate estimate of the BRB_{sbf}; with the reduction of a recorded signal with a typical length (10 s), the memory requirement is less in the OTD equipment and the condition of the cage rotor is more accurate. It is important to apply the considerations for Prony method application for RCIM OTD presented in Sections 3 and 4, so an accurate estimate and diagnostic can be achieved.

For future work and recommendations, it is suggested to consider its application and analysis for DC motor current signature. The scope of this work was to present and validate the application details and advantages to use Prony method to determine the condition of the cage of a RCIM using MCSA. Moreover, it should be mentioned that in comparison with conventional OTD equipment FFT analysis of current signal, Prony method requires at least 25 cycles of recorded data, which means that it leads to a great reduction in memory hardware requirement for the recorded current signal, so an integration of the Prony method in OTD equipment is recommended to increase the accuracy in the condition diagnostic of a cage rotor of a RCIM.

Author Contributions: Conceptualization, methodology, software, validation, resources, data curation, visualization, investigation, writing—original draft preparation, writing—review and editing, L.A.T.G.; software, resources, visualization, investigation, writing—review and editing, M.A.P.G., J.R.M., M.A.G.V., L.H.R.A. and F.S.S. All authors have read and agreed to the published version of the manuscript.

Funding: This research received no external funding.

Institutional Review Board Statement: Not applicable.

Informed Consent Statement: Not applicable.

Data Availability Statement: Not applicable.

Acknowledgments: The authors would like to thank Cuerpo Académico Diagnóstico, Análisis, Control y Medición de Sistemas Eléctricos y Electrónicos (UANL-CA-432) of the Ministry of Public Education of Mexico for allowing us to use the laboratory test system equipment for the corresponding analysis.

Conflicts of Interest: The authors declare no conflict of interest.

Abbreviations

RCIM	Rotor Cage Induction Motor
BRB	Broken Rotor Bar
BRB_{sbf}	Broken Rotor Bar sideband frequencies
Sbf+	Upper sideband frequency
Sbf−	Lower sideband frequency
CBER	Cracked/Broken End Ring
DOL	Direct-On-Line
MCSA	Motor Current Signature Analysis
AAGE	Abnormal Air Gap Eccentricity
OTD	Online Test Diagnostics
MSE	Mean Square Error
VFD	Variable Frequency Drive
FFT	Fast Fourier Transform
DSP	Digital Signal Processing

Appendix A

Table A1. RCIM nameplate data.

Motor Data		
Rated Current	1.98	A
Rated Voltage	208	V
Rated Power	0.5	HP
Temperature insulation class F	155	°C
Rated Frequency	60	Hz
Service factor	1.15	
Efficiency	72	%
Connection	YY	
Rotor cage material	Aluminum	
Rated Speed	1730	RPM

References

1. Singh, G.K.; Kazzaz, S.A.S.A. Induction machine drive condition monitoring and diagnostic research—A survey. *Electr. Power Syst. Res.* **2003**, *64*, 145–158. [CrossRef]
2. Toliyat, H.A.; Nandi, S.; Choi, S.; Meshgin-Kelk, H. *Electric Machines: Modeling, Condition Monitoring, and Fault Diagnosis*, 1st ed.; CRC Press: Boca Raton, FL, USA, 2013.
3. Bonnett, A.H.; Soukup, G.C. Cause and analysis of stator and rotor failures in three-phase squirrel-cage induction motors. *IEEE Trans. Ind. Appl.* **1992**, *28*, 921–937. [CrossRef]
4. Thomson, W.T.; Culbert, I. *Current Signature Analysis for Condition Monitoring of Cage Induction Motors: Industrial Applications and Case Histories*, 1st ed.; John Wiley & Sons, Inc.: Hoboken, NJ, USA, 2017.
5. Messaoudi, M.; Flah, A.; Alotaibi, A.A.; Althobaiti, A.; Sbita, L.; Ziad El-Bayeh, C. Diagnosis and Fault Detection of Rotor Bars in Squirrel Cage Induction Motors Using Combined Park's Vector and Extended Park's Vector Approaches. *Electronics* **2022**, *11*, 380. [CrossRef]
6. Mazouji, R.; Khaloozadeh, H.; Arasteh, M. Fault Diagnosis of Broken Rotor Bars in Induction Motors Using Finite Element Analysis. In Proceedings of the 2020 11th Power Electronics, Drive Systems, and Technologies Conference (PEDSTC), Tehran, Iran, 4–6 February 2020. [CrossRef]
7. Ferrucho-Alvarez, E.R.; Martinez-Herrera, A.L.; Cabal-Yepez, E.; Rodriguez-Donate, C.; Lopez-Ramirez, M.; Mata-Chavez, R.I. Broken Rotor Bar Detection in Induction Motors through Contrast Estimation. *Sensors* **2021**, *21*, 7446. [CrossRef] [PubMed]
8. Drakaki, M.; Karnavas, Y.L.; Karlis, D.A.; Chasiotis, I.D.; Tzionas, P. Study on fault diagnosis of broken rotor bars in squirrel cage induction motors: A multi-agent system approach using intelligent classifiers. *Inst. Eng. Technol.* **2020**, *14*, 245–255. [CrossRef]
9. Bazan, G.H.; Goedtel, A.; Duque-Perez, O.; Morinigo-Sotelo, D. Multi-Fault Diagnosis in Three-Phase Induction Motors Using Data Optimization and Machine Learning Techniques. *Electronics* **2021**, *10*, 1462. [CrossRef]
10. Martinez-Herrera, A.L.; Ferrucho-Alvarez, E.R.; Ledesma-Carrillo, L.M.; Mata-Chavez, R.I.; Lopez-Ramirez, M.; Cabal-Yepez, E. Multiple Fault Detection in Induction Motors through Homogeneity and Kurtosis Computation. *Energies* **2022**, *15*, 1541. [CrossRef]
11. Liu, X.; Yan, Y.; Hu, K.; Zhang, S.; Li, H.; Zhang, Z.; Shi, T. Fault Diagnosis of Rotor Broken Bar in Induction Motor Based on Successive Variational Mode Decomposition. *Energies* **2022**, *15*, 1196. [CrossRef]
12. Bonet-Jara, J.; Morinigo-Sotelo, D.; Duque-Perez, O.; Serrano-Iribarnegaray, L.; Pons-Llinares, J. End-ring wear in deep well submersible motor pumps. In Proceedings of the IEEE Transactions on Industry Applications, Greenfield, WI, USA, 12 April 2022. [CrossRef]
13. Elvira-Ortiz, D.A.; Morinigo-Sotelo, D.; Zorita-Lamadrid, A.L.; Osornio-Rios, R.A.; Romero-Troncoso, R.d.J. Fundamental Frequency Suppression for the Detection of Broken Bar in Induction Motors at Low Slip and Frequency. *Appl. Sci.* **2020**, *10*, 4160. [CrossRef]
14. Zamudio-Ramirez, R.A.; Osornio-Rios, J.A.; Antonino-Daviu, H.R.; Romero-Troncoso, R.d.J. Magnetic Flux Analysis for the Condition Monitoring of Electric Machines: A Review. In Proceedings of the IEEE Transactions on Industry Applications, Taipei, Taiwan, 2 April 2022; Volume 18, pp. 2895–2908. [CrossRef]
15. Garcia-Calva, T.A.; Morinigo-Sotelo, D.; Romero-Troncoso, R.D.J. Fundamental Frequency Normalization for Reliable Detection of Rotor and Load Defects in VSD-fed Induction Motors. In Proceedings of the IEEE Transactions on Industry Applications, Trieste, Italy, 13 November 2021. [CrossRef]
16. Fernandez-Cavero, V.; Pons-Llinares, J.; Duque-Perez, O.; Morinigo-Sotelo, D. Detection of Broken Rotor Bars in Nonlinear Startups of Inverter-Fed Induction Motors. In Proceedings of the IEEE Transactions on Industry Applications, Greenfield, WI, USA, 17 March 2021; Volume 57, pp. 2559–2568. [CrossRef]
17. Fernandez-Cavero, V.; Pons-Llinares, J.; Duque-Perez, O.; Morinigo-Sotelo, D. Detection and quantification of bar breakage harmonics evolutions in inverter-fed motors through the dragon transform. *ISA Trans.* **2021**, *109*, 352–367. [CrossRef]
18. Fernandez-Cavero, V.; García-Escudero, L.A.; Pons-Llinares, J.; Fernández-Temprano, M.A.; Duque-Perez, O.; Morinigo-Sotelo, D. Diagnosis of Broken Rotor Bars during the Startup of Inverter-Fed Induction Motors Using the Dragon Transform and Functional ANOVA. *Appl. Sci.* **2021**, *11*, 3769. [CrossRef]
19. Garcia-Calva, T.A.; Morinigo-Sotelo, D.; Fernandez-Cavero, V.; Garcia-Perez, A.; Romero-Troncoso, R.d.J. Early Detection of Broken Rotor Bars in Inverter-Fed Induction Motors Using Speed Analysis of Startup Transients. *Energies* **2021**, *14*, 1469. [CrossRef]
20. Pezzani, C.; Donolo, P.; Bossio, G.; Donolo, M.; Guzmán, A.; Zocholl, S.E. Detecting Broken Rotor Bars With Zero-Setting Protection. In Proceedings of the IEEE Transactions on Industry Applications, Greenfield, WI, USA, 31 July 2014; Volume 50, pp. 1373–1384. [CrossRef]
21. Ayhan, B.; Chow, M.-Y.; Trussell, H.J.; Song, M.-H. A case study on the comparison of non-parametric spectrum methods for broken rotor bar fault detection. In Proceedings of the 29th Annual Conference of the IEEE Industrial Electronics Society (IECON), Roanoke, VA, USA, 2–6 November 2003. [CrossRef]
22. Ribeiro, P.F.; Duque, C.A.; Ribeiro, P.M.; Cerqueira, A.S. *Power Systems Signal Processing for Smart Grids*, 1st ed.; Wiley: London, UK, 2014.
23. Trujillo-Guajardo, L.A.; Rodriguez-Maldonado, J.; Moonem, M.A.; Platas-Garza, M.A. A Multiresolution Taylor–Kalman Approach for Broken Rotor Bar Detection in Cage Induction Motors. *IEEE Trans. Inst. Meas.* **2018**, *67*, 1317–1328. [CrossRef]

24. Naha, A.; Samanta, A.K.; Routray, A.; Deb, A.K. A method for detecting half-broken rotor bar in lightly loaded induction motors using current. *IEEE Trans. Inst. Meas.* **2016**, *65*, 1614–1625. [CrossRef]

25. Valles-Novo, R.; Rangel-Magdaleno, J.; Ramirez-Cortes, J.M.; Peregrina-Barreto, H.; Morales-Caporal, R. Empirical mode decomposition analysis for broken-bar detection on squirrel cage induction motors. *IEEE Trans. Inst. Meas.* **2015**, *64*, 1118–1128. [CrossRef]

26. Picazo-Rdenas, M.J.; Antonino-Daviu, J.; Climente-Alarcon, V.; Royo-Pastor, R.; Mota-Villar, A. Combination of noninvasive approaches for general assessment of induction motors. *IEEE Trans. Ind. Appl.* **2015**, *51*, 2172–2180. [CrossRef]

27. Wang, J.; Gao, R.X.; Yan, R. Broken-Rotor-Bar Diagnosis for Induction Motors. In Proceedings of the 9th International Conference on Damage Assessment of Structures (DAMAS 2011), London, UK, 11–13 July 2011.

28. Bessam, B.; Menacer, A.; Boumehraz, M.; Cherif, H. Detection of broken rotor bar faults in induction motor at low load using neural network. *ISA Trans.* **2016**, *64*, 241–246. [CrossRef]

29. Sahraoui, M.; Cardoso, A.J.M.; Ghoggal, A. The Use of a Modified Prony Method to Track the Broken Rotor Bar Characteristic Frequencies and Amplitudes in Three-Phase Induction Motors. *IEEE Trans. Ind. Appl.* **2014**, *51*, 2136–2147. [CrossRef]

30. Jia, Z.; Zhu, H.; Liu, X.; Shang, H. Incipient Broken Rotor Bar Fault Diagnosis Based on Extended Prony Spectral Analysis Technique. In Proceedings of the 2018 37th Chinese Control Conference (CCC), Wuhan, China, 25–27 July 2018. [CrossRef]

31. Chen, S.; Zivanovic, R. Estimation of frequency components in stator current for the detection of broken rotor bars in induction machines. *Measurement* **2010**, *43*, 887–900. [CrossRef]

32. Xu, B.Q.; Tian, S.H. *A Detection Method for Broken Rotor Bar Fault in Induction Motors Based on SVD Combined MUSIC with Extended Prony*; Applied Mechanics and Materials Trans Tech Publications, Ltd.: Basel, Switzerland, 2014; Volume 707, pp. 333–337. [CrossRef]

33. Dehina, W.; Boumehraz, M.; Kratz, F. On-line detection and estimation of harmonics components in induction motors rotor fault through a modified Prony's method. *Int. Trans. Electr. Energ. Syst.* **2021**, *31*. [CrossRef]

34. Lobos, T.; Rezmer, J.; Schegner, J. Parameter estimation of distorted signals using Prony method. In Proceedings of the 2003 IEEE Bologna Power Tech Conference Proceedings, Bologna, Italy, 23–26 June 2003. [CrossRef]

35. Leonowicz, Z.; Lobos, T.; Rezmer, J. Advanced Spectrum Estimation Methods for Signal Analysis in Power Electronics. *IEEE Trans. Ind. Electr.* **2003**, *50*, 514–519. [CrossRef]

36. Wrocław University of Science and Technology Digital Library. Parametric Methods for Time–Frequency Analysis of Electric Signals. Available online: https://www.dbc.wroc.pl/dlibra/publication/1877/edition/2021?language=pl (accessed on 15 February 2022).

37. Qi, L.; Qian, L.; Woodruff, S.; Cartes, D. Prony Analysis for Power System Transients. *EURASIP J. Adv. Signal Processing* **2007**, *2007*, 048406. [CrossRef]

38. Meunier, M.; Brouaye, F. Fourier transform, Wavelets, Prony Analysis: Tools for Harmonics and Quality of Power. In Proceedings of the 8th Int. Conf. on Harmonics and Quality of Power ICHQP'98, Athens, Greece, 14–16 October 1998. [CrossRef]

39. Johnson, M.A.; Zarafonitis, I.P.; Calligaris, M. Prony analysis and power system stability-some recent theoretical and applications research. In Proceedings of the 2000 Power Engineering Society Summer Meeting, Seattle, WA, USA, 16–20 July 2000. [CrossRef]

40. Castillo, R.; Ramírez, J.R.; Alonso, G.; Ortiz-Villafuerte, J. Prony's method application for BWR instabilities characterization. *Nucl. Eng. Des. J.* **2014**, *284*, 67–73. [CrossRef]

41. Trujillo Guajardo, L.A. Relevador De Protección De Distancia Con Estimador Fasorial De Prony (MX Patent No. 351620 B). México Patent Office, Instituto Mexicano de la Propiedad Industrial, IMPI. 2017. Available online: https://vidoc.impi.gob.mx/visor?usr=SIGA&texp=SI&tdoc=E&id=MX/a/2014/012486 (accessed on 12 January 2022).

42. Trujillo Guajardo, L.A. Prony filter vs conventional filters for distance protection relays: An evaluation. *Electr. Power Syst. Res.* **2016**, *137*, 163–174. [CrossRef]

Article

A Negative Sequence Current Phasor Compensation Technique for the Accurate Detection of Stator Shorted Turn Faults in Induction Motors

Syaiful Bakhri and Nesimi Ertugrul *

Research Centre for Nuclear Fuel Cycle and Radioactive Waste Technology, National Research and Innovation Agency of Republic Indonesia, Puspiptek Complex, Building 20, Tangerang Selatan 15314, Banten, Indonesia; bakhrisy@batan.go.id or syai001@brin.go.id
* Correspondence: nesimi.ertugrul@adelaide.edu.au

Abstract: Stator faults are the most critical faults in induction motors as they develop quickly hence requiring fast online diagnostic methods. A number of online condition monitoring techniques are proposed in the literature to respond to such faults, including the signature analysis of the stator current, vibration monitoring, flux leakage monitoring, negative sequence components of voltage and current and sequence components monitoring based on the identification of asymmetrical behavior in a machine. Negative sequence components of voltage and current and sequence components monitoring are commonly considered for the identification of asymmetrical behavior of induction motors. Negative sequence current analysis is a sensitive technique for the detection of shorted turns as it directly measures the asymmetry produced by the fault. However, the technique is sensitive to other asymmetrical faults and disturbances, which can also produce negative sequence currents. These disturbances include sensor errors, inherent asymmetry and voltage unbalance. This paper provides a comprehensive investigation of the disturbances using a motor model along with experimental measurements under varying load conditions. Then, a new phasor compensation technique is explained to eliminate such disturbances effectively. This technique enables the accurate detection of even relatively small shorted turn faults, even at an early stage. The technique is tested experimentally, and a set of practical results are given to validate the methods developed.

Keywords: condition monitoring; induction machines; negative sequence currents; shorted turn faults; phasor compensation

Citation: Bakhri, S.; Ertugrul, N. A Negative Sequence Current Phasor Compensation Technique for the Accurate Detection of Stator Shorted Turn Faults in Induction Motors. *Energies* **2022**, *15*, 3100. https://doi.org/10.3390/en15093100

Academic Editors: Daniel Morinigo-Sotelo, Joan Pons-Llinares and Rene Romero-Troncoso

Received: 7 March 2022
Accepted: 7 April 2022
Published: 24 April 2022

1. Introduction

Electric motors consume about 45% of the world's electric energy. In total, 10.3% of these are medium size (0.75–375 kW), and 0.3% of them are large size (>375 kW) motors that consume a significant level of energy and usually operate in critical applications. Among the motor types, induction motors cover a greater portion of the applications, from direct-online (DOL) grid-connected motors to electric vehicles, primarily due to their robustness, reliability and low cost. However, the failure of induction motors has a significant impact on both their running cost and the efficiency of production.

Catastrophic failure of the motors usually develops over a period of time (from seconds to days), first as a low degree of fault, which is investigated intensively under "condition monitoring" involving various electrical quantities. The faults in induction motors can be classified into five groups (Figure 1b): bearing related faults (41%), stator related faults (37%), rotor faults (10%) and eccentricity related and mechanical faults (12% in total).

Online condition monitoring and diagnostic methods are preferred to predict any incipient failures in induction motors. As the most critical faults, stator faults are commonly detected using steady-state condition monitoring.

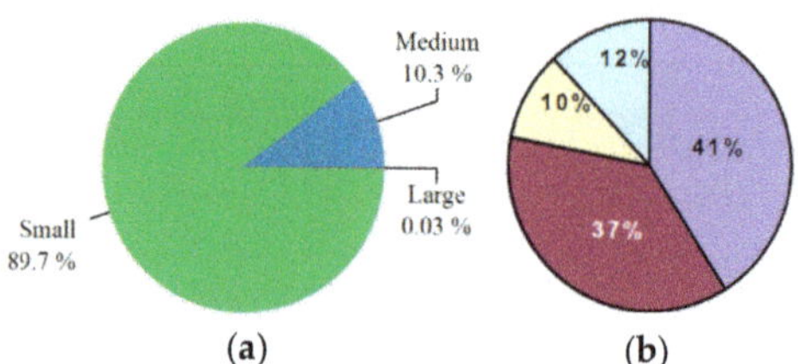

Figure 1. The breakdown of global motor usage by size (**a**) [1], and distribution of common faults of electric motors (**b**) according to EPRI survey results [2].

Various online diagnostic methods are proposed in the literature, including the signature analysis of the stator current, vibration monitoring, flux leakage monitoring, negative sequence components of voltage and current and sequence components monitoring that is based on the identification of asymmetrical behavior in a machine.

Among these, the sequence components method offers a fast and reliable solution in which any unidentified unbalanced three-phase voltage or current phasors are transformed into a set of three simple independent balanced component phasors: positive sequence, negative sequence and zero sequence phasors. Moreover, sequence component monitoring provides opportunities to increase the accuracy of the results through non-idealities and non-linearity compensation techniques.

Note that the positive sequence component always exists due to the supply voltage, but the negative sequence component exists only under asymmetrical voltage supply or under motor faults. Hence, the negative sequence component is utilized to monitor the health of the machine as well as can identify the supply voltage unbalances. For example, negative sequence current monitoring can detect one of the most critical faults, stator shorted turn faults in induction motors, as an alternative to other signature analysis techniques such as stator current, vibration and flux leakage [1,2], and this non-invasive method has low computational requirements [3,4].

However, the measured negative sequence may contain inherent non-idealities (such as asymmetries in real machines, saturations, inherent winding asymmetry) and is sensitive to external effects (such as load changes, supply and temperature variations). In order to eliminate such secondary fault effects, various compensation methods for voltage unbalance and other inherent non-idealities using look-up table databases, empirical formulas or neural networks are proposed in the literature. These studies are summarized below in Figure 2.

The negative sequence current monitoring method is based on understanding the sources of asymmetry in the three-phase machine using measurements on the line currents. As illustrated in Figure 3, the sources of negative sequence currents in induction machines can be classified into four main groups: inherent asymmetry, supply voltage unbalances, instrumentation asymmetry in measuring devices and actual motor faults. The figure also shows the complex interactions among the sources of negative sequence currents, specifically when thermal effects, supply voltage variations and load variations are considered in real machines, which have a significant effect on the sequence currents.

Although a significant amount of research is reported in the literature which utilizes the negative sequence currents for condition monitoring [5–10], the causes of these currents are not investigated comprehensively. These include the contribution of measurement related asymmetry, the complex interaction among the causes of the negative sequence currents, the angle of the negative sequence current in the analysis for inherent asymmetry, voltage unbalance and finally, shorted turn faults under load using phasor plots. It should be emphasized here that to identify the real faults of induction motors, the disturbances need to be eliminated from the measured negative sequence current. This can be conducted by compensation methods such as using simple look-up tables [6], a specific proportional integral (PI) negative sequence regulator [11] or using advanced neural networks [8,12].

Table 1 show the various compensation techniques reported in the literature that are proposed to detect stator shorted turn faults using the negative sequence component, which includes negative sequence currents [5–10,12,13], negative sequence impedance [14] and a matrix of impedances [15] modeling cross-coupling between the positive and negative sequence components. Three major disturbances, inherent asymmetry, voltage unbalances and load variations, are also assessed in the table for each given reference.

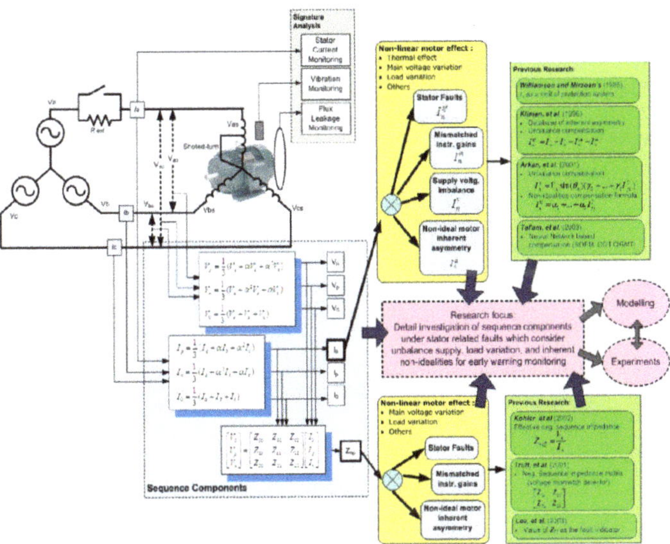

Figure 2. The summary of previous studies and the research opportunities for negative sequence components.

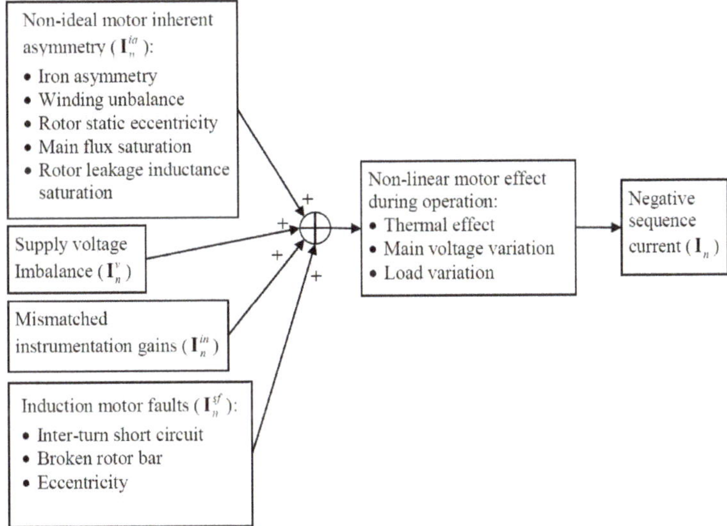

Figure 3. Main sources of negative sequence currents in induction motors.

Table 1. Features of Negative Sequence Current Compensation Techniques Reported in the Literature.

References	Inherent Asymmetry	Voltage Unbalance	Load Variation	Shorted Turn Fault
[6]	Compensation of non-linearity (including inherent asymmetry) using look-up table	Negative sequence impedance Z_n of motor (assumed independent load variation and turn faults)	Unaffected by load variation	Negative sequence current $I_{n\text{-}sf}$
[5]	Minimization of thermal effect by eliminating the current phase $I_{sn} = (V_n \sin \theta_n)/X_{hn}$	Semi-empirical quadratic function of healthy reactance (X_{hn}) $$X_{hn}^{-1} = \gamma_0 + \gamma_1 V_n + \gamma_2 \sin 2\phi_n + \dots$$ $$\gamma_3 \cos 2\phi_n + \gamma_4 I_{px} + \gamma_5 I_{py}^2$$	Semi-empirical quadratic function of stator current under load variations $$I_{mnlv} = \alpha_0 + \alpha_1 I_{px} +$$ $$\alpha_2 I_{px}^2 + \alpha_3 I_{py} + \alpha_4 I_{py}^2$$	Negative sequence current $I_{n\text{-}sf} = I_n - I_{sn} - I_{mnlv}$
[7]	Complex constants (k) $I_n = k_1 V_p + k_2 V_n$	Complex constants (k) $I_n = k_1 V_p + k_2 V_n$	k_1 and k_2 are load dependent	Negative sequence current $I_{n\text{-}sf}$
[10,13]	Calculated from two current sensors based-method I_a and I_b; Negative sequence due to uncalibrated sensor is also considered $I_{n-ia} = \frac{1}{3}(1 - a)(I_a - aI_b)$	Voltage unbalance supply is not considered	Tested under no-load, half load and full load	Negative sequence current $I_{n\text{-}sf}$
[8,12]	Neural network	Neural network	Neural network	Negative sequence current $I_{n\text{-}sf}$
[14]	Need to be perfectly balanced	Need a balanced voltage supply	Load is not affected	Effective negative sequence impedance: $Z_{n\text{-}eff} = V_n/I_n$
[15]	Independent of inherent asymmetry	Independent of voltage unbalance	Calculation under speed variation	Negative sequence impedance matrix Z_{np}
[11]	Proportional integral (PI) controller. The PI negative sequence regulator is not intended for monitoring	PI controller	PI controller	Not available

Notes: X: Magnitude of reactance; ϕ: Phase angle; I: Magnitude of current; h: Operator for symmetrical components; V: Magnitude of voltage; x, y: Real and imaginary parts of a phasor; s, r: Subscripts for the quantities of stator and rotor; 0, p, n: Subscripts for zero, positive and negative sequence components; f: Supply frequency; m, s, l, v: Subscripts denoting motor, supply, load and voltage.

As it is highlighted in the table, no literature is found to address and demonstrate the interaction between shorted turn faults and disturbances, which is critical for the accurate detection of faults. This paper aims to address this using a compensation technique based on phasor analysis for sensor calibration, supply voltage unbalances and inherent machine asymmetry, with a target aim of detecting small degrees of stator shorted turn faults.

The layout of the paper is as follows. In Section 2, the negative sequence current analysis is discussed in detail by expanding the research studies reported in [7,16]. The principles of the phasor compensation technique are also provided in the same section. Section 3 discusses the simulation model and the test machine. Section 4 describes the effects of sensor calibration, motor temperature and supply voltage unbalance. Section 5 presents test results demonstrating the effectiveness of the proposed negative sequence phasor compensation for supply voltage unbalances and inherent asymmetry. The effects of motor loadings are also considered. Finally, conclusions are drawn in Section 6.

2. Negative Sequence Current Analysis

2.1. The Sources of Negative Sequence Current

An unbalanced three-phase set of current phasors, I_a, I_b and I_c can be represented as the superposition of three sets of balanced symmetrical component phasors: positive sequence I_p, negative sequence I_n and zero sequence I_0.

Although balanced three-phase voltages applied to an ideal induction machine produces balanced currents with no negative sequence components, practical induction machines have some negative sequence current due to inherent asymmetries between the windings. In addition, even a perfectly balanced induction machine, when operating from an unbalanced supply, will have a negative sequence current which is equal to the negative sequence supply voltage divided by the negative sequence impedance, Z_n, of the machine. In light of these practical limitations and using the summary of disturbances given in

Figure 3, the sources of the negative sequence currents are discussed in detail below, which is critical for the phasor based compensation technique developed in the paper.

2.1.1. Measurement Asymmetry ($I_{n\text{-}in}$)

Given the negative sequence, components are associated with imbalances between the supply voltages/currents; even small differences in gain and phase between the voltage/current transducers in different phases can produce substantial errors in the results. Thus careful calibration of the transducers is crucial when seeking to accurately measure negative sequence currents.

2.1.2. Inherent Asymmetry ($I_{n\text{-}ia}$)

As indicated previously, due to the manufacturing limitations in machine production, negative sequence currents even occur in healthy motors as they contain inherent asymmetry. As summarized in Figure 3, inherent asymmetry may be due to iron asymmetry, stator winding unbalances and rotor static eccentricity.

2.1.3. Voltage Unbalances ($I_{n\text{-}v}$)

All practical ac supplies have some degree of voltage imbalance. Compensation of the effect of supply voltage imbalance is important to separate this effect from the stator winding faults.

2.1.4. Induction Motor Faults ($I_{n\text{-}sf}$)

Negative sequence currents are produced by faults that cause asymmetries in the induction machine. Three major contributing faults are listed in Figure 3. However, the shorted turn fault is considered to be the most critical one since it generally develops faster than the eccentricity and/or broken rotor bar faults, and these other faults can also be effectively detected by alternative methods.

2.2. Principle of Phasor Compensation Technique

The negative sequence input current I_n of an induction machine can be expressed as the phasor sum of four separate negative sequence currents: motor faults $I_{n\text{-}sf}$, inherent asymmetry $I_{n\text{-}ia}$, measurement errors $I_{n\text{-}in}$ and supply voltage unbalance $I_{n\text{-}v}$.

$$I_n = I_{n\text{-}sf} + I_{n\text{-}ia} + I_{n\text{-}in} + I_{n\text{-}v} \tag{1}$$

The above equation implies that negative sequence current components can exist even in healthy machines, and Figure 4 illustrate the implementation of the shorted turn fault extraction based on the same equation.

The negative sequence current component due to motor faults $I_{n\text{-}sf}$ can be obtained from the measured negative sequence phasor I_n by subtracting the negative sequence current components due to voltage unbalance $I_{n\text{-}v}$, the motor inherent asymmetry $I_{n\text{-}ia}$ and instrumentation asymmetry $I_{n\text{-}in}$. In addition, it can be assumed that the negative sequence current component due to shorted turn faults $I_{n\text{-}sf}$ is proportional to the fault severity. It is also important to emphasize that the negative sequence current components can be sensitive to motor load variations and changes in motor temperature during operation.

As it is listed in Table 1, the negative sequence current can be used for shorted turn fault detection by utilizing the key fault indicators (given in the rightmost column) after compensating for the inherent asymmetry, voltage unbalance and load variations. In [6], the negative sequence impedance Z_n due to voltage unbalance was assumed unaffected by load variations and hence was assumed independent of stator shorted-turn faults. Therefore, the negative sequence current for the voltage unbalance compensation can be obtained from the measured negative sequence voltage and a look-up table for Z_n. The complex admittances method reported in [7] assumed that the healthy negative sequence current was a function of the positive and negative sequence supply voltages.

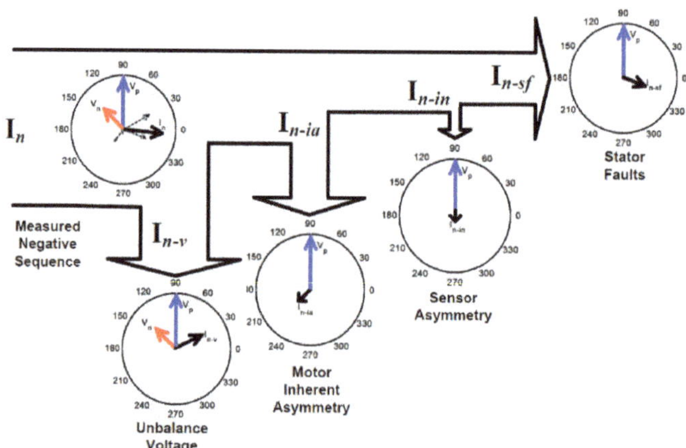

Figure 4. The compensation techniques using phasor calculations, where both the magnitude and the phase of the negative sequence current are illustrated to separate the stator faults.

In [5], the healthy negative sequence impedance was estimated based on empirical formulas taking into account the effect of the voltage unbalance, load variation and inherent asymmetry. In [10,13], the negative sequence of the online sensor and inherent asymmetry model and measurement was demonstrated. These two papers show the negative sequence current due to stator faults which were free from the inherent asymmetry and uncalibrated sensor effects. Furthermore, in [8,12], a neural network was applied to estimate the healthy negative sequence current.

However, as observed from these earlier studies, the previous techniques consider only a limited number of cases without separating the negative sequence of current phasor components related to each type of disturbance. Therefore, in the following section, the principle of the phasor compensation technique will be explained graphically to demonstrate its effectiveness as a winding short-circuit fault detection method.

3. Simulation Study and Test Setup

To be able to understand the behavior of the induction machine under various fault levels, a Simulink model was developed. The parameters of the commercial test machine are given in Table 2. The simulation was based on a dynamic machine coupled circuit model [6], which allowed the simulation of unequal numbers of winding turns in each phase as well as inter-turn short circuits. This provided a comprehensive understanding of disturbances at different fault levels for the phasor-based compensation technique.

Table 2. Equivalent Circuit Parameters of the Motor under Test.

Rated output power	2.2 kW	Referred rotor resistance	4.65 ohm
Rated frequency	50 Hz	Stator leakage inductance	14.8 mH
Line voltage	415 V	Referred rotor leakage inductance	14.8 mH
Rated stator current	4.9 A	Rotor inertia	0.05 kg m^2
Number of series turns/phase	282 turns	Magnetizing inductance	312 mH
No. of poles	4	Rated power factor	0.8 lag
Stator phase winding resistance	5.22 ohm	Rated speed	1415 rpm

The stator winding of the test machine was specially re-wound to allow the introduction of various levels of shorted turn faults via external connections. In the test motor, 5, 10, 15 or 20 shorted turns can be applied in either one or two phases of the motor (see Figure 5). This corresponds to shorted turn faults of 1.7%, 3.5%, 5.3% and 7.1% per phase, respectively.

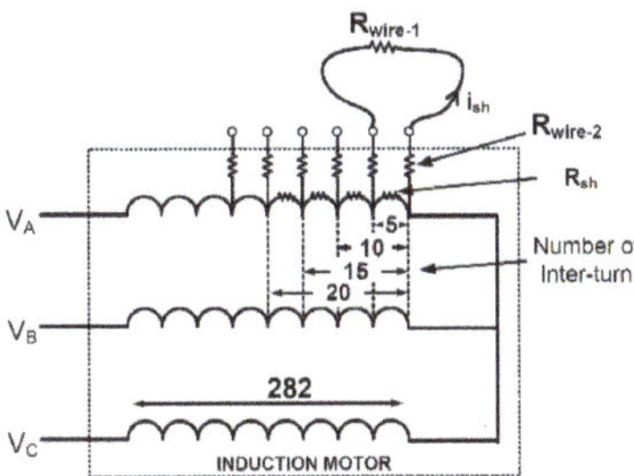

Figure 5. The modified stator winding configuration of the test motor and the external connection circuit used to introduce shorted turn faults (the motor windings have 282 series turns per phase).

Note that the stator winding resistance for 5 turns was about 93 mΩ while for 20 turns, it was about 370 mΩ, which was estimated using the total measured resistance of the stator winding. As illustrated in Figure 5, the short-circuit current path had an additional stray resistance of $R_{wire1} + 2R_{wire2}$, which was measured at about 125 mΩ. The machine model used in the simulation study included this stray resistance effect.

The measurement and the calibration system consisted of three voltage and three current transducers signal conditioning circuits, an eight-channel 8th order low-pass Butterworth analog filter and an eight-channel, 12-bit simultaneous sampling data acquisition card (see Figure 6). The current measurement resolution was 0.2 mA.

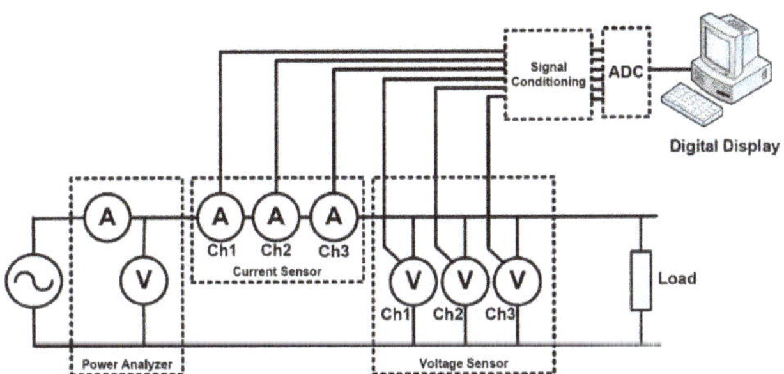

Figure 6. The measurement system used for the voltage and current sensor calibration.

The analysis of instrumentation asymmetry will be discussed in the following section.

4. System Calibration and Disturbances

4.1. Sensor Calibration

During the calibration process, three current and three voltage sensors are configured to measure the same line current and the same phase voltage, as it is shown in Figure 6. A single-phase AC supply is also connected to a load via a precision power analyzer to provide the reference measurement. Table 3 indicate the calibration constants for each voltage and current sensor, and Figure 7 summarize the results of the calibration tests.

Table 3. The Calibration of Voltage and Current Channels.

Channel	Voltage	Current
A	$y = 2.4 \times 10^{-5} + 8.84 \times 10^{-3}x$	$y = -4.93 \times 10^{-5} + 0.489x$
B	$y = 1.1 \times 10^{-3} + 8.88 \times 10^{-3}x$	$y = -8.19 \times 10^{-5} + 0.490x$
C	$y = 4.6 \times 10^{-4} + 8.93 \times 10^{-3}x$	$y = -1.07 \times 10^{-4} + 0.488x$

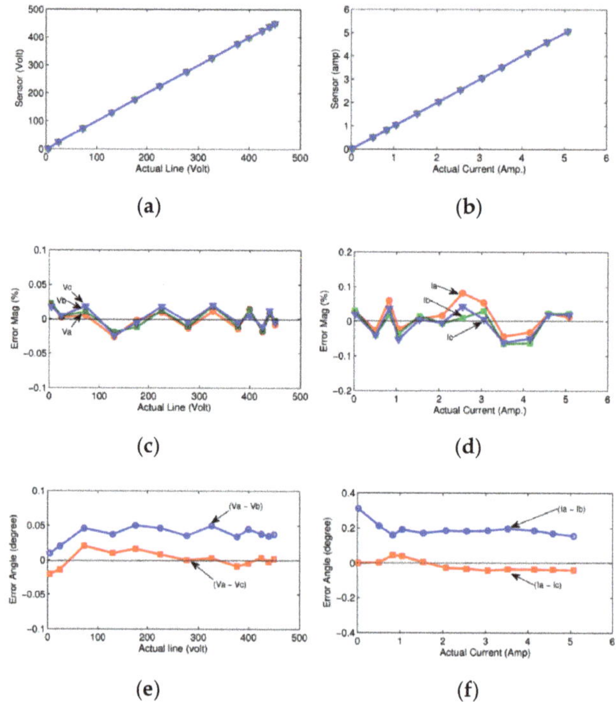

Figure 7. The measured and processed calibration characteristics: the calibrated values of voltage (**a**) and current (**b**), the residual magnitude errors in three voltage (**c**) and current sensors (**d**) and the relative angle errors for the three voltage (**e**) and current sensors (**f**).

Note that before the calibration process, the maximum gain error of the voltage channels was about 1% and for the current channels was about 0.4%. After the calibration process, all the sensor measurements are compared with the values measured by the power analyzer to obtain residual magnitude errors, as given in Figure 7. The percentage error is expressed as the ratio of the error value and the full-scale reading output. The results show that the percentage error is less than 0.05% for the voltage magnitude measurement, and it is less than 0.1% for the current magnitude measurement. Such errors can be considered acceptable for the negative sequence component analysis.

The angle error analysis of the voltage and the current measurement is also given in Figure 7, which is defined as the relative angle difference between two channels, i.e., between the V_a and V_b also the V_a and V_c voltage channels. Note that the average angle error of the worst pair of currents (about 0.2°) is much lower than for the worst pair of voltages (about 0.05°). To understand the impact of such discrepancies, the measurement asymmetry needs to be analyzed. Since, in this test, the three current and voltage channels are supplied from the same source, the three phasors should ideally create a zero sequence component. Figure 8 show the three measured current phasors showing small residual magnitude errors and uncompensated phase errors. In order to find the negative sequence component phasors due to the measurement asymmetry, the two phasors (I_b and I_c) are each rotated 120°, as shown in the rightmost figure.

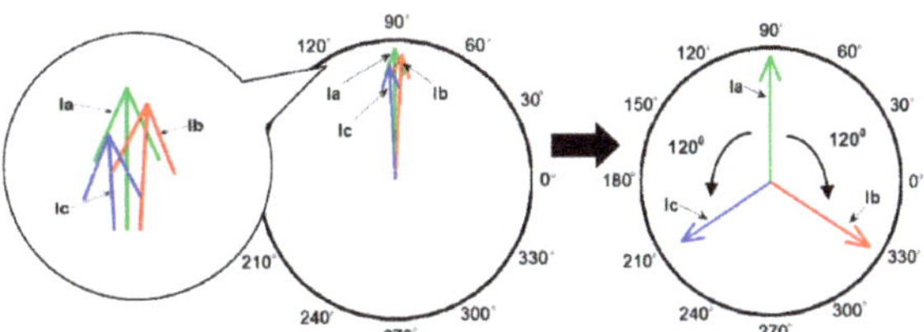

Figure 8. Illustration of the angle displacement for the sequence component analysis.

The magnitude of the negative sequence component due to measurement errors is given in Figure 9. This shows a 0.3% current component and 0.03% voltage component at rated current and voltage, respectively. This error can be reduced by subtracting a fixed value of angle offset between the channels of 0.2° for the current and 0.05° for the voltages. This angle offset correction reduces the negative sequence current error to lower than 0.1% of rated current and the negative sequence voltage error to 0.015% of rated voltage.

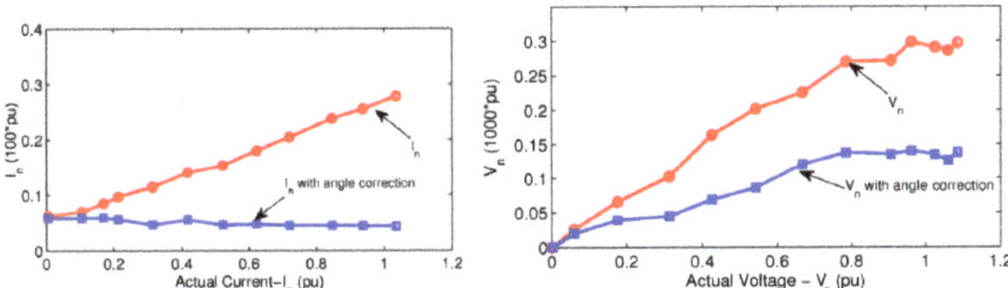

Figure 9. The magnitude of negative sequence components before and after offset angle correction, (**left**) for current and (**right**) for voltage.

4.2. Motor Temperature

The effect of temperature is investigated and presented in Figure 10 using the measurement system and the induction motor under test. The no-load cold data test is performed within the first 15 min after the motor is started. The no-load hot data is obtained after running the motor in a generator mode at the full load condition for half an hour. Then, the generator is decoupled from the electrical load and tested and measured under no-load conditions. In both cases, the measurements are performed over a range of supply voltages.

Figure 10 show that there is little difference between the hot and cold positive sequence currents, but there are significant differences between the hot and cold negative sequence currents and voltages. At the rated voltage, the negative sequence current is about 1%, and the negative sequence voltage is about 0.2% which is much larger than the residual measurement errors discussed previously.

Figure 10b show that the negative sequence current increases rapidly for voltages above about 0.7 pu. This may be due to asymmetries in the saturation of the three phases.

4.3. Voltage Unbalances

A supply voltage unbalance produces a negative sequence supply voltage which in turn produces a negative sequence current that is inversely proportional to the motor's negative sequence impedance. The supply voltage unbalance is measured using the voltage unbalance factor (VUF), which is defined as the ratio between the negative sequence V_n and positive sequence V_p voltage magnitudes:

$$\text{VUF} = V_n/V_p \tag{2}$$

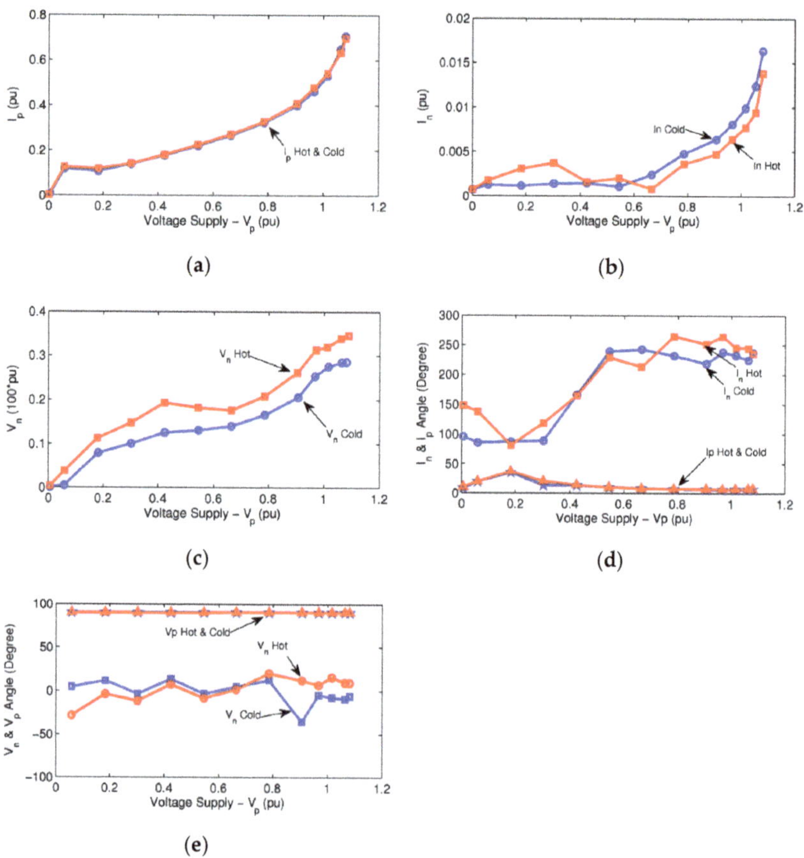

Figure 10. Hot and cold motor measurements: (**a**) magnitude of the stator positive sequence current I_p, (**b**) magnitude of the stator negative sequence current I_n, (**c**) magnitude of the negative sequence voltage V_n, (**d**) the angle of I_n and I_p, (**e**) the angle of V_n and V_p where the reference angle V_p is $90°$.

A variable voltage unbalance is introduced experimentally in the computer simulations by adding a variable external resistor in series with one of the supply phases of the motor while operating under no-load conditions.

Figure 11 highlights the variations of the negative sequence current magnitudes as a function of voltage unbalance (left column) and current phasor plots of the negative sequence current (right column).

Figure 11a,b show the measured negative sequence current using uncalibrated (squares) and calibrated (circles) voltage and current sensors, as discussed at the beginning of this section, which all demonstrate the significance of calibration.

Figure 11a show an approximately linear relationship between the negative sequence current component and the voltage unbalance while the positive sequence current component (crosses) remains almost constant. Note that the percentage change in the negative sequence current is approximately four times the percentage change in the negative sequence current. The results are consistent with up to 5% voltage unbalance, which is an acceptable level in practice [17]. The effect of inherent asymmetry in the machine is also observed in the same figures, where extrapolating the measured voltage unbalance results in zero unbalance (see squares in Figure 11c) and does not result in zero negative sequence current.

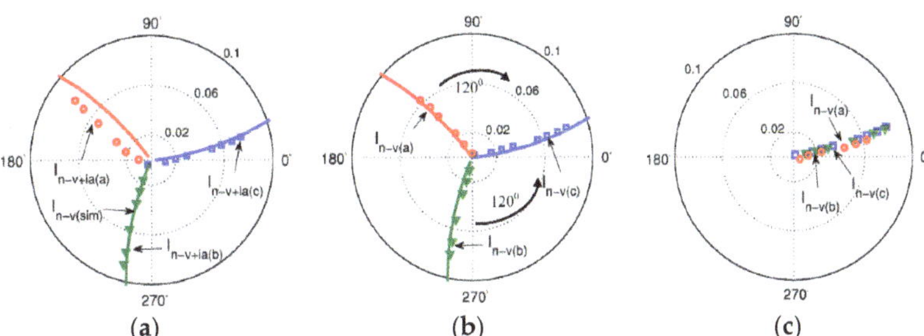

Figure 11. Three-phase phasor representation of the negative sequence current under a supply voltage unbalance: before (**a**) and after (**b**) inherent asymmetry compensation, and after 120° shift of two phases (**c**). The simulation results are shown by solid lines. The experimental results are indicated by symbols.

The simulation results are shown as solid lines in Figure 11a,b. Although the simulation results in Figure 11a show significant discrepancies as a function of voltage unbalance, the simulation model predicts the current phasor trajectory accurately, as illustrated in Figure 11b (except for the un-modelled inherent asymmetry). The corrected experimental results (triangles) are also given after the experimentally determined inherent asymmetry currents (squares) are subtracted. The results are now consistent with zero negative sequence current at zero voltage unbalance.

Note that the above-described procedure was repeated in each of the three phases of the machine. Figure 12 show the simulation and the experimental results of this study before (Figure 12a) and after (Figure 12b) inherent asymmetry compensation, at which point the results become centered on the origin. Figure 12c illustrate the effect of 120° phase shifting in two of the phases, which demonstrates that the effect of voltage unbalance is similar in each phase. This is the basis for compensating the negative sequence current that is due to the supply voltage unbalance.

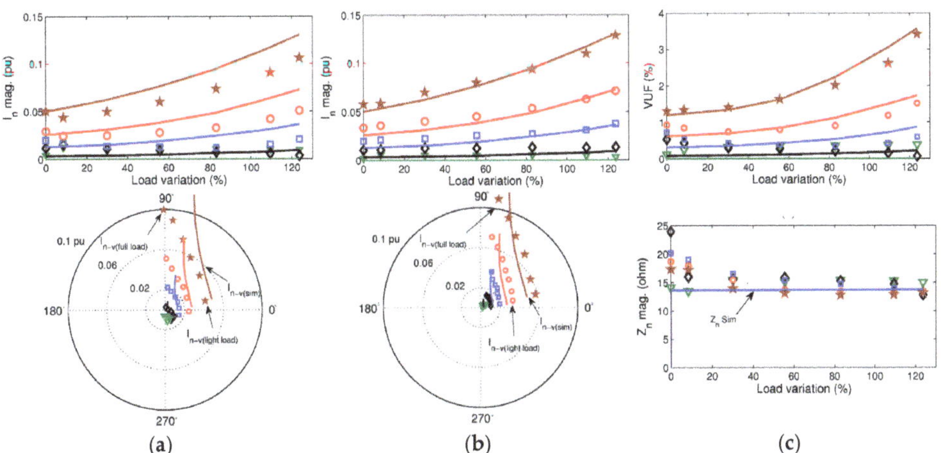

Figure 12. The experimental (symbols) and the simulated (solid lines) results of the current and voltage unbalance as a function of the load with a series resistor in one of the phases. The magnitude (**top**) and phasor trajectories (**bottom**) are also given before (**a**) and after (**b**) inherent asymmetry elimination, and the voltage unbalance factor (**top**) and negative sequence impedance (**bottom**) (**c**). The values of the external resistors were 0.25 Ω (triangle), 0.5 Ω (diamond), 1 Ω (square), 2 Ω (circle) and 4 Ω (star).

Figure 12 show the effect of voltage unbalance (using a resistor in series with one of the phase lines) on the negative sequence current under varying motor loads. The results are given before (Figure 12a) and after (Figure 12b) inherent asymmetry elimination, which shows that this substantially improves the correspondence between the simulated and test results.

The corresponding voltage unbalance factor and the negative sequence impedance are provided in Figure 12c. For a fixed supply unbalance resistor, increasing the load increases the voltage unbalance, which increases the negative sequence current and produces a phase angle change. The negative sequence impedances stay relatively constant.

5. Shorted Turn Motor Fault and Novel Phasor Compensation Technique

The effect of shorted turns was experimentally verified on the test machine using the tapped stator windings given in Figure 5 previously. This section explores the ability to detect such fault types and aims to estimate the severity of the fault.

5.1. Effect of Shorted Turn Fault

Figure 13 show the variation of the fault current as a function of the shorted turn (inter-turn) percentage and the voltage supply level. If the resistance of the external wires used to tap the stator winding (R_{wire} in Figure 5) is zero, then the fault current should be ideally constant. However, with a finite external resistance, the fault current increases with a number of shorted turns and asymptotes towards the true short-circuit current value. This current is also proportional to supply voltage.

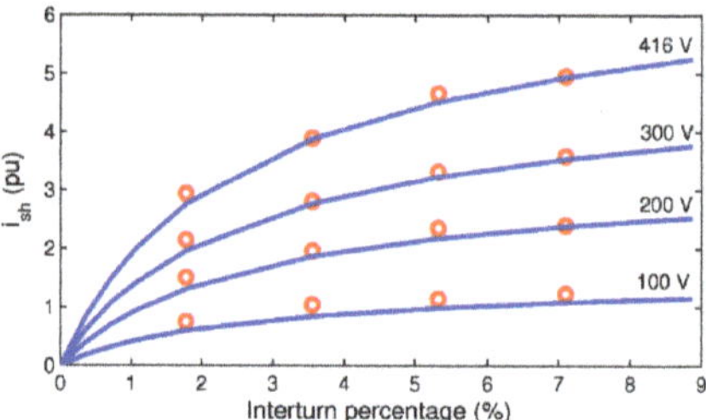

Figure 13. The fault current in the shorted-turns vs. the fraction of shorted-turns with different supply voltages. Simulation results (lines) and experimental results (circles).

Since the fault currents produced by shorted turns are large, they can cause the rapid heating of the shorted windings and consecutive catastrophic failure. For example, at the rated voltage, a fault current of five times the rated current is produced with only 7% of the shorted turns. Therefore, fast fault detection is crucial for early inference.

Figure 14a show the negative and positive sequence current magnitudes as a function of the short-turn fault percentage. The figure indicates that the negative sequence current is almost proportional to the fault level with a 2% negative sequence current, which corresponds to the 7% shorted turns. It can be observed in the same results that the short-turn fault does not affect the positive sequence current significantly.

Figure 14b show the negative sequence current phasor trajectory corresponding to the shorted-turn faults on two phases only, which are tested one phase at a time. The results show a good correspondence between the simulation and the experimental results.

To be able to investigate the practical operating scenarios, the shorted turn faults are also studied under the combination of voltage unbalance and motor loading. The results before and after inherent asymmetry elimination are given in Figure 15. The figure shows that shorted turn faults increase the magnitude of the negative sequence current as well as varying its phasor trajectory under voltage unbalances.

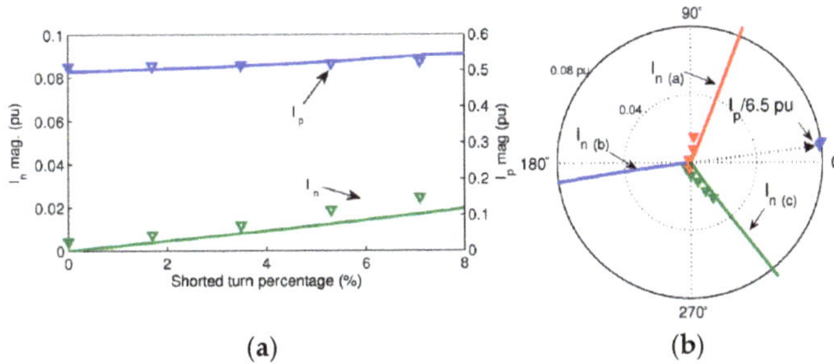

(a) **(b)**

Figure 14. The experimental (symbols) and simulated (solid lines) results at varying shorted turn fault levels of 1.7%, 3.1%, 5.3% and 7.1%: the magnitudes of the positive and the negative sequence currents (**a**) and the negative sequence phasor trajectory for faults in two phases (**b**).

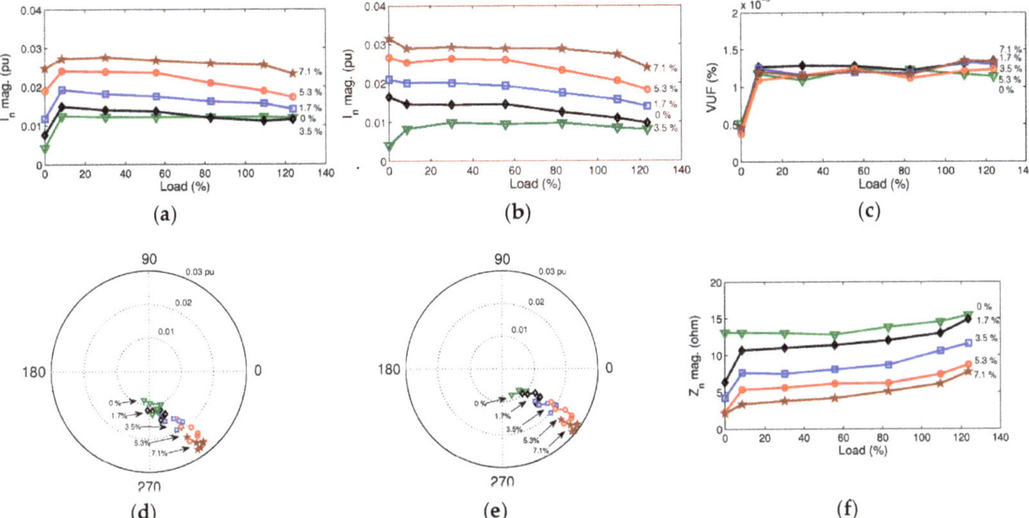

(a) **(b)** **(c)**

(d) **(e)** **(f)**

Figure 15. The test results at various shorted turn fault levels of 0% (triangle), 1.7% (diamond), 3.5% (square), 5.3% (circle) and 7.1% (star), before inherent asymmetry elimination of the magnitude of I_n (**a**), phasor diagram of I_n (**d**), after inherent asymmetry elimination of the magnitude of I_n (**b**), phasor diagram of I_n (**e**), and the corresponding voltage unbalance factors and (**c**) the negative sequence impedances (**f**).

Figure 15c,f show the percentage of voltage unbalance and the negative sequence impedance as a function of the motor load. As can be observed in the figures, the effect of the negative sequence current magnitude decreases slightly with increasing load (Figure 15a). This is caused by the negative sequence impedance (Figure 15f) that increases slightly with the load. However, the negative sequence voltage remains relatively constant, as illustrated in Figure 15c.

5.2. Shorted Turn Faults and Phasor Compensation

Figure 16 show the variation of the measured negative sequence current magnitude vs. the shorted-turn ratio (up to 7%) for different supply voltage unbalances (from 0.1% to 1.7%) under no-load conditions. Current trajectory plots are also given in the second row of the figure. The left column in the figure shows the measured results using the calibrated sensors; the middle column shows the results with compensation for supply voltage unbalance and the right column illustrates the results after both supply voltage imbalance and inherent asymmetry compensation.

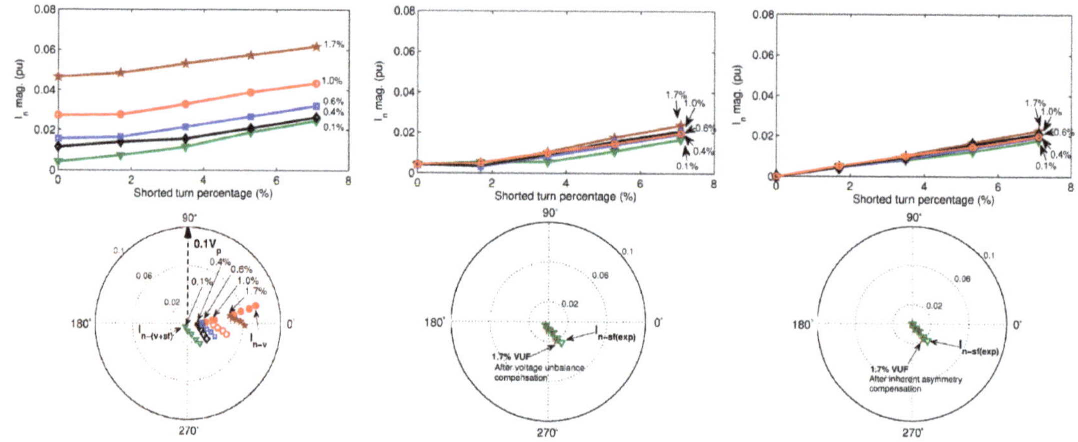

Figure 16. The test results (magnitude and phasor plots) of the negative sequence current magnitude as a function of shorted turn fault severity (up to 7.1%) with voltage unbalance factors between 0.1% to 1.7%: measured results (**left column**) after voltage unbalance compensation (**middle column**) and after both voltage unbalance and inherent asymmetry compensation (**right column**).

As can be seen in the figure, with near-zero supply voltage unbalance (0.1%), the negative sequence current is almost proportional to the number of shorted turns. For example, a 7% shorted turn fault produces slightly more than 0.02 pu negative sequence current. However, in a healthy motor (zero shorted turns), 1% supply voltage unbalance can produce a comparable magnitude of negative sequence current, which indicates the necessity of compensation for the supply voltage unbalance.

In addition, the phasor plot in the left column of Figure 16 show that the phase angle of the negative sequence current component due to supply voltage unbalance (about +15°) is substantially different than that due to the shorted turn fault (about −45°). These phase angles are dependent on the negative sequence voltage phasor and the phase of the supply in which the shorted-turn fault is present.

The negative sequence current magnitude depends on the phasor summation of the negative sequence current components of the fault level and the supply voltage unbalance. Therefore, any compensation technique must be based on a phasor calculation rather than just considering the magnitudes.

The results of the negative sequence supply voltage compensation are given in the middle column of Figure 16, which utilized the calibration curve given previously. Note that when the sensitivity of the negative sequence current to the supply voltage unbalance is eliminated, this can magnify the influence of the shorted turn fault. In the phasor plot of Figure 16, the close correspondence between the results at 0.1% and 1.8% VUF is given.

The presence of inherent asymmetry, hence non-zero negative sequence current, in a healthy machine makes it difficult to identify the small level of shorted turn faults (less than 2% or five shorted turns in the test machine as it is given in the middle column of Figure 16).

The results given in the right column of Figure 17 show the compensation of both voltages unbalance and inherent asymmetry. Note that the compensation reduces the negative sequence current to near zero for healthy machines, which allows even small faults of greater than, say, 1% (3 shorted turns) to be distinguished. Moreover, an almost ideal linear relationship between the fault severity and the negative sequence current is obtained, which reduced the sensitivity to supply voltage unbalance further. The phasor plot shows the result for the healthy motor now lies at the origin. This means that the negative sequence current for the stator shorted turn faults is now fully compensated.

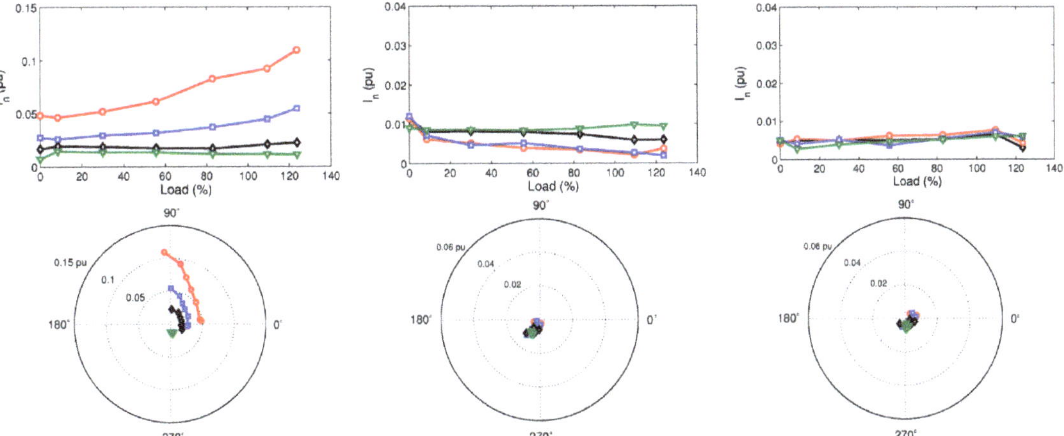

Figure 17. The test results (magnitude and phasor plots) of the negative sequence current at 1.7% shorted turn fault severity vs. motor load with various voltage unbalance factors: measured results (**left column**) after voltage unbalance compensation (**middle column**) and after voltage unbalance and inherent asymmetry compensation (**right column**).

The left column in Figure 17 show the negative sequence current test results with a 1.7% shorted turn fault combined with a varying supply voltage unbalance (produced by external resistors of 0 Ω, 1 Ω, 2 Ω and 4 Ω) under various motor loads. The significant effects of loading and voltage unbalance are also visible from the change in angle of the negative sequence current components, i.e., from −43° to 95°. At the light motor loading, the negative sequence phase angle is shifted from −43° to +12°, primarily due to the shorted turn fault.

The measured negative sequence current variation under the load given in the middle of Figure 17 shows a significant reduction after compensation for the voltage unbalance. Note the order of magnitude changes in the scales of the vertical axis in Figure 17.

The right column in Figure 17 show the processed result after the inherent asymmetry elimination. Note also that the supply unbalance and the load variation are reduced even further. This demonstrates that the compensation technique can successfully eliminate the voltage unbalance and the inherent asymmetry under a wide range of motor load variations, which is performed to gain the true magnitude of the negative sequence current due to the early and accurate detection of the shorted turn fault.

6. Conclusions

The stator faults in three-phase induction motors are primarily associated with the windings and are found to be the most critical faults as they develop quickly, requiring fast online diagnostic methods. This paper utilized the negative sequence components of voltage and current for the identification of the asymmetrical behavior of motor windings under stator faults.

In order to understand the causes of negative sequence currents for the detection of shorted-turn faults in the line-operated induction machines, this paper provided a comprehensive study of the effects of the major asymmetrical disturbances on the negative sequence currents. These included the effects of measurement errors, motor temperature, inherent machine asymmetry and supply voltage unbalance, all under varying motor load conditions.

A novel phasor compensation technique was described for inherent asymmetry and supply voltage unbalance. This relies on performing tests to determine the inherent asymmetry and identify the effective negative sequence impedance at the operating voltage.

The simulation and experimental results demonstrated that the proposed approach allows the use of the negative sequence current to detect even at small shorted turn faults (2% for the test motor). In addition, the fault severity was estimated accurately using the compensating method for sensor errors, inherent asymmetry and voltage unbalance, which are all present in practical machines. The compensation of supply voltage unbalance under varying motor loads was also investigated in detail.

It can be concluded that further works can be conducted while the motor is operating under multiple faults, which may be based on an online model of the machine.

Author Contributions: Data curation, S.B.; Formal analysis, S.B.; Funding acquisition, N.E.; Investigation, S.B. and N.E.; Methodology, S.B. and N.E.; Software, S.B.; Supervision, N.E.; Validation, S.B.; Writing—review & editing, S.B. and N.E. All authors have read and agreed to the published version of the manuscript.

Funding: This research received no external funding.

Conflicts of Interest: The authors declare no conflict of interest.

References

1. Haines, G.; Ertugrul, N. Application Sensorless State and Efficiency Estimation for Integrated Motor Systems. In Proceedings of the 13th IEEE International Conference on Power Electronics and Drive Systems (PEDS 2019), Toulouse, France, 9–12 July 2019.
2. Bakhri, S. Detailed Investigation of Negative Sequence Current Compensation Technique for Stator Shorted Turn Fault Detection of Induction Motors. Ph.D. Thesis, University of Adelaide, Adelaide, Australia, 2013.
3. Supangat, R.; Grieger, J.; Ertugrul, N.; Soong, W.L.; Gray, D.A.; Hansen, C. Investigation of static eccentricity fault frequencies using multiple sensors in induction motors and effects of loading. In Proceedings of the IECON 2006-32nd Annual Conference on IEEE Industrial Electronics, Paris, France, 6–10 November 2006; pp. 958–963.
4. Supangat, R.; Ertugrul, N.; Soong, W.L.; Gray, D.A.; Hansen, C.; Grieger, J. Broken rotor bar fault detection in induction motors using starting current analysis. In Proceedings of the 2005 European Conference on Power Electronics and Applications, Dresden, Germany, 11–14 September 2005.
5. Grubic, S.; Aller, J.M.; Lu, B.; Habetler, T.G. A survey on testing and monitoring methods for stator insulation systems of low-voltage induction machines focusing on turn insulation problems. *IEEE Trans. Ind. Electron.* **2008**, *55*, 4127–4136. [CrossRef]
6. Bakhri, S.; Ertugrul, N.; Soong, W.L.; Arkan, M. Investigation of negative sequence components for stator shorted turn detection in induction motors. In Proceedings of the Australasian Universities Power Engineering Conference (AUPEC), Christchurch, New Zealand, 5–8 December 2010; pp. 1–6.
7. Arkan, M.; Perovic, D.K.; Unsworth, P. Online stator fault diagnosis in induction motors. *IEE Proc. Electr. Power Appl.* **2001**, *148*, 537–547. [CrossRef]
8. Kliman, G.B.; Premerlani, W.J.; Koegl, R.A.; Hoeweler, D. Sensitive, on-line turn-to-turn fault detection in AC motors. *Electr. Mach. Power Syst.* **2000**, *28*, 915–927.
9. Tallam, R.M.; Habetler, T.G.; Harley, R.G. Transient model for induction machines with stator winding turn faults. *IEEE Trans. Ind. Appl.* **2002**, *38*, 632–637. [CrossRef]
10. Tallam, R.M.; Habetler, T.G.; Harley, R.G. Experimental testing of a neural-network-based turn-fault detection scheme for induction machines under accelerated insulation failure conditions. In Proceedings of the 4th IEEE International Symposium on Diagnostics for Electric Machines, Power Electronics and Drives 2003 (SDEMPED 2003), Atlanta, GA, USA, 24–26 August 2003; IEEE: Piscataway, NJ, USA, 2003; pp. 58–62.
11. Asfani, D.A.; Morel, J.; Purnomo, M.H.; Hiyama, T. Simulation and detection of temporary short circuit in induction motor windings using negative sequence current. In Proceedings of the 2011 International Conference on Advanced Power System Automation and Protection (APAP), Beijing, China, 16–20 October 2011; pp. 2304–2308.
12. Bouzid, M.B.K.; Champenois, G. New Expressions of Symmetrical Components of the Induction Motor Under Stator Faults. *IEEE Trans. Ind. Electron.* **2013**, *60*, 4093–4102. [CrossRef]
13. Mengoni, M.; Zarri, L.; Gritli, Y.; Tani, A.; Filippetti, F.; Lee, S.B. Online Detection of High-Resistance Connections with Negative-Sequence Regulators in Three-Phase Induction Motor Drives. *IEEE Trans. Ind. Appl.* **2015**, *51*, 1579–1586. [CrossRef]
14. Lashkari, N.; Poshtan, J. Detection and discrimination of stator interturn fault and unbalanced supply voltage fault in induction motor using neural network. In Proceedings of the 6th 2015 Power Electronics, Drives Systems & Technologies Conference (PEDSTC), Tehran, Iran, 3–4 February 2015; IEEE: Piscataway, NJ, USA, 2015; pp. 275–280.

15. Bouzid, M.; Champenois, G. Experimental compensation of the negative sequence current for accurate stator fault detection in induction motors. In Proceedings of the Industrial Electronics Society, IECON 2013-39th Annual Conference of the IEEE, Vienna, Austria, 10–13 November 2013; IEEE: Piscataway, NJ, USA, 2013; pp. 2804–2809.
16. Kohler, J.L.; Sottile, J.; Trutt, F.C. Condition monitoring of stator windings in induction motors: Part I-Experimental investigation of the effective negative-sequence impedance detector. *IEEE Trans. Ind. Appl.* **2002**, *38*, 1447–1453. [CrossRef]
17. Lee, S.B.; Tallam, R.M.; Habetler, T.G. A robust, on-line turn-fault detection technique for induction machines based on monitoring the sequence component impedance matrix. *IEEE Trans. Power Electron.* **2003**, *18*, 865–872.

energies

Article

Machine Learning-Based Fault Detection and Diagnosis of Faulty Power Connections of Induction Machines

David Gonzalez-Jimenez, Jon del-Olmo, Javier Poza *, Fernando Garramiola and Izaskun Sarasola

Faculty of Engineering, Mondragon Unibertsitatea, 20500 Arrasate-Mondragón, Gipuzkoa, Spain; dgonzalez@mondragon.edu (D.G.-J.); jdelolmo@mondragon.edu (J.d.-O.); fgarramiola@mondragon.edu (F.G.); isarasola@mondragon.edu (I.S.)
* Correspondence: jpoza@mondragon.edu; Tel.: +34-943794700

Citation: Gonzalez-Jimenez, D.; del-Olmo, J.; Poza, J.; Garramiola, F.; Sarasola, I. Machine Learning-Based Fault Detection and Diagnosis of Faulty Power Connections of Induction Machines. *Energies* **2021**, *14*, 4886. https://doi.org/10.3390/ en14164886

Academic Editors: Daniel Moriñigo-Sotelo, Rene Romero-Troncoso and Joan Pons-Llinares

Received: 20 June 2021
Accepted: 3 August 2021
Published: 10 August 2021

Publisher's Note: MDPI stays neutral with regard to jurisdictional claims in published maps and institutional affiliations.

Abstract: Induction machines have been key components in the industrial sector for decades, owing to different characteristics such as their simplicity, robustness, high energy efficiency and reliability. However, due to the stress and harsh working conditions they are subjected to in many applications, they are prone to suffering different breakdowns. Among the most common failure modes, bearing failures and stator winding failures can be found. To a lesser extent, High Resistance Connections (HRC) have also been investigated. Motor power connection failure mechanisms may be due to human errors while assembling the different parts of the system. Moreover, they are not only limited to HRC, there may also be cases of opposite wiring connections or open-phase faults in motor power terminals. Because of that, companies in industry are interested in diagnosing these failure modes in order to overcome human errors. This article presents a machine learning (ML) based fault diagnosis strategy to help maintenance assistants on identifying faults in the power connections of induction machines. Specifically, a strategy for failure modes such as high resistance connections, single phasing faults and opposite wiring connections has been designed. In this case, as field data under the aforementioned faulty events are scarce in industry, a simulation-driven ML-based fault diagnosis strategy has been implemented. Hence, training data for the ML algorithm has been generated via Software-in-the-Loop simulations, to train the machine learning models.

Keywords: fault diagnosis; fault detection; induction motor; electric machine; machine learning; supervised learning; data-driven; power connection failures

1. Introduction

Induction motors (IM), especially squirrel cage motors, constitute the core of many electric drives. They are widely used in industrial applications such as machining tools, electric vehicles and railway traction systems. Their simplicity, robustness and ease of maintenance have made them popular in industry. However, like any component, they are not totally free from failures. Therefore, in the last decades, numerous studies have been carried out analysing their failure modes, their probabilities of happening and proposing fault detection and diagnosis (FDD) strategies.

From the point of view of a generic electric drive, the electric machine can be defined as one of the main subsystems together with the invert block, the power source and the sensors. Bearing in mind this schema, it is worth summarizing the different failure modes that may appear in these subsystems due to the influence they can have on the behaviour of the induction machine. Regarding the inverter, the main failure modes can be summarized regarding power semiconductors (MOSFET, IGBT, Diode) and electrolytic filtering capacitors faults. When referring to semiconductors, the most typical faults are short- and open-circuit faults. The former is normally considered a destructive fault because of its high overcurrent effects, therefore, it typically requires the adoption of actions to shut down the drive immediately. The latter, which usually leads to complete or partial losses of the current at the exit of the inverter, is not usually classified as catastrophic. This means that these faults can remain undetected for a long time since the entire system can continue to operate in a degraded mode, so, it is interesting to develop health management strategies to detect the anomalies in advance [1–5]. In the case of the electrolytic capacitors that usually make up the DC link, the most frequent failure mode tends to be the ageing of the component because of operating in hard working conditions. This mainly leads to the variation of the capacitance (C) and the equivalent series resistance (ESR) which may cause not fulfilling the tasks of maintaining a constant DC voltage value, neither protecting power converters from over-voltages and sudden drops in the energy voltage source, nor presenting a high impedance against the harmonics generated by the inverter [6–8]. Concerning the sensors, they are usually used for control and protection tasks of the electric drive. However, when they operate in harsh working environments, they sometimes become prone to failure, causing abnormal operation of the electrical machine, reducing the efficiency of the traction force, or even causing an emergency stop. The most common failure modes can be summarized in gain or offset in the measurement or direct disconnection of the device, as they can be understood in [3,9–12].

Focussing on the electric machine subsystem, as summarised in [13–16], the failure modes of induction motors can be grouped into stator failures and rotor failures. Among the most common stator failures, stator winding short-circuits (in their different modes), vibration problems and phase connection failures can be found. The most common rotor failures are bar breakages, rotor misalignments or bearing problems, either due to bearing failures, lack of lubrication or misalignment. As a result, several FDD strategies have been proposed for these types of problems in different applications [17], such as electric vehicles, railway traction drives or renewable energy systems.

On the one hand, model-based techniques are usually used for the identification of motor parameters in order to monitor their deviation from the nominal values [18–21]. On the other hand, in the field of signal-based methods, there are alternatives such as the Park's Vector monitoring for stator short-circuit [22,23], stator imbalance [24] and rotor bar breakage [25] detection. However, the most widely used technique in induction motor FDD has been the frequency analysis of phase currents. This frequency analysis known as Motor Current Signature Analysis (MCSA) is the most popular one [26–29].

In recent years, the emergence of Industry 4.0 and the use of artificial intelligence methods, such as machine learning (ML) or deep learning (DL) have led to the development of data-driven FDD techniques. ML or DL have been used as a complement to the aforementioned techniques to help in the classification and prediction of failure modes. For example, there are many examples that use vibration or current measurements to diagnose stator and rotor failures [30–35]. Article [36] presents an extensive review of the application of data-driven methods for electric drives.

However, among all the IM failure modes, the one that has perhaps been studied the least is motor wiring or connection failures. It is worth mentioning that failure modes such as open-phase, High Resistance Connections or opposite-phase wiring in IM connections are usually catastrophic. In other words, although these failure modes are not frequent, when they occur, the maintenance tasks become very costly because most of the time the rolling stock must be stopped and the electric machine must be repaired. Furthermore, in the majority of cases, they is often a link to human errors during manufacture, resulting in incorrectly tightened terminals or poor wirings. In an industrial context, there are situations where due to manufacturing or maintenance mistakes, current imbalance and opposite-phase wiring problems can occur. For example, when a faulty inverter is replaced in a depot, the wiring can be deficiently installed and faulty equipment is put in service. As a result, they are considered high-cost, low-probability cases.

For example, the detection of HRC has been approached using different techniques. Thermal imaging can be a useful and effective technique for manual inspections [37]. However, it can be costly and difficult to automatise. During the last decades, online detection methods have been developed mainly based on resistance estimation or current sequence analysis. The first method consists of estimating the resistance by injecting voltage pulses with the inverter already installed in the drive, as proposed in [38]. The authors of this paper propose to measure the voltage between the neutral point of the motor and the negative of the DC-link. The main disadvantage of this method is the need for an additional sensor, as well as the fact that the neutral point is usually not accessible. In [39], the authors use the same method but without additional sensors. They inject two voltage vectors and with the measurements of their respective currents calculate the phase resistance. It is mentioned that with this method the effect of inverter nonlinearities on the estimation can be eliminated. Furthermore, ref. [40] proposes a similar method but while compensating for the effect of the inverter by pre-calculating the voltage drop across the semiconductors. The second method proposes to detect the negative sequence of currents due to system unbalance. In [41,42], the authors develop an induction motor model taking into account the effect of HRC and stator short circuits. From these models, a negative sequence of current and voltage due to unbalance can be estimated and used as a fault indicator. They also make an effort to be able to identify the specific failure mode (HRC or short-circuit). Furthermore, in [43] a similar technique is presented, but this time the drive control strategy is used to calculate the negative sequence and to implement a fault-tolerant control.

As far as open-circuit faults are concerned, few papers refer to the detection of this failure mode when it occurs at the motor connection. However, the effect of this failure mode is similar regardless of whether it occurs in the wiring, the inverter or the motor. Therefore, the techniques proposed in the literature could be used for all of them. Park's Vector Approach is one of the most widely used methods [44]. In [45–47], condition monitoring using this technique is proposed to detect open-circuit faults in inverters. The same fault is detected in [48] by calculating indicators from the mean value of the currents. Moreover, there are also model-based techniques, such as the one presented in [49], where a model is proposed and validated, which takes into account open-circuit faults in the phases and in the wiring.

All these strategies need to be executed at high frequency and usually embedded in the controller of the drive. This can be challenging in some applications such as electric transportation or renewable energy systems, where controller memory and computational capacity is limited. Furthermore, increasing the cost of a drive by adding FDD functionalities is not justified nowadays, especially in view of the rise of communication and cloud-based technologies. As mentioned in [36], FDD strategy trends show that data-driven methodologies based on ML or DL have emerged as a valid solution for electric drives. As an example, several publications show the application of ML or DL for the detection of faults in stator, rotor and bearings [32,50–52]. In the case of HRCs in electric motors, ref. [53] shows the training, validation and testing of an artificial neural network. It can be said that this is an evolution of the classical negative/zero sequence technique, where the neural network models the healthy state and classifies faulty states.

In applications such as a railway traction, fault detection and isolation is more difficult due to the composition of the system. Usually, a rolling stock is composed by several inverter boxes that can feed more than one induction motor in parallel (see Figure 1). As an example, a train can have six motors, each of them controlled in pairs by three controllers/inverters. Thus, when the driver sets a general torque command for the whole traction chain, the motors are controlled independently dividing the total command by the number of inverters. As a result, in some failure modes, this structure can be defined as catastrophic, because even if there is a faulty motor in the total 6, the rest ones will keep working. As the rotor of the motor and the axle of the boogie are coupled mechanically, even if the motor is faulty, the rotor will continue turning due to the train inertia. While the control unit tries to control the torque of the motor, high overcurrent and overvoltages can generate irreversible failures.

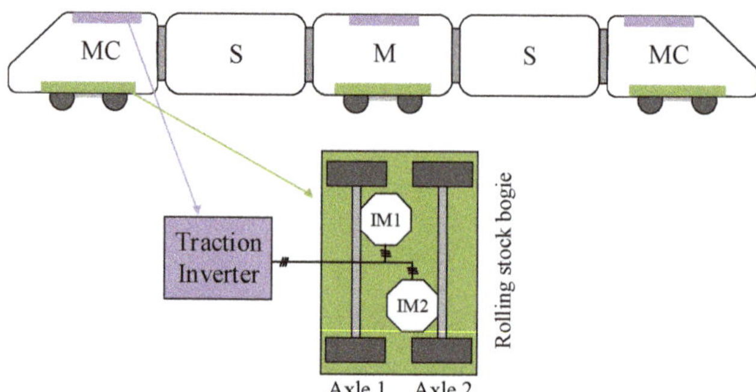

Figure 1. Schematic of the two induction motors in parallel structure from a railway application.

Seeing the potential that ML and DL techniques have had on other failure modes, this paper proposes a data-driven strategy for the detection and classification of HRC, open-phase and opposite-phase wiring faults in induction machines implemented in railway applications. For this, a Software-in-the-Loop (SiL) simulation platform is used in order to generate the data to train the ML models. Concretely, the SiL simulation replicates the behaviour of an electric drive from a tram traction application. The rest of the paper is organized as follows: Section 2 introduces the SiL platform used for data generation. Afterwards, Section 3 presents the development of the ML-based fault diagnosis strategy. Step-by-step data preprocessing, feature engineering and ML model training and testing are explained. Finally, Section 4 presents the main conclusions of the work.

2. SiL Simulation-Based Data Generation

As it has been mentioned before, one of the challenges in developing data-driven strategies for FDD is data availability. Electric machines are designed not to fail, so it is difficult to find a sufficient volume of information to allow reliable training, validation and testing of ML algorithms. Hence, data from healthy machine operation is available in abundance, while data from representative faulty operating conditions is limited. That is why, in many applications it is very common to deal with unbalanced datasets [54]. Furthermore, still nowadays, little field datasets from real industrial applications are available. Normally, implementing an effective data acquisition approach can be interpreted as expensive, as well as time-consuming. As a result, this data scarcity and imbalance has become an important drawback when trying to design data-driven condition monitoring strategies, specially those based on Machine Learning and Deep Learning.

In order to overcome these limitations, one of the used techniques is to generate the training dataset via simulations. In a digital environment, simulation-driven synthetic data generation is used to emulate conditions that are not easily available in existing field data, such as different working conditions, specific failure modes, etc. Therefore, in the present work, a SIL simulation platform developed in Matlab/Simulink platform, has been used to obtain synthetic data on the effects of deficient connections in induction motors. Specifically, it emulates the operation of a 160 kW electric traction drive from a railway traction application with two induction motors connected in parallel. It is worth mentioning that this platform has been validated by our industrial partner, using it in the development of railway traction systems. Furthermore, the results of the platform in healthy cases were previously compared with laboratory results. At the same time, both its nominal behaviour and the fault insertion block have been discussed in other publications of our research group [55–57].

The Matlab/Simulink platform consists of several blocks that simulate the operation of an electric drive (see Figure 2). The plant of this electric drive is composed of an input stage (contactors, filter and braking crowbar), a three-phase inverter and two induction motors fed in parallel. Moreover, the mechanical system is simplified to an inertia and a static load. It is worth mentioning that power electronics and traction motors can be simulated either using basic blocks from Simulink or Simscape blocks, depending on the required accuracy and simulation speed.

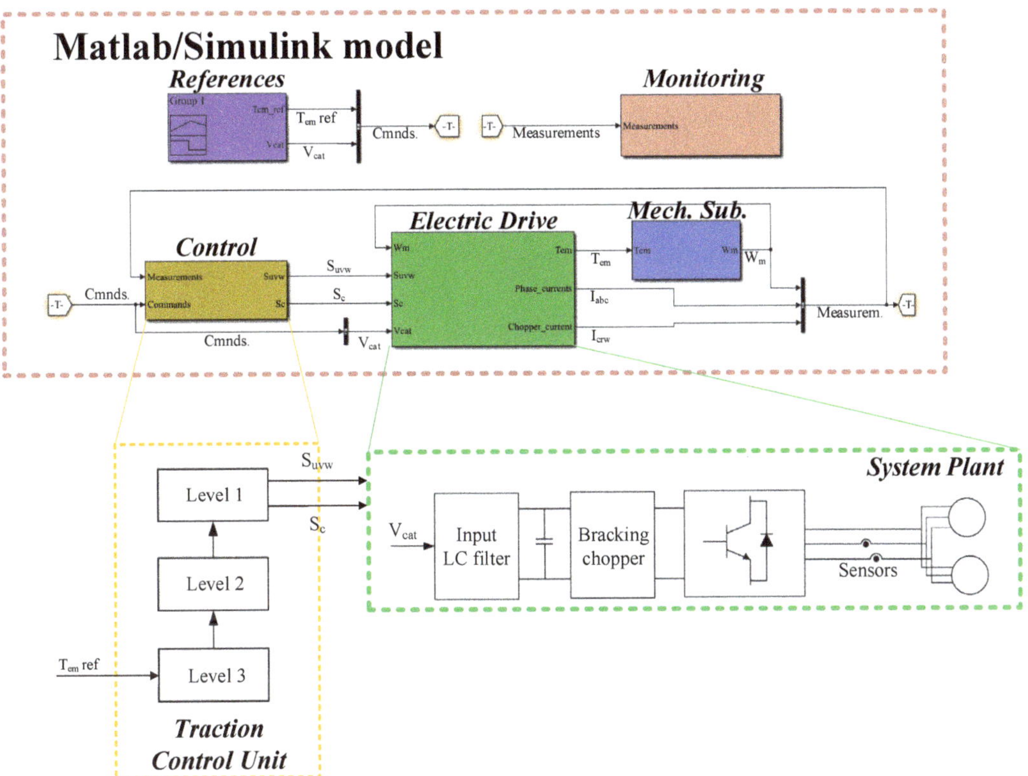

Figure 2. Schematic of the Matlab/Simulink based SiL simulation platform of a tram application for synthetic data generation.

The control functionalities are integrated following the Software-in-the-Loop strategy. Here, the control software used in the real device is embedded into the simulation so that its operation is as close as possible to the real application. Concretely, the TCU is built with three different control levels. In control level 3, the references for the control strategy of level 2 are calculated. Typically, level 2 implements some variant of vector control, so torque and flux references are obtained first in level 3. In this level, other control functionalities, such as bus voltage control or torque reference limitations, can be activated. Once level 2 obtains the voltage references for the inverter, in level 1, modulation strategies calculate the switching states for the inverter and the crowbar. It is worth mentioning that the vector control of the IM is an average control, because two motors are fed in parallel with only one inverter and one current sensor per inverter phase. Therefore, the measured current is the total current flowing from the inverter.

This platform allows the control strategy to be validated in different scenarios. As a result, and with the aim of analysing the effects of power connection faults, first a set of baseline healthy simulations has been defined. Torque is controlled following a predefined profile in order to obtain the desired acceleration and deceleration rates, as well as target speed. In this case, 3 target speed profiles have been simulated in the conditions shown in Table 1.

Table 1. Simulation scenarios for faulty data generation.

Speed Ref. [rpm]	Load Torque [Nm]
900	50
	100
2400	50
	100
4500	50
	100

Figure 3 shows the torque, phase currents, speed and DC-link voltage for a 2400 rpm target speed and 50 Nm load torque simulation environment. Furthermore, Figure 4 shows the detail of the phase currents.

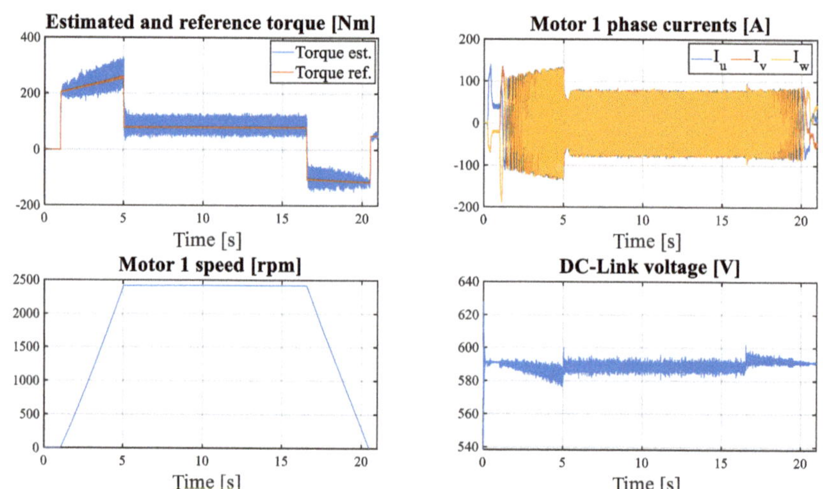

Figure 3. Torque, speed, phase currents and DC-link voltage signals at 2400 rpm and 50 Nm load (healthy state).

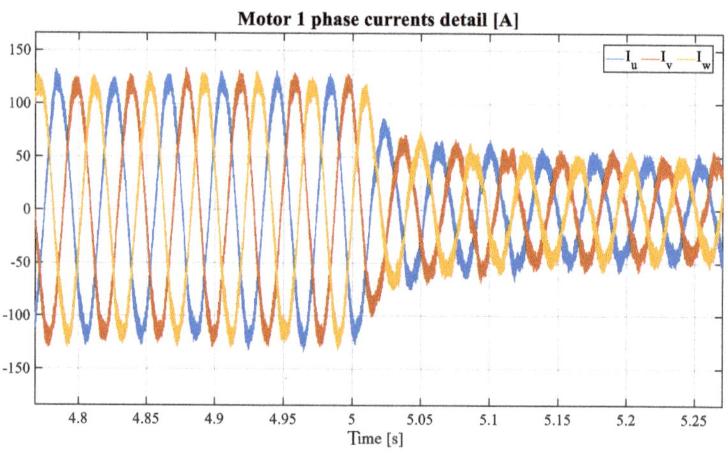

Figure 4. IM phase currents detail at 2400 rpm and 50 Nm load (healthy state).

Using the baseline simulations shown previously, power connection failures have been injected in the plant model. In particular, HRC faults, open-phase faults and opposite-phase wiring connection faults were simulated. Thanks to the use of Simulink's Simscape toolbox, these faults can be easily injected in the simulation. As it is shown in Figure 5, the HRC was emulated connecting a series resistor in a phase of the motor, while the other two faults were provoked by changing directly the motor connections.

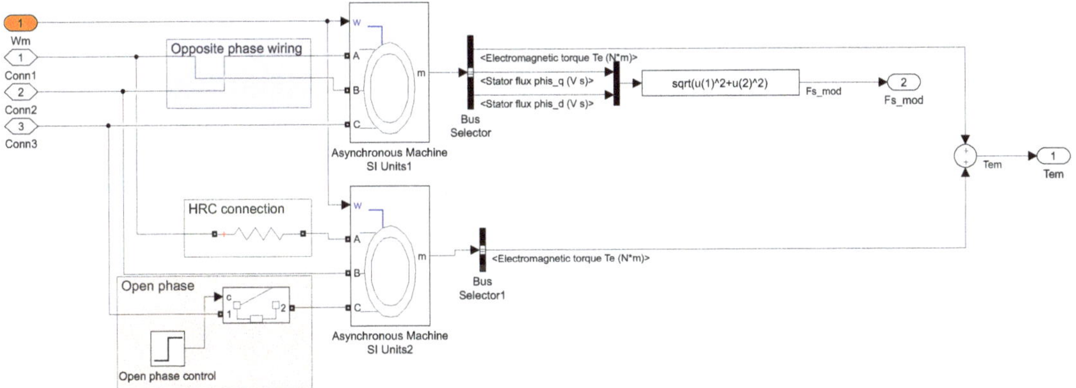

Figure 5. IM power connection faults modelled in the SiL platform.

Using this modified model and the operation conditions described in Table 1, faulty operation scenarios were simulated. In total, 72 simulations were launched for the generation of faulty data (6 scenarios, with 3 fault modes injected at 4 different instants). In this way, a database of 355 million samples at 50 µs was created.

In the following lines, some of the simulation results are presented. Figure 6 shows the different signals obtained from the simulation while causing an open-phase fault in the induction motor number 1. Looking at the IM-1 phase currents, phase A is disconnected and, as a consequence, the rest of the phase currents increase. As it was mentioned before, the vector control of the motors is an average control as two motors are fed in parallel with only one inverter. Hence, any failure in one of the motors causes the abnormal operation of the other one. At the same time, current imbalance causes torque oscillations due to the current measurement feedback and the vector control structure. It is important to remark that the high value of the load inertia filters torque oscillations mitigating them in the speed (bottom right graph). Therefore, as the simulation does not implement any speed control loop, when the fault occurs, the average value of the real torque and the speed decrease. In the real application, this speed loss would be compensated by the user increasing the torque command.

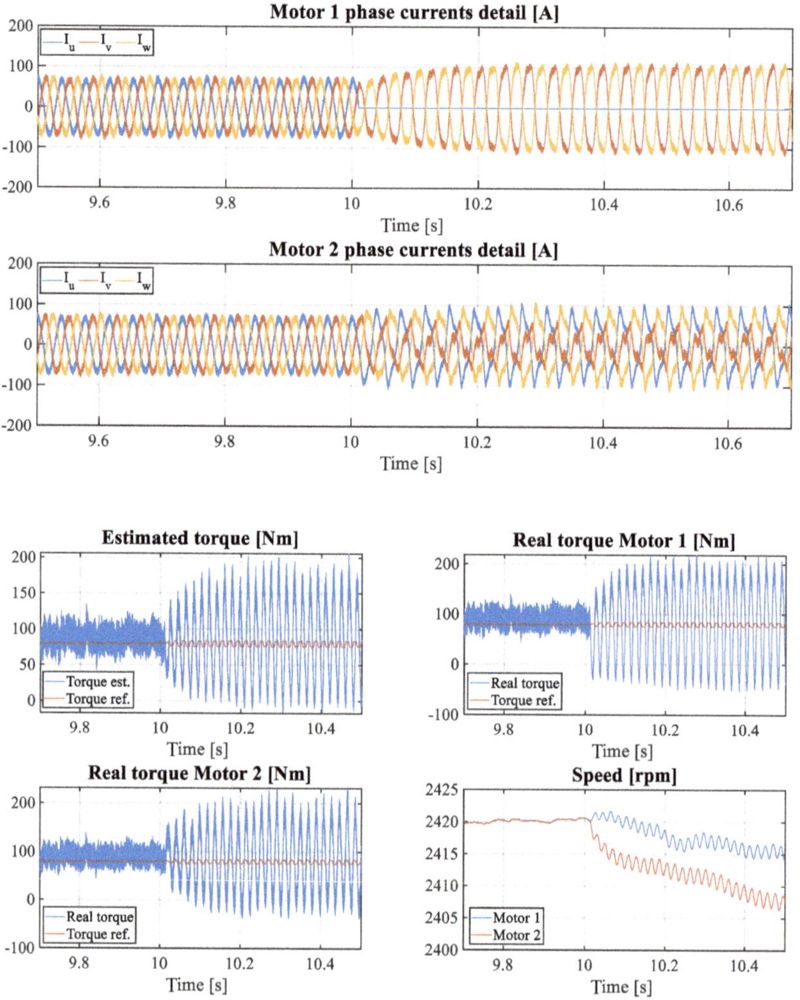

Figure 6. Phase currents of motor 1 and motor 2 at 2400 rpm and 50 Nm load (open-phase fault at t = 10 s).

In the case of HRC, it can be said that it is a less severe version of the open-phase failure. Current imbalance is translated to the torque as low frequency oscillations, as it is presented in Figure 7.

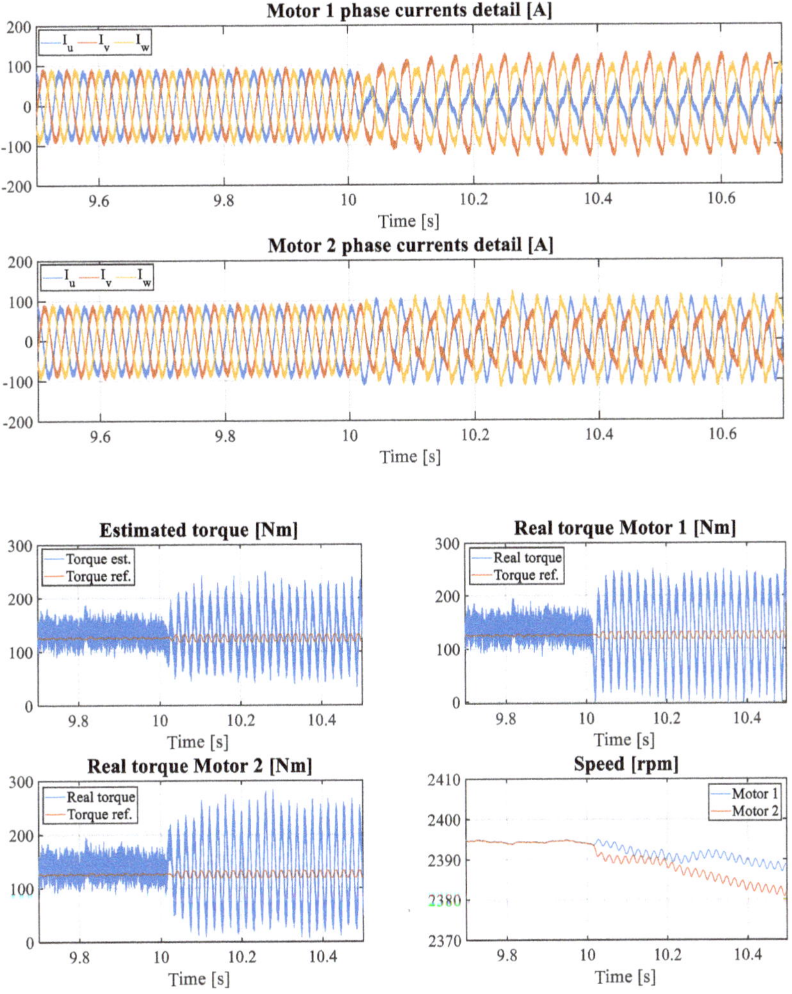

Figure 7. Phase currents of motor 1 and motor 2 at 2400 rpm and 100 Nm load (HRC/unbalanced fault at t = 10 s).

Finally, in the opposite-phase wiring mode (see Figure 8), since there is a closed torque control loop, it sets the current necessary for the estimated torque to follow the reference. However, as one of the motors is wired incorrectly, the actual torques of the motors do not follow the reference and the target speed is not achieved.

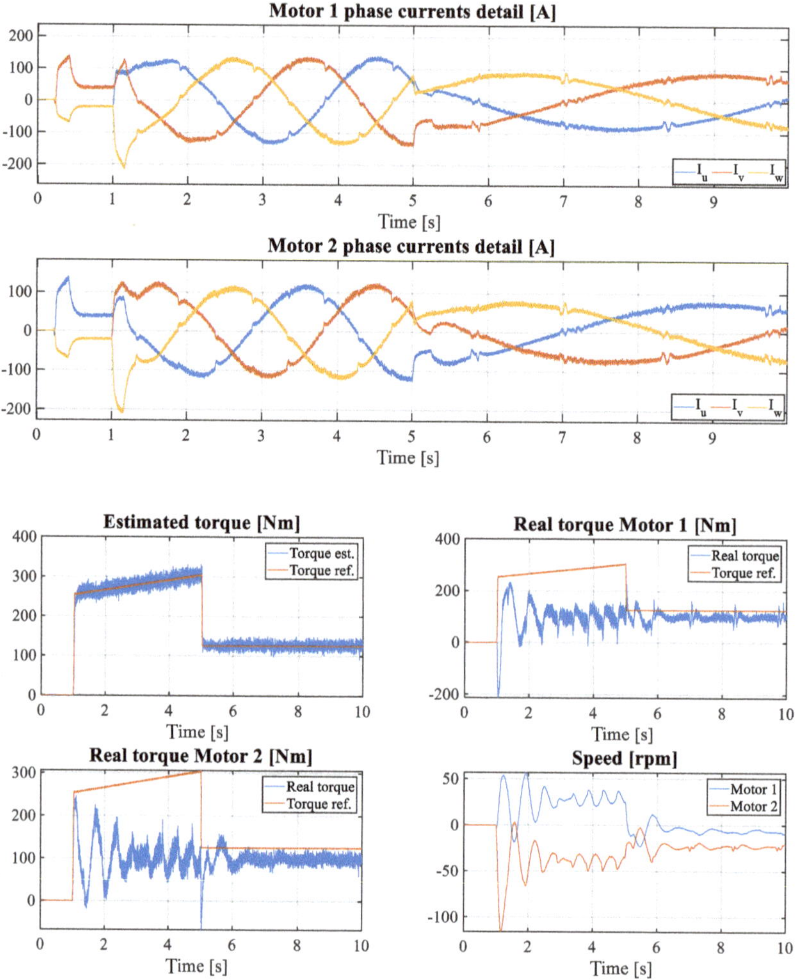

Figure 8. Phase currents of motor 1 and motor 2 at 2400 rpm and 100 Nm load (opposite-phase wiring fault at t = 0 s).

In the following sections, the development of the data-driven FDD strategy for the detection and classification of these fault modes will be presented. However, as it was shown previously, the HRC and the open-phase faults have similar effects in terms of torque oscillations, speed deviations and current in the healthy motor, therefore, they will be grouped in the same cluster, labelled as current imbalance. Therefore, the main task of the FDD strategy is to distinguish current imbalance and opposite-phase wiring anomalies from the healthy behaviour.

3. ML-Based Fault Diagnosis Strategy

Once the synthetic data have been generated, it is time to develop the machine learning-based fault diagnosis strategy for induction machines power connection failures. As mentioned in Section 1, these approaches developed via data-driven strategies seek to generate computer systems capable of performing tasks that normally require human intelligence, through artificial intelligence. In this research, the analysis of the health status of induction machine power connexions is proposed by differentiating the aforementioned failure modes from the healthy behaviour. For this, the synthetic data acquired from the simulation platform were used to train and validate ML classification algorithms, in order to categorise the different health status in groups.

In this way, it is important to mention that in order to implement these data-driven solutions efficiently, a specific and standardized ML workflow is generally put into practice. As it can be seen in Figure 9, it is not only based on selecting and optimizing the ML algorithm, but it also consists of carrying out different tasks, such as the acquisition and organization of raw data, the raw data preprocessing and the implementation and integration of the algorithm in the application [58–60].

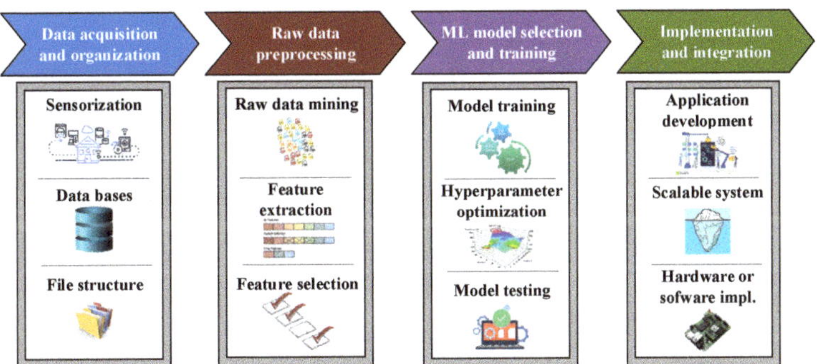

Figure 9. Standardized workflow to apply effectively machine learning approaches.

Thus, in this section, the main steps of this workflow have been developed and optimized taking into account the application requirements. The solution should be able to distinguish the healthy behaviour from the opposite-phase wiring faults and the current imbalance faults. Furthermore, false positives should be avoided. It has to be taken into account that a false positive could cause an unnecessary maintenance shutdown of the equipment, which in applications such as railway could cause important availability and economic losses.

In addition, it is important to mention that one of the main advantages of simulations is their flexibility to create faulty environments, since different failure modes can be injected. Therefore, the supervised ML method becomes an effective alternative to face this classification problem. In these cases, the task of labelling data samples becomes much easier, owing to the fact that the exact failure injection time and even its characteristics are known. In a real application environment, the labelling task is much more time-consuming, as it demands considerable expert knowledge. Figure 10 shows schematically the Supervised ML approach.

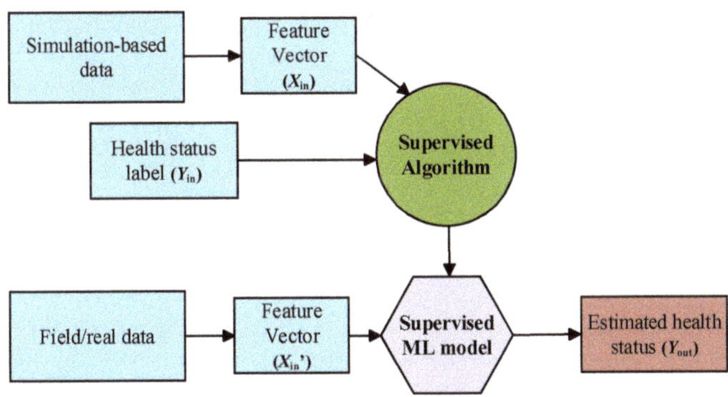

Figure 10. Schematic of the supervised ML methodology.

In supervised ML methods, a certain label (Y_{in}) is attached to each training dataset sample (X_{in}) with information about their momentary state of health. Therefore, it is easier to interpret the output predictions (Y_{out}) of the ML model from the new unseen dataset (X'_{in}). Specifically, in this article, a three-class classification ML algorithm was trained. Although in Section 1 more than three health statuses are explained, in a first approximation the open-phase fault and the HRC fault are unified due to their similar effects in phase current imbalance. Therefore, these are the different health status labels that the supervised ML algorithm should differentiate: healthy (H), current imbalance (CI) and opposite-phase wiring fault (OPW).

Moreover, before starting with each of the stages from the ML workflow, it is worth mentioning that Amazon Web Services (AWS) is the cloud service platform where the fault diagnosis approach was developed. Apart from the development of the FDD strategy, a secondary objective of the work has been to use commercial cloud-based tools. The use of these tools has several advantages: the management of big datasets is easier (than with software such as Matlab) and the proposed solution will be closer to a future industrial implementation. Figure 11 shows the architecture of the platform for development of the data-driven FDD strategy.

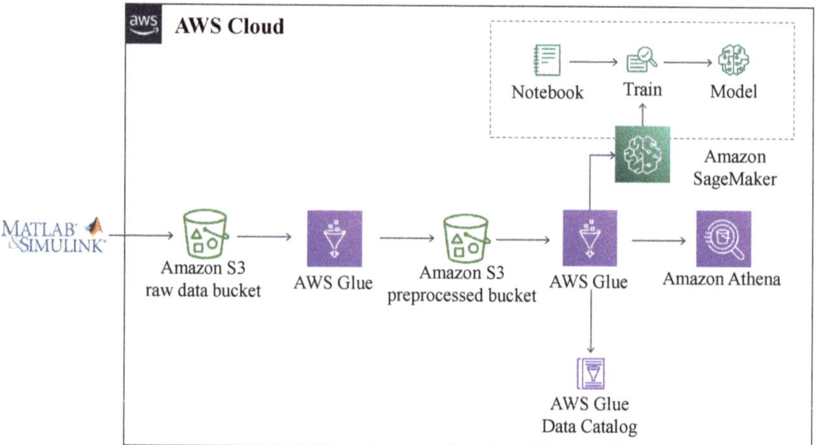

Figure 11. Data pipeline for the ML-based FDD strategy developed in Amazon Web Services.

3.1. Data Acquisition and Organization

As mentioned before, simulated data should be generated in the a way that is as similar as possible to how it is acquired in the real application, in terms of recorded variables, sampling frequency and acquisition mode (average, RMS, etc.) Therefore, output data from simulation must be modified to replicate a real application environment.

As an example, the simulation explained previously runs at 50 µs. Therefore, it provides many different variables with a 20 kHz sampling rate. However, in a real application, not all the variables are accessible, nor is the acquisition frequency that high. In this research, 11 variables that can be recorded in real applications were downsampled at a 64 ms rate, replicating the limitation of sensors installed in real applications. Table 2 shows a summary of the recorded variables from the simulation. Consequently, the complete raw dataset contains approximately 27,700 samples per each variable.

Finally, all the signals were saved in .csv files. In each of these files, an acceleration/deceleration profile with different health status and fault injection times was recorded. These .csv files were uploaded to AWS platform, specifically to the AWS S3 service which is an object storage service that offers industry-leading scalability, data availability, security and performance.

Table 2. Recorded variables from the SiL simulation platform.

Variable Name	Acq. Freq [ms]	Acq. Mode	Explanation
Torque_Motores	64	Inst.	Sum of the two parallel IM torques [Nm]
Tem_ref	64	Inst.	Torque reference for each of the IMs [Nm]
Tem_ref_TL	64	Inst.	Torque reference after control limitations [Nm]
wm1	64	Inst.	Speed of IM1 [Hz]
wm2	64	Inst.	Speed of IM2 [Hz]
Ia_medida	64	RMS	Total output current of the inverter in phase A [A]
Ib_medida	64	RMS	Total output current of the inverter in phase B [A]
Ic_medida	64	RMS	Total output current of the inverter in phase C [A]
Icat	64	Inst.	Input measured current to the system [A]
Icrw	64	Inst.	Crowbar current [A]
Vbus	64	Inst.	BUS voltage [V]

3.2. Raw Data Preprocessing

After acquiring and organizing the raw data in AWS S3 service, the next step is preprocessing it. That means cleaning and manipulating the raw data to train different machine learning algorithms. This stage is normally divided into two levels of preprocessing—on the one hand, the general preprocessing and, on the other hand, the feature engineering. To do this, the raw dataset was exploited with the AWS SageMaker service.

As for general preprocessing, it involves data cleaning, which consists of filtering messy data, detecting outliers and missing values, applying standardization [61,62] and even segmentation [63,64]. However, since the raw data source for this research is a simulation platform, it can be said that cleaning tasks are not as necessary as they are for the data from a real application environment.

As an example of the general preprocessing, a search for outliers was performed. Therefore, different samples that can distort the training dataset were removed, as can be seen in Figure 12.

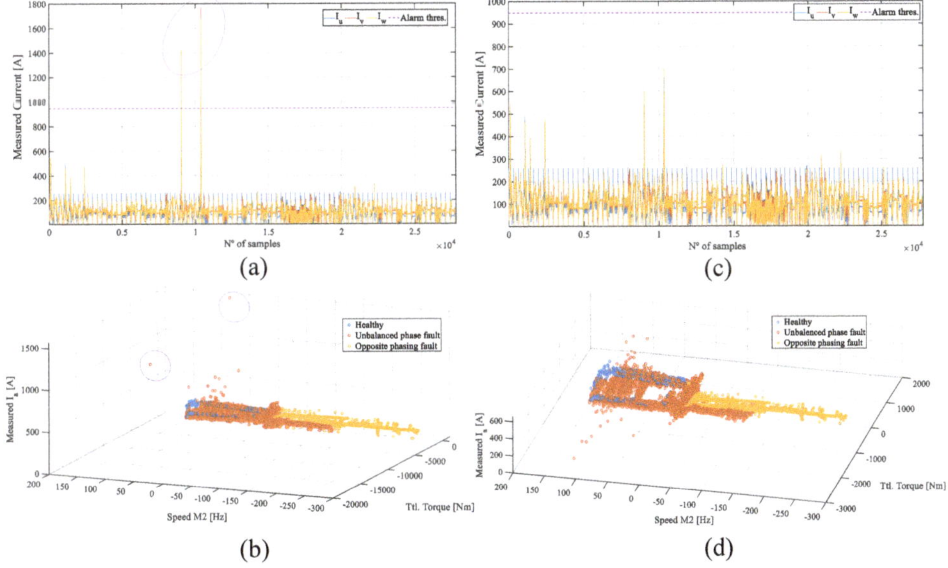

Figure 12. Example of outlier detection tasks. The outliers circled in purple were removed. (**a**) IM phase currents plot with outliers. (**b**) Scatterplot of the raw dataset with outliers. (**c**) IM phase currents plot without outliers. (**d**) Scatterplot of the raw dataset without outliers.

Furthermore, the entire independent set of variables was normalized, using *Z*-score normalization [61,62], which rescales independent variables with a zero mean and unit-variance range, as shown in Equation (1):

$$Z\text{-score} = \frac{x_i^{(t)} - \mu_i}{\sigma_i} \tag{1}$$

where *Z*-score is the normalized instance, μ_i and σ_i are the mean value and standard deviation of the *i*th acquired variable, respectively.

After cleaning the raw data, feature engineering is applied to extract important information from the dataset, in order to efficiently feed the ML algorithms. In particular, feature engineering can be divided into two main tasks, feature extraction (FE) and feature selection (FS).

The main goal of feature extraction is to transform raw data into numerical features, while preserving important information from the original data set. This can be done manually by calculating features in domains such as time, frequency or time-frequency, or automatically by applying modifications such as principal component analysis (PCA), linear discriminant analysis (LDA), etc. In this research, 5 time-domain features were extracted per each of the 11 initially recorded variables with a dynamic window, jumping each 10 samples. These new statistical features are maximum, minimum, mean, variance and standard deviation. As a result, from having a raw dataset matrix of 11 variables with 27,700 samples each, now we have 55 time-domain features with 2770 samples. This FE operation is explained in Figure 13.

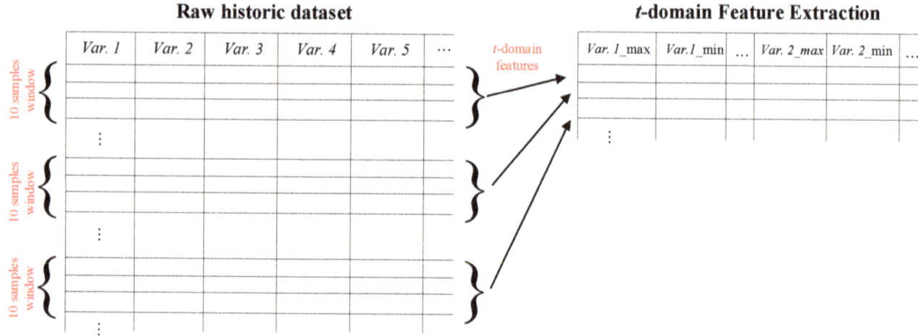

Figure 13. Schematic of the time-domain feature extraction method.

Then, feature selection is implemented, which means ranking the importance of the extracted features by applying certain evaluation criteria, while discarding the less important ones. In this article, the *SelectKBest* function with the *F-test* filter method from *sklearn* library in Python was implemented. Therefore, the best 14 statistical features were selected to train the ML algorithms. Figure 14 shows a list of the selected feature names, as well as a barplot with the different scores of the features.

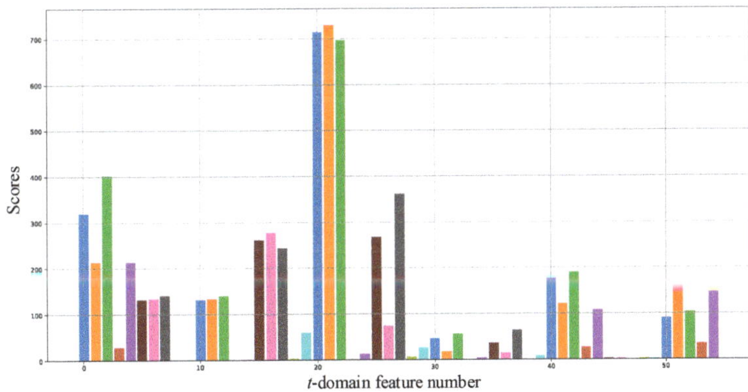

Figure 14. Scores of the feature selection step and list of *t*-domain features.

At the end of this ML workflow step, the definitive dataset which will be used to train and test the ML algorithms is obtained. Figure 15 shows a 3D scatterplot of the distribution of three important features from the definitive dataset. In blue the healthy samples are shown, in red the instances with unbalanced phase fault, and finally, in yellow the samples with opposite-phase faults. If a physical interpretation of the different clusters is created, it can be seen that while the samples with nominal health status (blue) have positive speed values and stable ranges of torque and phase current, for both the samples with failure due to imbalance and opposite connection phasing this is not the case. Regarding the first fault mode (red), it can be seen that the variance of both the torque and the current is considerable. This is due to the effect of the appearance of the phase current ripple that translates into torque vibrations due to the control strategy. In the case of the opposite connection phasing fault (yellow), the clearest effect can be seen in the speed, which in the majority of the samples contains negative values without normalizing, as well as in the mechanical torque, which never reaches the reference.

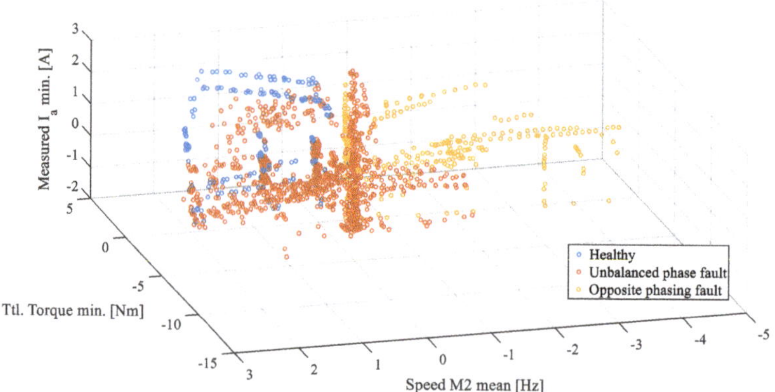

Figure 15. 3D scatterplot after applying raw dataset preprocessing step from the ML workflow.

Although the different classes can be differentiated visually by colours, the classification task of new unseen instances should be performed by a previously trained ML algorithm. Therefore, the main objective of this preprocessing step is to improve the separability of classes as much as possible to facilitate the training process.

3.3. ML Model Selection and Training

The third step of this workflow consists of choosing the Machine Learning topology, as well as training and validating the algorithm to leave it ready to be integrated into the required application. For that, it is helpful to rely on the specific procedure shown in Figure 16.

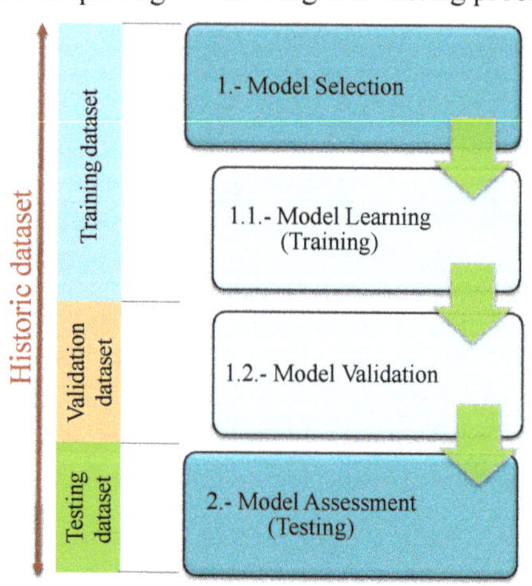

Figure 16. Training/testing process for a machine learning algorithm.

In the Model Selection phase, an empirical comparison of different algorithms from the same topology is carried out and the one with the best results is selected. This phase is divided into two basic tasks. On the one hand, in the former (Model Learning), algorithms with similar characteristics (supervised, unsupervised, semi-supervised) are trained with the training sub-dataset. That is to say, adjusting the internal parameters of each algorithm to efficiently estimate the outputs. In this way, four different algorithms have been trained in these research: logistic regression (LR), support vector machine (SVM), random forest (RF), and k-nearest neighbors (k-NN). On the other hand, the second task (Model Validation) requires optimizing the hyperparameters, as well as validating the different algorithms with the validation sub-dataset. When we refer to validation, we think of evaluating the performance of the algorithms by different criteria. In this research, as we have been working with supervised classification algorithms, the evaluation criterion applied is the confusion matrix. From here, accuracies and precision values were analysed to select the best algorithm. When speaking about accuracy, we refer to the percentage of the correct classified values over the total classified samples. Regarding precision, it quantifies the ability to classify the real positive samples correctly. It is worth mentioning that, given that the analysed failure modes are considered catastrophic (when one of these failures occurs, the entire system must be stopped to guarantee safety), within the misclassified samples those classified as false negative are more important than the false positives.

Finally, in the Model Assessment phase, the trained and selected algorithm is tested with new unseen data. If this last evaluation is positive, the ML model is supposed to be ready to implement in the respective application.

In addition, it is important to have in mind that the most efficient way to perform this training and testing process is to use independent sub-datasets at each stage. Therefore, in this research, the dataset obtained from the preprocessing step of the workflow has been split into three sub-datasets, namely, training, validation and testing sub-datasets. As a result, 70% of the initial dataset was used for the Model Selection step and the remaining 30%—for the Model Assessment step.

The results obtained during the whole process of the ML workflow are collected in Table 3.

Table 3. Accuracy and Precision results of the different algorithms during different steps of the ML workflow.

Steps	Accuracy [(TP + TN)/(Total Samples)]				Precision [TP/(TP + FP)]			
	LR	*SVM*	*RF*	*k-NN*	*LR*	*SVM*	*RF*	*k-NN*
Training with raw dataset	0.719	0.806	0.967	0.938	0.753	0.831	0.978	0.932
Training with *t*-domain features	0.848	0.921	0.975	0.933	0.872	0.923	0.982	0.933
Optimized algorithms testing	0.923	0.967	0.985	0.942	0.911	0.953	0.976	0.934

Regarding the accuracy of the algorithms, it is clear that this increases, on the one hand, when preprocessing the raw data and, on the other hand, when optimizing the hyperparameters of the algorithms. As a result, best values while classifying healthy, current unbalanced and opposite-phase wiring health status have been obtained with the random forest algorithm, with 98.5% accuracy. In terms of precision, something similar happens. The algorithm which classified fewer samples as false positive is also random forest, with 97.6% accuracy. Therefore, the trained, validated and tested RF model was selected at the end of the training/testing process.

4. Conclusions

This article presents a ML-based FDD strategy for induction motor High Resistance Connection faults, open-phase faults and opposite-phase wiring faults. In order to develop this strategy, and due to the lack of faulty samples from field working conditions, these data were generated using a Matlab/Simulink-based Software-in-the-Loop simulation. To this end, these synthetic samples were used for training, validating and testing different algorithms, such as logistic regression, support vector machine, random forest and k-nearest neighbours. Previously, raw data preprocessing tasks, as well as feature extraction and selection methods have been performed to improve the efficiency of the ML workflow. The best results were obtained by optimizing the random forest ML algorithm, reaching values of 98.5% for accuracy, and 97.6% for precision. Among all the available metrics for the evaluation of the ML algorithms, the false positive rate was prioritized, taking into account the cost of maintenance shutdowns which can occur in industry.

As it was shown, the proposed method is capable of distinguishing the unbalanced operation of the motor from opposite wiring problems. This will improve future maintenance tasks, since the algorithm could guide the process of failure detection and isolation, even preventing further damages. A data-driven approach has been applied in failure modes that were previously approached using model-based or signal-based methods. Moreover, the proposed solution was designed and implemented using the Amazon Web Services cloud service, reducing the adaptation time for future industrial applications. With regard to future lines of research, firstly, work must be done to improve the separability of classes by means of advanced feature engineering techniques. In addition, the validation of this method with Hardware-in-the-Loop, laboratory and field data will be performed.

Author Contributions: Conceptualization: D.G.-J. and J.d.-O.; investigation: D.G.-J. and J.d.-O.; methodology: D.G.-J. and J.d.-O.; formal analysis: J.P. and F.G.; project administration: J.P.; validation: D.G.-J. and J.d.-O.; writing—original draft: D.G.-J. and J.d.-O.; writing—review and editing: J.d.-O., J.P., I.S. and F.G. All authors have read and agreed to the published version of the manuscript.

Funding: This research received no external funding.

Institutional Review Board Statement: Not applicable.

Informed Consent Statement: Not applicable.

Data Availability Statement: Not applicable.

Acknowledgments: This research work was supported by CAF Power & Automation. The authors are thankful to the colleagues from CAF P&A, who provided material and expertise that greatly assisted the research.

Conflicts of Interest: The authors declare no conflict of interest.

Abbreviations

Abbreviations

The following abbreviations are used in this manuscript:

IM	Induction Machine
FDD	Fault Detection and Diagnosis
ML	Machine Learning
DL	Deep Learning
HRC	High Resistive Connection
SiL	Software-in-the-Loop
TCU	Traction Control Unit
AWS	Amazon Web Services
RMS	Root Mean Square
LR	Logistic Regression
SVM	Support Vector Machine
*k*NN	*k*-Nearest Neighbours
RF	Random Forest

References

1. Riera-Guasp, M.; Antonino-Daviu, J.A.; Capolino, G.A. Advances in electrical machine, power electronic, and drive condition monitoring and fault detection: State of the art. *IEEE Trans. Ind. Electron.* **2015**, *62*, 1746–1759. [CrossRef]
2. Li, Z.; Gao, Y.; Zhang, X.; Wang, B.; Ma, H. A Model-Data-Hybrid-Driven Diagnosis Method for Open-Switch Faults in Power Converters. *IEEE Trans. Power Electron.* **2020**, *36*, 4965–4970. [CrossRef]
3. Gou, B.; Xu, Y.; Xia, Y.; Deng, Q.; Ge, X. An Online Data-driven Method for Simultaneous Diagnosis of IGBT and Current Sensor Fault of 3-Phase PWM Inverter in Induction Motor Drives. *IEEE Trans. Power Electron.* **2020**, *35*, 13281–13294. [CrossRef]
4. Oh, H.; Han, B.; Mccluskey, P.; Han, C.; Youn, B.D. Physics-of-Failure, Condition Monitoring, and Prognostics of IGBT Modules: A Review. *IEEE Trans. Power Electron.* **2015**, *30*, 2413–2426. [CrossRef]

5. Wang, H.; Blaabjerg, F. Reliability of capacitors for DC-link applications in power electronic converters—An overview. *IEEE Trans. Ind. Appl.* **2014**, *50*, 3569–3578. [CrossRef]

6. Liao, L.; Gao, H.; He, Y.; Xu, X.; Lin, Z.; Chen, Y.; You, F. Fault Diagnosis of Capacitance Aging in DC Link Capacitors of Voltage Source Inverters Using Evidence Reasoning Rule. *Math. Probl. Eng.* **2020**. [CrossRef]

7. Khelif, M.A.; Bendiabdellah, A.; Cherif, B.D.E. Short-circuit fault diagnosis of the DC-Link capacitor and its impact on an electrical drive system. *Int. J. Electr. Comput. Eng.* **2020**, *10*, 2807–2814. [CrossRef]

8. Imam, A.M.; Divan, D.M.; Harley, R.G.; Habetler, T.G. Real-time condition monitoring of the electrolytic capacitors for power electronics applications. In Proceedings of the IEEE Applied Power Electronics Conference and Exposition—APEC, Anaheim, CA, USA, 25 February–1 March 2007; doi:10.1109/APEX.2007.357646. [CrossRef]

9. Dybkowski, M.; Klimkowski, K. Artificial neural network application for current sensors fault detection in the vector controlled induction motor drive. *Sensors* **2019**, *19*, 571. [CrossRef] [PubMed]

10. Gou, B.; Xu, Y.; Xia, Y.; Wilson, G.; Liu, S. An Intelligent Time-adaptive Data-driven Method for Sensor Fault Diagnosis in Induction Motor Drive System. *IEEE Trans. Ind. Electron.* **2018**, *66*, 9817–9827. [CrossRef]

11. Chen, H.; Jiang, B.; Lu, N. Data-Driven Incipient Sensor Fault Estimation with Application in Inverter of High-Speed Railway. *Math. Probl. Eng.* **2017**. [CrossRef]

12. Mohammadhassani, A.; Teymouri, A.; Mehrizi-Sani, A.; Tehrani, K. Performance Evaluation of an Inverter-Based Microgrid under Cyberattacks. In Proceedings of the SOSE 2020—IEEE 15th International Conference of System of Systems Engineering, Budapest, Hungary, 2–4 June 2020; doi:10.1109/SoSE50414.2020.9130524. [CrossRef]

13. Singh, G.K.; Al Kazzaz, S.A.S. Induction machine drive condition monitoring and diagnostic research—A survey. *Electr. Power Syst. Res.* **2003**, *64*, 145–158. [CrossRef]

14. Henao, H.; Capolino, G.A.; Fernandez-Cabanas, M.; Filippetti, F.; Bruzzese, C.; Strangas, E.; Pusca, R.; Estima, J.; Riera-Guasp, M.; Hedayati-Kia, S. Trends in fault diagnosis for electrical machines: A review of diagnostic techniques. *IEEE Ind. Electron. Mag.* **2014**, *8*, 31–42. [CrossRef]

15. Liang, X. Condition monitoring techniques for induction motors. In Proceedings of the 2017 IEEE Industry Applications Society Annual Meeting, IAS 2017, Cincinnati, OH, USA, 1–5 October 2017; doi:10.1109/IAS.2017.8101860. [CrossRef]

16. Basak, D.; Tiwari, A.; Das, S.P. Fault diagnosis and condition monitoring of electrical machines—A review. In Proceedings of the IEEE International Conference on Industrial Technology, Mumbai, India, 15–17 December 2006; doi:10.1109/ICIT.2006.372719. [CrossRef]

17. Garramiola, F.; Poza, J.; Madina, P.; del Olmo, J.; Almandoz, G. A Review in fault diagnosis and Health Assessment for Railway traction drives. *Appl. Sci.* **2018**, *8*, 2475. [CrossRef]

18. Isermann, R. Model-based fault detection and diagnosis: Status and applications. *IFAC Proc. Vol.* **2004**, *37*, 49–60. [CrossRef]

19. Duvvuri, S.S. Model-Based Bearing Fault Detection in Induction Motors under Speed Varying Conditions. In Proceedings of the India International Conference on Power Electronics, IICPE, Jaipur, India, 13–15 December 2018; doi:10.1109/IICPE.2018.8709523. [CrossRef]

20. Duvvuri, S.S.; Detroja, K. Model-based stator interturn short-circuit fault detection and diagnosis in induction motors. In Proceedings of the 2015 7th International Conference on Information Technology and Electrical Engineering: Envisioning the Trend of Computer, Information and Engineering, ICITEE 2015, Chiang Mai, Thailand, 29–30 October 2015; doi:10.1109/ICITEED.2015.7408935. [CrossRef]

21. Karami, F.; Poshtan, J.; Poshtan, M. Model-based fault detection in induction Motors. In Proceedings of the IEEE International Conference on Control Applications, Yokohama, Japan, 8–10 September 2010; doi:10.1109/CCA.2010.5611214. [CrossRef]

22. Cruz, S.; Cardoso, A. Stator winding fault diagnosis in three-phase synchronous and asynchronous motors, by the extended Park's vector approach. *IEEE Trans. Ind. Appl.* **2001**, *37*, 1227–1233. [CrossRef]

23. Marques Cardoso, A.; Cruz, S.; Fonseca, D. Inter-turn stator winding fault diagnosis in three-phase induction motors, by Park's vector approach. *IEEE Trans. Energy Convers.* **1999**, *14*, 595–598. [CrossRef]

24. Nejjari, H.; Benbouzid, M. Monitoring and diagnosis of induction motors electrical faults using a current Park's vector pattern learning approach. *IEEE Trans. Ind. Appl.* **2000**, *36*, 730–735. [CrossRef]

25. Acosta, G.; Verucchi, C.; Gelso, E. A current monitoring system for diagnosing electrical failures in induction motors. *Mech. Syst. Signal Process.* **2006**, *20*, 953–965. [CrossRef]

26. Bruzzese, C.; Honorati, O.; Santini, E. Rotor bars breakage in railway traction squirrel cage induction motors and diagnosis by MCSA technique Part I: Accurate fault simulations and spectral analyses. In Proceedings of the 2005 5th IEEE International Symposium on Diagnostics for Electric Machines, Power Electronics and Drives, Vienna, Austria, 7–9 September 2005; pp. 1–6. [CrossRef]

27. Boccaletti, C.; Bruzzese, C.; Honorati, O.; Santini, E. Rotor bars breakage in railway traction squirrel cage induction motors and diagnosis by MCSA technique Part II: Theoretical arrangements for fault-related current sidebands. In Proceedings of the 2005 5th IEEE International Symposium on Diagnostics for Electric Machines, Power Electronics and Drives, Vienna, Austria, 7–9 September 2005; pp. 1–6. [CrossRef]

28. Thomson, W.; Rankin, D.; Dorrell, D. On-line current monitoring to diagnose airgap eccentricity in large three-phase induction motors-industrial case histories verify the predictions. *IEEE Trans. Energy Convers.* **1999**, *14*, 1372–1378. [CrossRef]

29. Zhang, P.; Du, Y.; Habetler, T.G.; Lu, B. A Survey of Condition Monitoring and Protection Methods for Medium-Voltage Induction Motors. *IEEE Trans. Ind. Appl.* **2011**, *47*, 34–46. [CrossRef]

30. Gou, X.; Bian, C.; Zeng, F.; Xu, Q.; Wang, W.; Yang, S. A Data-Driven Smart Fault Diagnosis Method for Electric Motor. In Proceedings of the 2018 IEEE International Conference on Software Quality, Reliability and Security Companion (QRS-C), Lisbon, Portugal, 16–20 July 2018; pp. 250–257. [CrossRef]

31. Shao, S.Y.; Sun, W.J.; Yan, R.Q.; Wang, P.; Gao, R.X. A Deep Learning Approach for Fault Diagnosis of Induction Motors in Manufacturing. *Chin. J. Mech. Eng.* **2017**, *30*, 1347–1356. [CrossRef]

32. Saucedo-Dorantes, J.J.; Delgado-Prieto, M.; Osornio-Rios, R.A.; De Jesus Romero-Troncoso, R. Multifault Diagnosis Method Applied to an Electric Machine Based on High-Dimensional Feature Reduction. *IEEE Trans. Ind. Appl.* **2017**, *53*, 3086–3097. [CrossRef]

33. Langarica, S.; Ruffelmacher, C.; Nunez, F. An Industrial Internet Application for Real-Time Fault Diagnosis in Industrial Motors. *IEEE Trans. Autom. Sci. Eng.* **2019**, *17*, 284–295. [CrossRef]

34. Zhang, W.; Li, C.; Peng, G.; Chen, Y.; Zhang, Z. A deep convolutional neural network with new training methods for bearing fault diagnosis under noisy environment and different working load. *Mech. Syst. Signal Process.* **2018**, *100*, 439–453. [CrossRef]

35. Alwan, H.O.; Farhan, N.M.; Sabbagh, Q.S.A. Detection of Static Air-Gap Eccentricity in Three Phase induction Motor by Using Artificial Neural Network (ANN). *Int. J. Eng. Res. Appl.* **2017**, *7*, 15–23. [CrossRef]

36. Gonzalez-Jimenez, D.; del-Olmo, J.; Poza, J.; Garramiola, F.; Madina, P. Data-Driven Fault Diagnosis for Electric Drives: A Review. *Sensors* **2021**, *21*, 4024. [CrossRef] [PubMed]

37. Barański, M.; Polak, M. Thermal diagnostic in electrical machines. *Prz. Elektrotech.* **2011**, *33*, 305–308.

38. de la Barrera, P.M.; Bossio, G.R.; Solsona, J.A. High-Resistance Connection Detection in Induction Motor Drives Using Signal Injection. *IEEE Trans. Ind. Electron.* **2014**, *61*, 3563–3573. [CrossRef]

39. Stojčić, G.; Wolbank, T.M. Detecting high-resistance connection asymmetries in inverter fed AC drive systems. In Proceedings of the 2013 9th IEEE International Symposium on Diagnostics for Electric Machines, Power Electronics and Drives (SDEMPED), Valencia, Spain, 27–30 August 2013; pp. 227–232. [CrossRef]

40. Lee, S.B.; Yang, J.; Hong, J.; Yoo, J.Y.; Kim, B.; Lee, K.; Yun, J.; Kim, M.; Lee, K.W.; Wiedenbrug, E.J.; et al. A New Strategy for Condition Monitoring of Adjustable Speed Induction Machine Drive Systems. *IEEE Trans. Power Electron.* **2011**, *26*, 389–398. [CrossRef]

41. Yun, J.; Cho, J.; Lee, S.B.; Yoo, J.Y. Online Detection of High-Resistance Connections in the Incoming Electrical Circuit for Induction Motors. *IEEE Trans. Ind. Appl.* **2009**, *45*, 694–702. [CrossRef]

42. Yun, J.; Lee, K.; Lee, K.W.; Lee, S.B.; Yoo, J.Y. Detection and Classification of Stator Turn Faults and High-Resistance Electrical Connections for Induction Machines. *IEEE Trans. Ind. Appl.* **2009**, *45*, 666–675. [CrossRef]

43. Mengoni, M.; Zarri, L.; Tani, A.; Gritli, Y.; Serra, G.; Filippetti, F.; Casadei, D. Online Detection of High-Resistance Connections in Multiphase Induction Machines. *IEEE Trans. Power Electron.* **2015**, *30*, 4505–4513. [CrossRef]

44. Leandro, E.; de Lacerda de Oliveira, L.E.; da Silva, J.G.B.; Lambert-Torresm, G.; da Silv, L.E.B. Predictive Maintenance by Electrical Signature Analysis to Induction Motors. *Induction Mot. Model. Control* **2012**. [CrossRef]

45. Rothenhagen, K.; Fuchs, F. Performance of diagnosis methods for IGBT open circuit faults in three phase voltage source inverters for AC variable speed drives. In Proceedings of the 2005 European Conference on Power Electronics and Applications, Dresden, Germany, 11–14 September 2005. doi:10.1109/EPE.2005.219426. [CrossRef]

46. Guan, Y.; Sun, D.; He, Y. Mean Current Vector Based Online Real-Time Fault Diagnosis for Voltage Source Inverter fed Induction Motor Drives. In Proceedings of the 2007 IEEE International Electric Machines Drives Conference, Antalya, Turkey, 3–5 May 2007; Volume 2, pp. 1114–1118. [CrossRef]

47. Trabelsi, M.; Boussak, M.; Gossa, M. Multiple IGBTs open circuit faults diagnosis in voltage source inverter fed induction motor using modified slope method. In Proceedings of the XIX International Conference on Electrical Machines—ICEM 2010, Rome, Italy, 6–8 September 2010; pp. 1–6. [CrossRef]

48. Estima, J.O.; Marques Cardoso, A.J. A New Algorithm for Real-Time Multiple Open-Circuit Fault Diagnosis in Voltage-Fed PWM Motor Drives by the Reference Current Errors. *IEEE Trans. Ind. Electron.* **2013**, *60*, 3496–3505. [CrossRef]

49. Thomsen, J.; Kallesoe, C. Stator fault modeling in induction motors. In Proceedings of the International Symposium on Power Electronics, Electrical Drives, Automation and Motion, 2006. SPEEDAM 2006, Taormina, Italy, 23–26 May 2006; pp. 1275–1280. [CrossRef]

50. Martin-Diaz, I.; Morinigo-Sotelo, D.; Duque-Perez, O.; Romero-Troncoso, R.J. An Experimental Comparative Evaluation of Machine Learning Techniques for Motor Fault Diagnosis Under Various Operating Conditions. *IEEE Trans. Ind. Appl.* **2018**, *54*, 2215–2224. [CrossRef]

51. Godoy, W.; Silva, I.; Goedtel, A.; Palácios, R.H.C.; Lopes, T.D. Application of intelligent tools to detect and classify broken rotor bars in three-phase induction motors fed by an inverter. *IET Electr. Power Appl.* **2016**, *10*, 430–439. [CrossRef]

52. Martins, J.F.; Ferno Pires, V.; Pires, A.J. Unsupervised Neural-Network-Based Algorithm for an On-Line Diagnosis of Three-Phase Induction Motor Stator Fault. *IEEE Trans. Ind. Electron.* **2007**, *54*, 259–264. [CrossRef]

53. Colby, R. Detection of high-resistance motor connections using symmetrical component analysis and neural network models. In Proceedings of the 4th IEEE International Symposium on Diagnostics for Electric Machines, Power Electronics and Drives, SDEMPED 2003, Atlanta, GA, USA, 24–26 August 2003; pp. 2–6. [CrossRef]

54. Lee, J.; Singh, J.; Azamfar, M.; Pandhare, V. Chapter 8—Industrial AI and predictive analytics for smart manufacturing systems. In *Smart Manufacturing*; Soroush, M., Baldea, M., Edgar, T.F., Eds.; Elsevier: Amsterdam, The Netherlands, 2020; pp. 213–244. [CrossRef]
55. Garramiola, F.; del-Olmo, J.; Poza, J.; Madina, P.; Almandoz, G. Integral sensor fault detection and isolation for railway traction drive. *Sensors* **2018**, *18*, 1543. [CrossRef] [PubMed]
56. Garramiola, F.; Poza, J.; Madina, P.; del-Olmo, J.; Ugalde, G. A Hybrid Sensor Fault Diagnosis for Maintenance in Railway Traction Drives. *Sensors* **2020**, *20*, 962. [CrossRef] [PubMed]
57. del Olmo, J.; Garramiola, F.; Poza, J.; Almandoz, G. Model-Based Fault Analysis for Railway Traction Systems. *Mod. Railw. Eng.* **2018**. [CrossRef]
58. MathWorks. *Mastering Machine Learning A Step-by-Step Guide with MATLAB*; Technical Report; MathWorks: Natick, MA, USA, 2018.
59. Zhong, K.; Han, M.; Han, B. Data-driven based fault prognosis for industrial systems: A concise overview. *IEEE/CAA J. Autom. Sin.* **2019**, *7*, 330–345. [CrossRef]
60. Bikov, E.; Boyko, P.; Sokolov, E.; Yarotsky, D. Railway incident ranking with machine learning. In Proceedings of the 16th IEEE International Conference on Machine Learning and Applications, ICMLA 2017, Cancun, Mexico, 18–21 December 2017; pp. 601–606. [CrossRef]
61. Melendez, I.; Doelling, R.; Bringmann, O. Self-supervised Multi-stage Estimation of Remaining Useful Life for Electric Drive Units. In Proceedings of the 2019 IEEE International Conference on Big Data (Big Data), Los Angeles, CA, USA, 9–12 December 2019; pp. 4402–4411. [CrossRef]
62. Xue, Z.Y.; Li, M.S.; Xiahou, K.S.; Ji, T.Y.; Wu, Q.H. A Data-Driven Diagnosis Method of Open-Circuit Switch Faults for PMSG-Based Wind Generation System. In Proceedings of the 2019 IEEE 12th International Symposium on Diagnostics for Electrical Machines, Power Electronics and Drives, SDEMPED 2019, Toulouse, France, 27–30 August 2019. doi:10.1109/DEMPED.2019.8864922. [CrossRef]
63. Xu, Z.; Hu, J.; Hu, C.; Nadarajan, S.; Goh, C.K.; Gupta, A. Data-Driven Fault Detection of Electrical Machine. In Proceedings of the 2018 15th International Conference on Control, Automation, Robotics and Vision, ICARCV 2018, Singapore, 18–21 November 2018; pp. 515–520. [CrossRef]
64. Shi, W.; Lu, N.; Jiang, B.; Zhi, Y.; Xu, Z. An Unsupervised Anomaly Detection Method Based on Density Peak Clustering for Rail Vehicle Door System. In Proceedings of the 2019 Chinese Control And Decision Conference (CCDC), Nanchang, China, 3–5 June 2019; pp. 1954–1959.

Article

Current-Based Bearing Fault Diagnosis Using Deep Learning Algorithms

Andre S. Barcelos * and Antonio J. Marques Cardoso

CISE—Electromechatronic Systems Research Centre, University of Beira Interior, Calçada Fonte do Lameiro, P-6201-001 Covilhã, Portugal; ajmcardoso@ieee.org
* Correspondence: andre.s.barcelos@ubi.pt; Tel.: +351-275-319-700

Abstract: Artificial intelligence algorithms and vibration signature monitoring are recurrent approaches to perform early bearing damage identification in induction motors. This approach is unfeasible in most industrial applications because these machines are unable to perform their nominal functions under damaged conditions. In addition, many machines are installed at inaccessible sites or their housing prevents the setting of new sensors. Otherwise, current signature monitoring is available in most industrial machines because the devices that control, supply and protect these systems use the stator current. Another significant advantage is that the stator phases lose symmetry in bearing damaged conditions and, therefore, are multiple independent sources. Thus, this paper introduces a new approach based on fractional wavelet denoising and a deep learning algorithm to perform a bearing damage diagnosis from stator currents. Several convolutional neural networks extract features from multiple sources to perform supervised learning. An information fusion (IF) algorithm then creates a new feature set and performs the classification. Furthermore, this paper introduces a new method to achieve positive unlabeled learning. The flattened layer of several feature maps inputs the fuzzy c-means algorithm to perform a novelty detection instead of clusterization in a dynamic IF context. Experimental and on-site tests are reported with promising results.

Keywords: bearing diagnosis; early damage detection; unlabeled learning; deep learning; dynamic information fusion

Citation: Barcelos, A.S.; Cardoso, A.J.M. Current-Based Bearing Fault Diagnosis Using Deep Learning Algorithms. *Energies* **2021**, *14*, 2509. https://doi.org/10.3390/en14092509

Academic Editors: Daniel Morinigo-Sotelo, Rene Romero-Troncoso and Joan Pons-Llinares

Received: 24 March 2021
Accepted: 23 April 2021
Published: 27 April 2021

Publisher's Note: MDPI stays neutral with regard to jurisdictional claims in published maps and institutional affiliations.

1. Introduction

Induction motors are present in most industrial processes because of their versatility for many applications, efficiency and robustness to operate in severe conditions. Recent studies have reported that 40% of operational failures caused in these machines are related to damage from bearings, which can be separated into two categories. The first category is the punctual damages that appear on a delimited bearing surface producing an impulsive mechanical vibration. The second category is the distributed damages that produce continuous mechanical vibrations with low magnitude harmonics [1,2].

The most recurrent approach to achieving bearing condition monitoring and damage diagnosis is to acquire vibration-based signals from accelerometers and perform supervised learning algorithms. However, many industrial motors are unable to provide vibration signals because they are installed at inaccessible locations or their housings are inadequate to install new devices. The vibration data acquisition is also expensive, demanding new sensors and devices to transduce, transmit and process [3]. Otherwise, current-based data acquisition is available in most industrial electric motors because the stator current is monitored for control, supply and protection purposes. Consequently, each motor phase is a distinct data source for the machine status because the phases lose symmetry due to disturbances, interferences, noises, intrinsic conditions, bearing damage or other reasons [4,5].

In this context, the wavelet transform (WT) becomes a recurrent tool for signal processing because of its advantage of a multiresolution analysis. The fractional wavelet transforms (FWTs) generalize the WT to represent signals in the fractional time-fractional frequency domain, preserving sparsity, removing redundancy, denoising and narrowing the resolution [6,7]. Therefore, appropriate denoising and reconstruction proceedings can generate a database with multiple data sources from each motor phase.

Consequently, deep learning (DL) algorithms with a data-driven approach can extract abstract features from multiple sources to improve the classification task [8,9]. Autoencoders [10], variational autoencoders [11], convolutional neural networks (CNN) [12,13], generative models [14], recurrent neural networks (RNN) [15] and extreme learning [16], among others, construct feature maps with deep levels of data abstraction. Moreover, the DL approaches can outperform traditional signal processing techniques and feature extraction methods to perform a bearing damage diagnosis [8,17].

Thus, this paper introduces a new deep learning approach based on an FWT, an independent CNN and long short-term memory (LSTM) to perform bearing damage identification from multiple sources. The raw and denoised signals become the multiple inputs of the maps of several features with different receptive fields. Each feature map inputs an independent flattened layer (FL) that feeds several artificial neural networks (ANN) to perform supervised learning and a SoftMax classification. LSTM cells are introduced in the feature maps with a large receptive field to improve the long time-dependency feature extraction. An information fusion (IF) algorithm unifies each SoftMax output into a novel feature set, reducing multiple source redundancy and preserving the bearing condition monitoring. With this new approach, the information fusion problem is replaced by a supervised classification task performed by a support vector machine (SVM).

Otherwise, in most industrial applications, the electric motors with damaged bearings are prevented from performing their nominal functions because of industrial process safety reasons. The bearing damage produces vibration signatures that propagate to pumps, compressors, pipelines or other types of loads that affect processes or subsystems [18]. Therefore, the acquisition of a labelled database with several damages is unrealistic or impracticable for most industrial facilities. However, one-class positive unlabeled learning (PUL) algorithms can detect novelty in unlabeled databases. The positive class, in this case, is the healthy bearing signals [19,20].

Thus, this paper introduces a second new approach that uses the feature maps from multiple phases (raw and denoised sources) to input several fuzzy c-means (FCM) algorithms in a paradigm to detect novelty in PUL instead of clustering data. Assuming that a healthy class is available and two clusters are present in each source, the Fisher discriminant ratio (FDR) and the Kullback–Leibler divergence (KLD) can monitoring the center and distribution behavior. A similar IF approach unifies the FDR and KLD of each FCM into a new feature set, preserving the bearing condition monitoring while inputting a SoftMax classifier to perform the bearing damage identification. The results of experimental and on-site tests with both approaches (supervised and PUL) are promising.

The sequence of this paper presents the theoretical background in Section 2. The test rig, setup and configurations are described in Section 3. Section 4 describes experimental and on-site tests. Section 5 presents the conclusion.

2. Theoretical Background

2.1. Fractional B-Spline Wavelet Transform

Fractional B-splines (FS) are the extended version of splines with order $\alpha > -1$ defined as:

$$\beta_{\pm}^{\alpha}(x) = \frac{1}{\Gamma(\alpha+1)} \Delta_{\pm}^{\alpha+1} x_{\pm}^{\alpha} \tag{1}$$

where $k \in \mathbb{Z}$ and $\Gamma(\alpha+1)$ is the gamma function. The one-side causal function (+) is $x_{+}^{\alpha} = x^{\alpha}$ and the anti-causal (−) is $x_{-}^{\alpha} = (-x)_{+}^{\alpha}$. The fractional finite difference operator is:

$$\Delta_{\pm}^{\alpha} f(x) = \sum_{k=0}^{\infty} (-1)^k \binom{\alpha}{k} f(x \mp k). \tag{2}$$

The FS is obtained by interpolating polynomial B-splines. The centered fractional B-splines of degree α are defined by the convolution operator $\star$ as:

$$\beta_\star^\alpha(x) = \beta_+^{\frac{\alpha-1}{2}} \star \beta_-^{\frac{\alpha-1}{2}} \tag{3}$$

and $\Delta_\star^\alpha \leftrightarrow |1 - e^{-jw}|^\alpha$ denotes the symmetric fractional finite difference operator. The Fourier transform of the one-side causal and centered fractional splines are calculated as follows:

$$\hat{\beta}_+^\alpha(w) = \left(\frac{1 - e^{-jw}}{jw}\right)^{\alpha+1} \qquad \hat{\beta}_\star^\alpha(w) = \left(\frac{\sin(w/2)}{w/2}\right)^{\alpha+1}. \tag{4}$$

Fractional B-splines satisfy all requirements to construct a wavelet basis for an $\alpha > -0.5$ given by:

$$\beta^\alpha(x/2) = \sum_{k\in\mathbb{Z}} h^\alpha(k)\beta^\alpha(x - k) \tag{5}$$

where the filters $h_+^\alpha(k)$, $h_-^\alpha(k)$ and $h_\star^\alpha(k)$ are defined as:

$$h_+^\alpha(k) = \frac{1}{2^\alpha}\begin{pmatrix} \alpha+1 \\ k \end{pmatrix} \leftrightarrow \hat{h}_+^\alpha(w) = 2\left(\frac{1+e^{-jw}}{2}\right)^{\alpha+1}$$

$$h_\star^\alpha(k) = \frac{1}{2^\alpha}\begin{pmatrix} \alpha+1 \\ k \end{pmatrix} \leftrightarrow \hat{h}_\star^\alpha(w) = 2\left(\frac{1+e^{-jw}}{2}\right)^{\alpha+1}.$$

The anti-causal (–) is obtained by substituting $h_-^\alpha(k) = h_+^\alpha(-k)$. The general approach to orthonormalize the fractional splines generates the scaling function given by:

$$\phi(x) = \sum_{k\in\mathbb{Z}} \left(a_\varphi^\alpha(k)\right)^{-\frac{1}{2}} \beta^\alpha(x - k) \tag{6}$$

where $(a_\varphi^\alpha(k))^{-1/2}$ is the convolution of the FS sequence. The Fourier transform $A_\varphi^\alpha(w)$ is defined as:

$$a_\varphi^\alpha = \beta_\star^{2\alpha+1}(k) \qquad A_\varphi^\alpha(w) = \sum_{k\in\mathbb{Z}} \beta_\star^{2\alpha+1}(n)e^{-jwn}.$$

Leading the corresponding two-relation:

$$\phi(x/2) = \sum_{k\in\mathbb{Z}} h_\perp^\alpha(k)\phi(x - k).$$

The low-pass filter and high-pass filter can be written as:

$$H_\perp^\alpha(w) = \hat{h}^\alpha(w)\sqrt{\frac{A_\varphi^\alpha(w)}{A_\varphi^\alpha(2w)}} \qquad G_\perp^\alpha(w) = e^{-jw}\overline{H_\perp^\alpha(w + \pi)}.$$

Thereby, the behavior of the filter tends to an ideal low-pass and high-pass filter as $\alpha \to 0$ [6,21]. The overlapping group shrinkage (OGS) algorithm reconstructs the denoised signal by observing the wavelet coefficients and performing a convex regularization while minimizing a cost function [22,23]. Therefore, in this paper, the equivalent filter bank denoises the raw signals to input the feature maps.

2.2. Long Short-Term Memory

A recurrent neural network (RNN) is a class of ANN that identifies patterns in sequential data. However, an RNN has a few drawbacks for most applications including gradient vanishing and gradient explosion in the backpropagation. Long short-term memory (LSTM) solves the gradient vanishing problem, using a memory cell that improves the RNN units [24,25]. Figure 1 shows a typical LSTM cell.

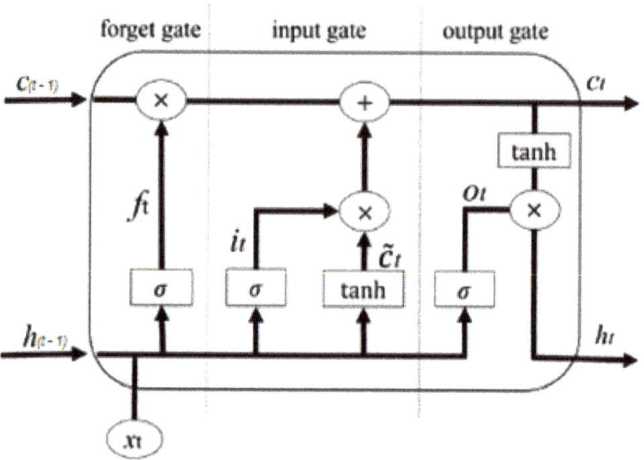

Figure 1. Long short-term memory cell [25].

The LSTM gates control the information flow of the current state, the input gate (i) and the output gate (o) [25]. The forget gate (f_t) determines how much previous information should be removed or saved as follows:

$$f_t = \sigma\left(w_{fz}z_t + w_{hf}h_{(t-1)} + b_f\right)$$

where σ is a sigmoid function, w is the weight, z_t is the current input, $h_{(t-1)}$ is the output of the previous cell and b_f is the bias. The input gate determines the behavior of x_t, the previous layer $h_{(t-1)}$ and the current state (C_t) as follows:

$$i_t = \sigma\left(w_{zi}z_t + w_{hi}h_{(t-1)} + b_i\right)$$
$$\hat{C}_t = \tan h\left(w_{xc}z_t + w_{hc}h_{(t-1)} + b_i\right).$$

The output gate controls the cell information and state as follows:

$$c_t = c_{t-1}f_t + i_t\hat{C}_t$$
$$o_t = \sigma\left(w_{zo}z_t + w_{ho}h_{(t-1)} + b_o\right)$$
$$h_t = o_t \times \tan(c_t).$$

Thus, the LSTM can memorize relevant time-dependent features to discriminate long-time delay events with overlapping low-frequency components [26].

2.3. Convolutional Neural Networks

The basic CNN contains three structures that provide feature extraction, perform classification and represent the decision with a probabilistic function. The convolution layer (CL) performs dot products, preserving the spatial structure of the previous layer and output abstract features [27,28]. The convolutional process is described as follows:

$$x_j^l = f\left(\sum_{i\,\in M_j} x_i^{l-1} * k_j^l + b_j^l\right) \tag{7}$$

where $*$ is the convolution operation and x^{l-1} denotes the input data of the previous layer. Each layer consists of n^l kernels with a weight matrix k_j^l and a bias vector b_j^l. The output of the nonlinear active function $f(*)$ is the n^l matrices x_j^l where $j = 1 : n^l$ corresponds with n kernels from the layer l. The activation function leaky-Relu has a linear identity for positive values and a slope for negative values to avoid gradient problems.

The pooling layer (PL) down-samples the previous CL to control the feature map size and save abstract information. The function $\max\left(x_j^l\right) = x_j^{l+1}$ is the most recurrent down-sampling operation in CNN models. Therefore, the feature map is an independent structure containing successive CLs and PLs that control the size and depth to extract more abstract features. The last PL can input LSTM cells to extract time-dependent features or be transformed into a flattened layer (FL) to input an ANN or other classifier [27,28]. The third structure is the SoftMax probabilistic distribution operator that transforms the classifier output z_i into a normalized vector as follows:

$$p_i = \frac{e^{(z_i)}}{\sum_{j=1}^{N} e^{(z_j)}} \qquad p_i \in [0,1]. \tag{8}$$

These three structures are used in different configurations to extract abstract features and improve accuracy classification. The feature maps can also be arranged in parallel with profound and shallow receptive fields to extract features from multiple sources [29].

2.4. Fuzzy C-Means Algorithm

The one-dimensional FL with data $X = \{x_1, x_2, \ldots x_k\}$ inputs the FCM algorithm to divide X into several clusters. The objective function of the FCM is defined as follows:

$$J = \sum_{i=1}^{c} \sum_{j=1}^{k} u_{i,j}^m ||x_j - v_i||^2$$

$$s.t. \quad \sum_{i=1}^{c} u_{ij} = 1, \qquad 0 \le u_{ij} \le 1$$

where $U = [u_{ij}]_{c \times k}$ is a membership matrix, $m > 1$ is the fuzzifier, v_i are the prototypes and c is the number of clusters. The solution for updating the partition matrix and the prototypes is given by:

$$u_{i,j} = \frac{\left(||x_j - v_i||^2\right)^{-\frac{1}{m-1}}}{\sum_{q=1}^{c} \left(||x_j - v_q||^2\right)^{-\frac{1}{m-1}}} \qquad v_i = \frac{\sum_{j=1}^{k} u_{i,j}^m x_j}{\sum_{j=1}^{k} u_{i,j}^m}.$$

A recurrent approach to stop criteria is a threshold between two successive partitions [30]. The fuzzy C-means can perform bearing detection and classification in an unlabeled context, presenting remarkable results with vibration-based signals. The main advantage is that the FCM algorithm allows changes in the regularization, cluster shape, cost function and membership function to improve the performance. The intra-cluster variance can also be minimized by adjusts in the fuzzifiers, keeping an adequate boundary. Indeed, the support vector data description (SVDD) and the one-class SVM present boundary problems including loose boundaries, data rejection and outlier misclassification among others that increase the complexity of the distribution interpretation and classification.

In this paper, the Fisher discriminant ratio (FDR) and Kullback–Leibler divergence (KLD) perform center monitoring and distribution behavior in the PUL context. Initially, an FL from a healthy source inputs the FCM algorithm, configurated in a paradigm to identify two clusters. Therefore, if the input remains healthy, the difference between the centers and distribution divergence must remain constant after successive batches.

2.5. Information Fusion

Multiple sources can merge into a new feature set throughout information fusion algorithms (IF) [31,32]. Assuming that s samples, c classes and n independent sources and classifiers are available, the features set from each source X^n is represented as follows:

$$
\begin{aligned}
X^1 &= \left[x_1^1, x_2^1, \ldots, x_s^1\right] \\
X^2 &= \left[x_1^2, x_2^2, \ldots, x_s^2\right] \\
X^n &= \left[x_1^n, x_2^n, \ldots, x_s^n\right].
\end{aligned}
$$

Therefore, n feature sets input n CNNs to classify C_i classes in each feature set X^n. The conditional probability $P(\cdot|\cdot,\cdot)$ of the class i based on the observation of k CNN on the sample x_j^k is defined as:

$$
P_i^{k,j} = P\left(C_i \middle| x_j^k, CNN^k\right) \qquad i = 1:c \quad j = 1:s \quad k = 1:n. \tag{9}
$$

All combinations of CNN^k are rearranged in the matrix P^k with the size $c \times s$. The output of all classifiers is then merged to the $(c \times n) \times s$ matrix P as follows:

$$
P^k = \begin{bmatrix} P_1^{k,1} & \cdots & P_1^{k,s} \\ \vdots & \ddots & \vdots \\ P_m^{k,1} & \cdots & P_m^{k,s} \end{bmatrix} \qquad P = \begin{bmatrix} P^1 \\ \vdots \\ P^n \end{bmatrix}. \tag{10}
$$

The task of analyzing multiple sources becomes a task of classifying the new feature set P. Consequently, the IF approach for the FCM is similar. The flattened vector from n feature maps input n FCM. The s samples are replaced by batches with measures m_j (KLD and FDR) to form the feature sets:

$$
\begin{aligned}
X^1 &= \left[FDR_1^1, KLD_1^1, \ldots, FDR_s^1, KLD_s^1\right] \\
X^2 &= \left[FDR_1^2, KLD_1^2, \ldots, FDR_1^2, KLD_1^2\right] \\
X^n &= \left[FDR_1^n, KLD_1^n, \ldots, FDR_1^n, KLD_1^n\right].
\end{aligned}
$$

Therefore, n FCM identifies the healthy class $(c = 1)$ or detects the novelty $(c = 2)$ of X^n. The conditional probability $P(\cdot|\cdot)$ of the class C_i based on the observation of k FCM on the measure m_j^k is defined as:

$$
P_i^{k,j} = P\left(C_i \middle| m_j^k, FCM^k\right) \qquad i = 1:c \quad j = 1:s \quad k = 1:n. \tag{11}
$$

This new approach leads to similar P^k and P matrices of Equation (10), which depends on the FCM performance, batch size and number measures.

3. Datasets

The tests were performed with every current-based signal developed by the Chair of Design and Drive Technology from the University of Paderborn in Germany containing the current-based signals from an induction motor. Two current probes acquired signals from the test rig with a sampling frequency of 64 kHz, a rotor speed of 900 rpm and 1500 rpm (N09 and N15) and loading conditions of 0.1 Nm and 0.7 Nm (M01 and M07). The classes of these damages were healthy and incipient, distributed and punctual damages [33]. In this test rig, the bearings were located externally from the induction motor to extract more sensitive vibration-based information. However, the internal bearings produced a few effects in data distribution that might cause misclassification in machine learning algorithms that learn from external bearing damage.

This work also used the test rig available at CISE, Electromechatronic Systems Research Centre at the University of Beira Interior in Portugal, to acquire bearing damaged current-based signals. The test rig consisted of an inverter-fed three-phase squirrel-cage induction motor, a programmable AC power source of 0~300 V, 12 kVA, 192 Amps, 15~1.2 kHz (Chroma), a data acquisition device USB-6366 (National Instruments) and a mechanical system that provided a stable load with speed control. Two current probes sent the stator currents to the acquisition board with a sampling frequency of 44 kHz, producing samples with a rotor speed of 1800 rpm and loading conditions of 0.1 Nm and 0.7 Nm. The first CISE damaged bearing had an incipient punctual damage in both rings caused by electrical discharges. The inner ring damage diameter was 1.5 mm and the spheres and cage remained intact. The outer ring had two opposite damages with diameters of 2.0 mm and 1.5 mm. This type of damage is common in industrial machinery but it was absent in the Padeborn dataset. The second CISE damaged bearing had punctual damage (hole) of 2.0 mm in the outer ring. Different from the Padeborn test rig, the CISE test rig inserted the damaged bearing at the fan on the drive-end side of the induction motor.

Pre-Processing

All stator phases (R1 and R2) of each bearing damage from both datasets were de-noised with a FWT and reconstructed with the OGS algorithm to generate F1 and F2 signals. Several signal segments were then rearranged in a square matrix $t \times t$ to convert 1-D signals into grayscale images (base2). In this work, the segment t was defined by the lower motor speed that produced $(64{,}000 \times 60)/900$ samples per revolution ($\simeq$4266). Therefore, $t^2 = 4096$ samples produced the normalized gray images with a size of 64 $\times$ 64. The sets of gray images from R1, R2, F1 and F2 sources inputted two independent arranges (A1 and A2) of feature maps with a profound and shallow receptive field. Table 1 summarizes the profound configuration, which was a conventional feature map with four CLs and PLs.

Table 1. Feature map architecture with a profound receptive field (A1).

Layer	K. Size	K. Number	Input	Output
CL 1	9 $\times$ 9	4	64 $\times$ 64	64 $\times$ 64
PL 1	2 $\times$ 2	4	64 $\times$ 64	32 $\times$ 32
CL 2	7 $\times$ 7	6	32 $\times$ 32	32 $\times$ 32
PL 2	2 $\times$ 2	6	32 $\times$ 32	16 $\times$ 16
CL 3	5 $\times$ 5	8	16 $\times$ 16	16 $\times$ 16
PL 3	2 $\times$ 2	8	16 $\times$ 16	8 $\times$ 8
CL 4	3 $\times$ 3	10	8 $\times$ 8	8 $\times$ 8
PL 4	2 $\times$ 2	10	8 $\times$ 8	4 $\times$ 4
FL			10 $\times$ 16	1 $\times$ 160

The feature map with a profound receptive field (A1) consisted of successive CLs and PLs to extract deeper abstract features. The kernel size of the CLs reduced, concentrating the abstract information into more compact structures, increasing the number of kernels. This procedure allowed the extraction of more abstract features with different kernel configurations. The PL controlled the output size through down-sampling operations while grouping relevant information, allowing the CL to increase the kernel number. The last PL inputted an FL with a 1 $\times$ 160 dimension. Thus, this feature map was capable of extracting abstract information at each CL, increasing the number of kernels to diversify the feature type.

The shallow receptive field (A2) was a feature map with two successive CLs and PLs and LSTM cells to extract long time-dependent features. In this configuration, the last cell of LSTM 2 also inputted an FL. Table 2 summarizes the shallow configuration.

This feature map consisted of a particular arrangement (A2) to extract shallow abstract features within a large receptive field. The CL and PL controlled the feature map size, avoiding deep features and allowing diversity in the kernels. The PLs reduced the feature map size, revealing inner relations in each kernel. PL 2 inputted the LSTM cells that behaved as recurrent neural networks, saving relevant time-dependent features to input an

FL. Indeed, the main advantage of the LSTM cells was that the internal structures controlled the flow of relevant information, keeping long-time relevant features.

Table 2. Feature map architecture with a shallow receptive field and LSTM cells (A2).

Layer	K. Size	K. Number	Input	Output
CL 1	9×9	5	64×64	64×64
PL 1	4×4	5	64×64	16×16
CL 2	7×7	10	16×16	16×16
PL 2	4×4	10	16×16	4×4
LSTM 1		32 cells	10×16	10×32
LSTM 2		16 cells	10×32	1×16

In summary, the multiple sources (4) inputted each feature map arrangement (2) to generate eight independent FLs. The supervised learning was performed by eight ANNs with a stochastic gradient descent, a learning rate at 0.0015, momentum at 0.5 and L2 regularization. The training set contained 1000 samples for each class (4000 in total) while the test set had 250 samples for each class. Each SoftMax had four outputs corresponding with each class. The IF unified the output from each SoftMax into the P matrix and a support vector machine performed the classification task. All possible four class combinations with different severity indexes were performed and the results were presented in terms of average accuracy.

Otherwise, in the PUL context, the objective was to identify incipient bearing damage using the KLD and FDR measures from FCM algorithms. Therefore, all possible combinations for healthy versus damaged signals with A1 and A2 arranges were performed.

4. Experimental and On-Site Tests

4.1. Supervised Learning

This research compared R1 and R2 and F1 and F2 performance to verify the effectiveness of the IF. The average accuracy of three operation conditions (N15M01, N09M07, N15M07) is summarized in Table 3.

Table 3. Average accuracy of supervised learning.

Source	Earlier	Punctual	Distributed
R1	91.38	94.82	91.87
R2	90.26	94.76	91.74
F1	91.70	95.21	92.24
F2	91.92	96.43	92.18
R1 and R2	93.56	95.97	92.65
F1 and F2	93.82	96.65	93.32
IF	94.11	97.02	93.55

The performance of F1 and F2 reached a similar accuracy of the IF with the four signals. Considering the implementation aspect, one can choose to fuse the F1 and F2 sources instead of performing the four sources (IF) to reduce the computational efforts, keeping a high accuracy performance. However, all tests performed in this paper were conducted with four sources in an IF context. Condition N15M07 improved the accuracy for each type of damage while the other two conditions decreased the average. An SVM with a linear kernel and a soft margin approach performed the P matrix classification. The setup of hyperparameters, training, stop criteria and kernel configurations were omitted for the sake of brevity. The polynomial and Gaussian kernels performed similar results although with more convergence time.

4.2. Unlabeled Learning

Five recent FCM algorithms were performed to identify novelty in a one-class PU context. The first was the FCM with a genetic optimization (FCM-GO) algorithm that searched for a suboptimal solution [34]. The Gustafson–Kessel (GK) clustering algorithm employed the Mahalanobis distance to update centers and proto-clusters. The FN-DBSC could be characterized by a convex function with a particular set of hyperparameters [35]. The FCM with a focal point (FCMFP) introduced a regularization term into the loss function [35]. Lastly, the Gath-Geva (GG) clustering is an extended version of the FCM that performed the previous detection of sizes and densities of clusters [36]. These fuzzy-based algorithms could perform novelty detection in the PUL context with an appropriate initialization method.

Assuming that two clusters were present in the PU data distribution, a previous batch $(-\tau, t_0)$ defined the centers and boundaries of these pseudo-clusters. In parallel, the KLD and FDR measured the distribution and the center behavior. A successive batch with current data $(t_0, -\tau,)$ was then used to calculate two new pseudo-clusters. The comparison between the KLD and FDR of previous and current batches identified the changes in the PU data. Figure 2 resumes the cluster behavior of the FCM algorithms in a one-class PU novelty detection paradigm.

The FDR to monitor the centers

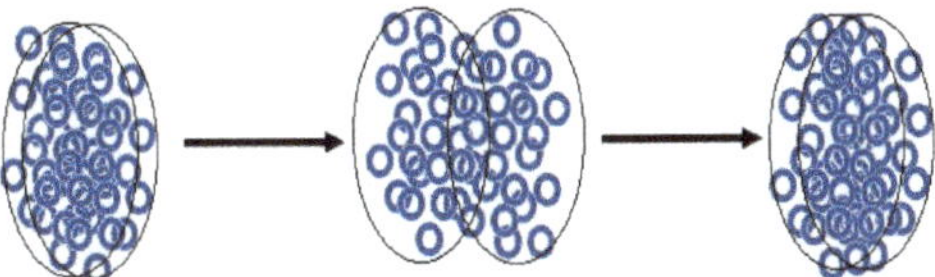

The KLD divergence to monitor the distribution

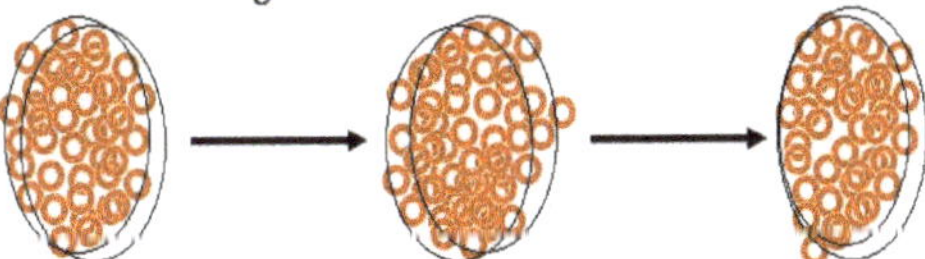

Figure 2. Center and distribution behavior of clusters.

Thus, the KLD and FDR measures could identify changes in the data distribution to perform novelty detection. In this case, a healthy bearing signal produced small changes in these measures because their outliers and noises were uncorrelated with bearing damage. Consequently, when damage arose, the previous cluster contained data from the healthy bearing $(-\tau, t_0)$ while the current cluster contained data from the damaged bearing. This discrepancy produced the center movement and divergence in distributions because the clusters acquired data from the same signal in different conditions.

Indeed, healthy bearing distributions can be described as symmetric alpha-stable probability density functions (PDFs) and damaged bearing distributions can be described as non-symmetrical alpha-stable PDFs with elongated, exponential or dense tails, which depend on the damage type and location. That is the principal advantage of KLD, which can monitor this complex distribution computing the PDF with numerical methods. Furthermore, the severity of the failure induced more significant changes in the distribution and center behavior. The relative distance between centers provided a measure to monitor the bearing damage evolution, quantifying the severity. Therefore, early bearing damage detection can be extended to damage severity monitoring.

In this paper, the multiple sources and arranges (A1 and A2) created the FLs that inputted eight FCM algorithms to measure the KLD and FDR and create the P matrix. The cumulative summation of each KLD and FDR, combined with changes in the P, performed earlier bearing damage detection in PUL. Table 4 presents the average performance of algorithms from healthy versus earlier bearing damage identification under different load and speed conditions.

Table 4. Results of earlier damage identification with FCM-IF.

Source	N15M01	N09M07	N15M07
FCM-GO	86.15	89.65	91.60
GK	87.32	88.91	92.13
FN-DBSC	86.31	89.00	91.93
FCMFP	87.45	89.72	91.23
GG	88.06	89.96	91.15

These algorithms presented a similar performance in a one-class PU context, confirming that it was challenging to identify incipient bearing damage with a stator current under several operating conditions. Indeed, the identification of distributed and punctual damage was performed with superior accuracy in the N15M07 condition but the results were omitted for the sake of brevity. This research also performed these algorithms with time, frequency and time-frequency features and the accuracy reached 88% in the best-case scenarios. Furthermore, the performance of these FCM algorithms was similar to the supervised learning approach of Table 3, attesting that the experimental tests presented a promising result.

In this context, both approaches (CNN-IF and FCM-IF) achieved a high accuracy in condition monitoring and bearing damage identification because of the FWT and LSTM, allowing that conventional techniques (e.g., a kurtogram and spectral envelope) provided the damage location (inner ring, outer ring or spheres). Indeed, well-known methods could predict the location of the punctual and distributed bearing damage with a high accuracy by a vibration signal analysis [37,38]. However, considering current-based signals, it was non-trivial to extract the relevant information without performing an adequate denoise technique (FWT) or monitoring relevant long-time behavior (LSTM). A remarkable example is that the FCM-IF could detect a novelty in current-based signals (e.g., a change in distribution) with insufficient information (e.g., harmonics buried in noises) to predict the location with a kurtogram or spectral envelope.

4.3. On-Site Tests

On-site tests were conducted in a wastewater pump driven by an electric motor at a gas processing facility (Figure 3). Initially, the supervised learning was achieved with the historical data, allowing the training and testing of the CNN-IF algorithm with two incipient bearing damage samples caused by wear and pitting, three punctual damages (electrical discharge, scratches and pitting with low severity) and two distributed damages.

This motor operated in two predominant speed conditions of 1500 rpm and 1800 rpm with a variable load that depended on process demand without vibration condition monitoring. The training accuracy reached around 92.15%, 95.26% and 93.08% for incipient, punctual and distributed damage identification, presenting similar results according to Table 3. In these tests, the data acquisition avoided the load transient, interrupting the training until the process (wastewater process) reached a more stable and stationary regime. This approach reduced the misclassification of the supervised algorithm.

The CNN-IF algorithm ran in real-time for sixteen weeks, performing bearing damage monitoring in both speed conditions with a variable load until the detection of an incipient distributed damage caused by wear. The kurtogram and the spectral envelope analysis using the R1 and R2 current-based signals were able to identify the same damage 48 h later. Indeed, the low magnitude harmonics, the poor SNR and the loss of information in the magnetic field reduced the performance of these approaches. Thus, the CNN-IF could perform transfer learning from test benches to on-site historical data (target source), saving the relevant inner structure to retrain partially with on-site data if available. It was also possible to perform transfer learning between similar on-site machines.

Figure 3. Industrial electric motor driving a wastewater pump.

The real-time test in a one-class PU context was then conducted with FCM-IF algorithms to perform early bearing damage detection in a centrifugal pump driven by the electric motor presented in Figure 4. In this case, the CNN-IF method was inviable because only two bearing damages caused by wear were reported in two years of historical data. This industrial motor pump was the main machine at this facility, running at 1800 rpm with variable loading that depended on processing demand.

Figure 4. Industrial electric motor driving the principal centrifugal pump.

The motor condition monitoring was performed by current envelope signatures while an automatized protection system prevented high levels of vibration and current. Therefore, there was no vibration-based condition monitoring or other dedicated systems to perform an independent bearing damage analysis. Figure 5 present the behavior of the most sensitive KLD, FDR and FDR moving average (FDR-MA) of the F1 source and the FCM-GO algorithm. In this case, the bearing damage was caused by wear and the early detection occurred at sample 240 by either the KLD or FDR. The current envelope signature identified the same damage at sample 282, approximately 50 h of difference.

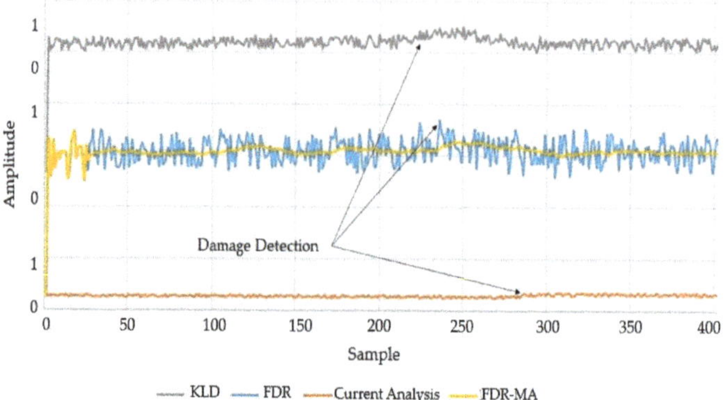

Figure 5. Early bearing damage detection caused by wear.

Indeed, the most sensitive FDR and KLD presented a drastic change around 200 samples, indicating that the distribution was becoming different and that the centers were moving in a new pattern. It was possible to identify these changes with the FDR and KLD because the FWT extracted relevant information and the LSTM saved the abstract long-time behavior from the healthy signal. Moreover, it was difficult to detect incipient wear by analyzing current-based signals with a kurtogram or spectral envelope. The distributed damage information produced low magnitude harmonics and energy information that were buried into noise due to a poor SNR.

Furthermore, every FCM related to this work was performed in real-time with this electric motor. The results were similar to Figure 5, surpassing the current envelope signature performance with an average difference of 50 h. Thus, the performance of this approach was independent of the FCM-IF choice but depended on sources, feature maps, measures and the initialization method. After the bearing damage detection (novelty detection), the clusters moved apart gradually because the successive data (damage versus damage evolving) produced a similar center and distribution. This effect occurred after 300 samples. Furthermore, a few slight variations in the KLD and FDR indexes might indicate that the severity evolved. Both on-site motors were driven by inverters but this methodology could be also applied in line-connected motors.

5. Conclusions

This research introduced the challenges of current-based condition monitoring and an early bearing damage diagnosis. Classic methods in supervised learning context that extract features in time, frequency and the time-frequency domain provided a high accuracy in a vibration-based analysis. However, these methods were insufficient for current-based approaches due to a poor SNR and low magnitude harmonics. The principal drawbacks for current-based bearing condition monitoring are the poor SNR, the loss of information in the magnetic field, saturation harmonics, electrical faults, interference and indirect measures, among others. Consequently, the traditional signal processing techniques that denoise and extract information from vibration-based signals had a lower performance in the current-based analysis. Current-based bearing condition monitoring has less available information (e.g., indirect measure) and more feature extraction complexity (e.g., a poor SNR). Thus, this paper introduced two new approaches with denoise methods and machine learning to detect incipient bearing damage by current-based signals with a high accuracy.

Therefore, the first contribution of this paper was the development of the fractional wavelet B-spline to denoise two phases of the stator current, taking advantage of multiple source analyses. The feature maps of CNNs then extracted profound and shallow features from each source while the shallow map contained LSTM cells that identified long time-dependent behavior. The ANN and SoftMax performed the classification and the information fusion algorithm merged each SoftMax classification into a new matrix. This approach addressed the multiple source information fusion problem to a supervised classification task. Indeed, this contribution improved the accuracy of current-based approaches because two arrangements of feature maps extracted more relevant and abstract features with different receptive fields from multiple sources.

The acquisition of a labelled database is unfeasible in most industrial applications because industrial motors are prevented from performing their functions under damaged conditions. Therefore, the second contribution of this work used multiple sources in two arrangements of feature maps and several FCM algorithms to perform bearing damage identification in a one-class positive unlabeled context. This new approach calculated the KLD and FDR from successive FCM batches to input an information fusion algorithm that merged these measures into a new matrix to perform bearing condition monitoring and early damage identification.

Experimental tests with Paderborn and CISE datasets were performed with the most representative type of damage and severity under several operation conditions with FW-CNN-IF and FW-FCM-IF algorithms. Both contributions presented remarkable results for incipient and distributed damage detection by current-based signals. Furthermore, on-site tests were performed in a gas processing facility and these algorithms surpassed the harmonic and envelope spectrum analysis every time.

Author Contributions: Conceptualization, A.S.B. and A.J.M.C.; methodology, A.S.B. and A.J.M.C.; software, A.S.B.; validation, A.S.B. and A.J.M.C.; formal analysis, A.J.M.C. and A.S.B.; investigation, A.S.B. and A.J.M.C.; resources, A.J.M.C.; data curation, A.S.B.; writing—original draft preparation, A.S.B.; writing—review and editing, A.J.M.C.; visualization, A.S.B. and A.J.M.C.; supervision, A.J.M.C.; project administration, A.J.M.C.; funding acquisition, A.J.M.C. All authors have read and agreed to the published version of the manuscript.

Funding: This work was supported by the European Regional Development Fund (ERDF) through the Operational Programme for Competitiveness and Internationalization (COMPETE 2020) under Project POCI-01-0145-FEDER-029494 and by National Funds through the FCT, Portuguese Foundation for Science and Technology under Projects PTDC/EEI-EEE/29494/2017, UIDB/04131/2020 and UIDP/04131/2020.

Conflicts of Interest: The authors declare no conflict of interest.

References

1. Cardoso, A.J.M. *Diagnosis and Fault Tolerance of Electrical Machines, Power Electronics and Drives*; IET: London, UK, 2018.
2. Merizalde, Y.; Hernández-Callejo, L.; Duque-Perez, O. State of the Art and Trends in the Monitoring, Detection and Diagnosis of Failures in Electric Induction Motors. *Energies* **2017**, *10*, 1056. [CrossRef]
3. Leite, V.C.M.N.; Da Silva, J.G.B.; Veloso, G.F.C.; Da Silva, L.E.B.; Lambert-Torres, G.; Bonaldi, E.L.; Oliveira, L.E.D.L.D. Detection of Localized Bearing Faults in Induction Machines by Spectral Kurtosis and Envelope Analysis of Stator Current. *IEEE Trans. Ind. Electron.* **2014**, *62*, 1855–1865. [CrossRef]
4. Cardoso, A.M.; Cruz, S.; Fonseca, D. Inter-turn stator winding fault diagnosis in three-phase induction motors, by Park's vector approach. *IEEE Trans. Energy Convers.* **1999**, *14*, 595–598. [CrossRef]
5. Xia, M.; Li, T.; Xu, L.; Liu, L.; De Silva, C.W. Fault Diagnosis for Rotating Machinery Using Multiple Sensors and Convolutional Neural Networks. *IEEE/Asme Trans. Mechatron.* **2018**, *23*, 101–110. [CrossRef]
6. Dai, H.; Zheng, Z.; Wang, W. A new fractional wavelet transform. *Commun. Nonlinear Sci. Numer. Simul.* **2017**, *44*, 19–36. [CrossRef]
7. Wang, L.; Zhang, X.; Liu, Z.; Wang, J. Sparsity-based fractional spline wavelet denoising via overlapping group shrinkage with non-convex regularization and convex optimization for bearing fault diagnosis. *Meas. Sci. Technol.* **2020**, *31*, 055003. [CrossRef]
8. Hoang, D.-T.; Kang, H.-J. A survey on Deep Learning based bearing fault diagnosis. *Neurocomputing* **2019**, *335*, 327–335. [CrossRef]
9. Perera, P.; Patel, V.M. Learning Deep Features for One-Class Classification. *IEEE Trans. Image Process.* **2019**, *28*, 5450–5463. [CrossRef]
10. Jiang, G.; Xie, P.; He, H.; Yan, J. Wind Turbine Fault Detection Using a Denoising Autoencoder With Temporal Information. *IEEE/ASME Trans. Mechatron.* **2018**, *23*, 89–100. [CrossRef]
11. San Martin, G.; López, D.E.; Meruane, V.; das Chagas, M.M. Deep variational auto-encoders: A promising tool for dimensionality reduction and ball bearing elements fault diagnosis. *Struct. Health Monit.* **2019**, *18*, 1092–1128. [CrossRef]

12. Yuan, L.; Lian, D.; Kang, X.; Chen, Y.; Zhai, K. Rolling bearing fault diagnosis based on convolutional neural network and support vector machine. *IEEE Access* **2020**, *8*, 137395–137406. [CrossRef]

13. Zhang, W.; Li, C.; Peng, G.; Chen, Y.; Zhang, Z. A deep convolutional neural network with new training methods for bearing fault diagnosis under noisy environment and different working load. *Mech. Syst. Signal Process.* **2018**, *100*, 439–453. [CrossRef]

14. Guo, Q.; Li, Y.; Song, Y.; Wang, D.; Chen, W. Intelligent fault diagnosis method based on full 1-D convolutional generative adversarial network. *IEEE Trans. Ind. Inform.* **2019**, *16*, 2044–2053. [CrossRef]

15. Yu, Y.; Si, X.; Hu, C.; Zhang, J. A review of recurrent neural networks: LSTM cells and network architectures. *Neural Comput.* **2019**, *31*, 1235–1270. [CrossRef]

16. Zhao, X.; Jia, M.; Ding, P.; Yang, C.; She, D.; Liu, Z. Intelligent fault diagnosis of multichannel motor–rotor system based on multimanifold deep extreme learning machine. *IEEE/ASME Trans. Mechatron.* **2020**, *25*, 2177–2187. [CrossRef]

17. Chen, S.; Meng, Y.; Tang, H.; Tian, Y.; He, N.; Shao, C. Robust deep learning-based diagnosis of mixed faults in rotating machinery. *IEEE/Asme Trans. Mechatron.* **2020**, *25*, 2167–2176. [CrossRef]

18. Randall, R.B. *Vibration-Based Condition Monitoring: Industrial, Aerospace, and Automotive Applications*; John Wiley & Sons: New Jersey, NJ, USA, 2011.

19. Zhang, J.; Wang, Z.; Meng, J.; Tan, Y.P.; Yuan, J. Boosting positive and unlabeled learning for anomaly detection with multi-features. *IEEE Trans. Multimed.* **2018**, *21*, 1332–1344. [CrossRef]

20. Gong, C.; Liu, T.; Yang, J.; Tao, D. Large-margin label-calibrated support vector machines for positive and unlabeled learning. *IEEE Trans. Neural Netw. Learn. Syst.* **2019**, *30*, 3471–3483. [CrossRef]

21. Pitolli, F.A. fractional B-spline collocation method for the numerical solution of fractional predator-prey models. *Fractal Fract.* **2018**, *2*, 13. [CrossRef]

22. Debarre, T.; Fageot, J.; Gupta, H.; Unser, M. B-spline-based exact discretization of continuous-domain inverse problems with generalized TV regularization. *IEEE Trans. Inf. Theory* **2019**, *65*, 4457–4470. [CrossRef]

23. Chen, P.Y.; Selesnick, I.W. Group-sparse signal denoising: Non-convex regularization, convex optimization. *IEEE Trans. Signal Process.* **2014**, *62*, 3464–3478. [CrossRef]

24. Zhang, T.; Song, S.; Li, S.; Ma, L.; Pan, S.; Han, L. Research on gas concentration prediction models based on LSTM multidimensional time series. *Energies* **2019**, *12*, 161. [CrossRef]

25. Pan, H.; He, X.; Tang, S.; Meng, F. An improved bearing fault diagnosis method using one-dimensional CNN and LSTM. *J. Mech. Eng.* **2018**, *64*, 443–452.

26. Qian, P.; Tian, X.; Kanfoud, J.; Lee, J.L.Y.; Gan, T.H. A novel condition monitoring method of wind turbines based on long short-term memory neural network. *Energies* **2019**, *12*, 3411. [CrossRef]

27. Hsueh, Y.M.; Ittangihal, V.R.; Wu, W.B.; Chang, H.C.; Kuo, C.C. Fault diagnosis system for induction motors by CNN using empirical wavelet transform. *Symmetry* **2019**, *11*, 1212. [CrossRef]

28. Esakimuthu, P.S.; Mizuno, Y.; Nakamura, H. A comparative study between machine learning algorithm and artificial intelligence neural network in detecting minor bearing fault of induction motors. *Energies* **2019**, *12*, 2105. [CrossRef]

29. Guo, S.; Zhang, B.; Yang, T.; Lyu, D.; Gao, W. Multitask convolutional neural network with information fusion for bearing fault diagnosis and localization. *IEEE Trans. Ind. Electron.* **2019**, *67*, 8005–8015. [CrossRef]

30. Arora, J.; Khatter, K.; Tushir, M. Fuzzy c-means clustering strategies: A review of distance measures. *Softw. Eng.* **2019**, *731*, 153–162.

31. Duan, Z.; Wu, T.; Guo, S.; Shao, T.; Malekian, R.; Li, Z. Development and trend of condition monitoring and fault diagnosis of multi-sensors information fusion for rolling bearings: A review. *Int. J. Adv. Manuf. Technol.* **2018**, *96*, 803–819. [CrossRef]

32. Wang, J.; Fu, P.; Zhang, L.; Gao, R.X.; Zhao, R. Multilevel information fusion for induction motor fault diagnosis. *IEEE/ASME Trans. Mechatron.* **2019**, *24*, 2139–2150. [CrossRef]

33. Lessmeier, C.; Kimotho, J.K.; Zimmer, D.; Sextro, W. Condition monitoring of bearing damage in electromechanical drive systems by using motor current signals of electric motors: A benchmark data set for data-driven classification. In Proceedings of the European Conference of the Prognostics and Health Management Society, Bilbao, Spain, 5–8 July 2016.

34. Das, S.; De, S. A modified genetic algorithm based FCM clustering algorithm for magnetic resonance image segmentation. In *Proceedings of the 5th International Conference on Frontiers in Intelligent Computing: Theory and Applications*; Springer: Bhubaneswar, India, 16–17 September 2016; pp. 435–443.

35. Li, C.; Cerrada, M.; Cabrera, D.; Sanchez, R.V.; Pacheco, F.; Ulutagay, G.; Valente de Oliveira, J. A comparison of fuzzy clustering algorithms for bearing fault diagnosis. *J. Intell. Fuzzy Syst.* **2018**, *34*, 3565–3580. [CrossRef]

36. Hou, J.; Wu, Y.; Gong, H.; Ahmad, A.S.; Liu, L. A novel intelligent method for bearing fault diagnosis based on EEMD permutation entropy and GG clustering. *Appl. Sci.* **2020**, *10*, 386. [CrossRef]

37. Bessous, N.; Zouzou, S.; Bentrah, W.; Sbaa, S.; Sahraoui, M. Diagnosis of bearing defects in induction motors using discrete wavelet transform. *Int. J. Syst. Assur. Eng. Manag.* **2018**, *9*, 335–343. [CrossRef]

38. Irfan, M.; Saad, N.; Ali, A.; Kumar, K.; Sheikh, M.; and Awais, M. A condition monitoring system for the analysis of bearing distributed faults. In *Proceedings of the 10th Annual Ubiquitous Computing Electronics & Mobile Communication Conference*; IEEE: New York, NY, USA, 2019; pp. 911–915.

Article

Early Detection of Broken Rotor Bars in Inverter-Fed Induction Motors Using Speed Analysis of Startup Transients

Tomas A. Garcia-Calva [1], Daniel Morinigo-Sotelo [2,*], Vanessa Fernandez-Cavero [3], Arturo Garcia-Perez [1] and Rene de J. Romero-Troncoso [4]

[1] HSPdigital-Electronics Department, University of Guanajuato, Salamanca 36700, Mexico; ta.garciacalva@ugto.mx (T.A.G.-C.); arturo@ugto.mx (A.G.-P.)
[2] Research Group HSPdigital-ADIRE, Institute of Advanced Production Technologies (ITAP), University of Valladolid, 47011 Valladolid, Spain
[3] Miguel de Cervantes European University, 47012 Valladolid, Spain; vfernandez@uemc.es
[4] HSPdigital-Mechatronics Department, Autonomous University of Queretaro, San Juan del Rio 76806, Mexico; troncoso@hspdigital.org
* Correspondence: daniel.morinigo@eii.uva.es

Abstract: The fault diagnosis of electrical machines during startup transients has received increasing attention regarding the possibility of detecting faults early. Induction motors are no exception, and motor current signature analysis has become one of the most popular techniques for determining the condition of various motor components. However, in the case of inverter powered systems, the condition of a motor is difficult to determine from the stator current because fault signatures could overlap with other signatures produced by the inverter, low-slip operation, load oscillations, and other non-stationary conditions. This paper presents a speed signature analysis methodology for a reliable broken rotor bar diagnosis in inverter-fed induction motors. The proposed fault detection is based on tracking the speed fault signature in the time-frequency domain. As a result, different fault severity levels and load oscillations can be identified. The promising results show that this technique can be a good complement to the classic analysis of current signature analysis and reveals a high potential to overcome some of its drawbacks.

Keywords: fault detection; fault diagnosis; frequency analysis; induction motors; rotating machines; signal processing; spectral analysis; time-frequency decompositions

Citation: Garcia-Calva, T.A.; Morinigo-Sotelo, D.; Fernandez-Cavero, V.; Garcia-Perez, A.; Romero-Troncoso, R.J. Early Detection of Broken Rotor Bars in Inverter-Fed Induction Motors Using Speed Analysis of Startup Transients. *Energies* **2021**, *14*, 1469. https://doi.org/ 10.3390/en14051469

Academic Editor: Mario Marchesoni

Received: 2 February 2021
Accepted: 1 March 2021
Published: 8 March 2021

Publisher's Note: MDPI stays neutral with regard to jurisdictional claims in published maps and institutional affiliations.

1. Introduction

The use of induction motors (IMs) in industrial applications with variable speed systems is widespread because they are more reliable, versatile, and efficient than line-fed machines [1]. High-performance IMs are considered robust machines, but they require reliable condition monitoring systems to avoid unprogrammed stops in production lines and reduce maintenance costs. The rotor cages of IMs are usually made from copper bars; these bars are exposed to failures in applications with variable operating conditions because of the excessive mechanical stresses involved [2]. Much attention has been directed to the study of broken rotor bar (BRB) fault, because if undetected it can develop into catastrophic machine breakdown [3]. Many papers in the literature have studied the diagnosis and condition monitoring of the IM rotor, but most only deal with machines operating at a constant speed and whole broken rotor bars. Nowadays, variable speed systems where the IM is driven by voltage source inverters (VSIs) are more common; they are used in a wide range of applications—namely, material handling, lifting, textile, compressors, pumps, mills, winders, and lifts [4]. These VSIs are usually pulse width modulators (PWM) that produce non-sinusoidal voltages using a solid-state inverter by rapidly switching the output voltage on and off. It is known that no matter how accurate the switching of the PWM is, they are an inherent source of harmonic distortion in IM voltages and currents, which has a negative impact on the efficiency of fault detection techniques [5].

Motor current signature analysis (MCSA) is the most popular method in preventive maintenance and it is considered the reference technique for broken bar fault diagnosis in squirrel-cage motors. The stator current signature analysis of motors under transient operation (such as startup) has received special attention in the last decade as an alternative to improve the reliability of stationary analysis and reduce the rate of false alarms in the classic MCSA [6,7]. To do this, detection and diagnosis methodologies have been proposed based on time-frequency (t, f) decompositions capable of identifying fault signatures and complementing the analysis of the stationary current signals [8–12]. These works achieve the identification of fault signatures and extract condition indicators from the single-phase electrical current in the startup transient. Nevertheless, most of these methods assume a line-fed induction machine, and the analysis of transients, such as startups, in inverter-fed IMs is still an active field and an open question.

Different approaches for the condition monitoring of inverter-fed IMs under startup transient regimes have been studied recently [13,14]. BRB detection under non-stationary conditions consists of tracking fault-related signatures called sideband harmonics. Unlike motors powered directly from the grid, VSIs introduce several harmonic components produced by the PWM that overlap with these signatures, obstructing their recognition and the accuracy of the fault diagnosis. Different tools have been proposed to overcome these issues and improve the reliability of fault diagnostic methods based on transient analysis. These tools include adaptive transforms [15,16], non-linear signal decompositions [17,18], demodulation schemes [19–21], statistical methods [22–24], intelligent algorithms [25–28], and combined techniques. However, all these methodologies focus on the analysis of the stator current signal.

This paper proposes the analysis of the rotor speed signal for the detection of broken rotor bars at incipient states in inverter-fed induction motors during the startup transient. This technique is also used to distinguish load oscillations to avoid false positives. It is demonstrated that the speed analysis of induction motors in the time-frequency domain offers a reliable detection of broken rotor bars when the stator current analysis fails due to low-slip operation, load oscillations, overlapped signatures, and other non-stationary conditions. The proposed technique is used to obtain and evaluate speed fault pattern variations and evolutions along the startup transient. The method is applied to real speed signals from laboratory experiments and compared to the analysis of the stator current of an induction motors subjected to different fault severities and load oscillations. This proposal can be a valuable and attractive complement to other techniques based on stator current, vibrations, or thermal analysis [29–31]. Results demonstrate that the analysis of the motor speed during startups in VSI-fed IMs can detect broken rotor bars, even at low fault severities, and distinguish this fault from load oscillations to avoid false positives. This proposal can be a valuable and attractive complement to other techniques based on stator current analysis and can help to avoid false alarms in VSI-fed IM systems. The capability of speed analysis is examined in this work, which is an extended contribution of the conference paper presented in [32].

2. Influence of Rotor Fault on Motor Speed

2.1. Theoretical Background

Electrical current analysis techniques for detecting BRB in IM are based on how the power spectral density (PSD) of the stator current is affected. This fault type increases the bar resistance, produces asymmetry in the airgap's electromagnetic field, and gives rise to sideband harmonics [33] located at:

$$f_{rf} = f_s(1 \pm 2ks), \; k = 1, 2, 3..., \tag{1}$$

where f_s is the power supply fundamental frequency, s is the rotor slip, and k is an integer. The low-order components (when $(k = 1)$) are of special interest for fault diagnosis because they have larger amplitudes than high-order components. The $(1 - 2s)f_s$ component is known as the left side-band harmonic (LSH) and the frequency component at $(1 + 2s)f_s$ is known as the right side-band harmonic (RSH), since they are located to the left and the right of f_s in steady-state analysis. In a faulty IM, the air-gap torque is affected by the flux linkages and the stator currents. The linkages fluxes are given by:

$$\psi = \int (v_a - R_s i_a) dt, \tag{2}$$

where v_a is the stator voltage, R_s is the stator resistance, and i_a is the stator current. The interaction of the $(1 \pm 2s)f_s$ components with the fundamental magnetic flux produces an oscillatory torque at frequency $2sf_s$ in the total torque, which is given by:

$$\sum \Gamma(t) = \Gamma_0 + 3P\psi \sum_{k=1}^{N} I'_k \sin(2ksf_st + \alpha_\psi - \alpha_k), \tag{3}$$

where Γ_0 is the dominant torque component produced by the fundamental component of the stator current, α_k is the ripple phase, and P is the pole-pair number. The fault components at the torque ripple produce a low-frequency modulation on the motor speed, with twice the slip frequency when the rotor is damaged [34]. The content of the angular speed for a faulty motor can be modeled in (rad/sec) by [35]:

$$\omega_r(t) = \omega_m(t) + \sum_{k=1}^{N} \frac{3P\psi}{J2s\omega} I'_k \cos(2ksf_st + \alpha_\psi - \alpha_k), \tag{4}$$

This can be expressed in (r.p.m.) by:

$$n_r(t) = \frac{120\omega_r(t)}{4\pi}, \tag{5}$$

where $2ksf_s$ are the speed oscillations due to the faulty rotor, J is the inertia, and ω_m is the angular speed fundamental component for a healthy induction motor in r.p.m., given by:

$$n_m(t) = \frac{120f_s(t)}{P}(1 - 2s(t)). \tag{6}$$

The most used method to detect fault-related oscillations in induction motors is the steady-state analysis of the stator current based on frequency analysis by Fourier transform (FT). Spectral leakage around the first harmonic is the main drawback of this analysis, whose causes are non-strict stationary conditions, non-integral digital frequencies, and the inherent finite time window of the analysis [36]. A comparison of the stator current and the speed analysis performed on healthy and faulty conditions of the motor operating under steady state is shown in Figure 1. The fundamental supply frequency at 50 Hz is noticed in Figure 1a, where a concentration of leaked energy is present in adjacent spectral bins. The LSH is observed at 44.86 Hz with an amplitude of -35.7 dB in the current spectrum of the motor with one BRB. Its amplitude is higher than the LSH of the healthy motor with -38.58 dB, which may be undetectable without a speed reference. Figure 1a,b shows the spectra of the stator current and the speed of the same tests of an induction motor in healthy and faulty states (one broken rotor bar). For the latter, the spectrum shows an increase in the speed oscillation at a $2sf_s$ frequency of 11.3 dB compared to the same component for the healthy case. The spectral leakage around the dominant frequency f_s (50 Hz) in the stator current spectra is significant when compared to the low level of leakage around the fundamental speed component n_r (0 Hz).

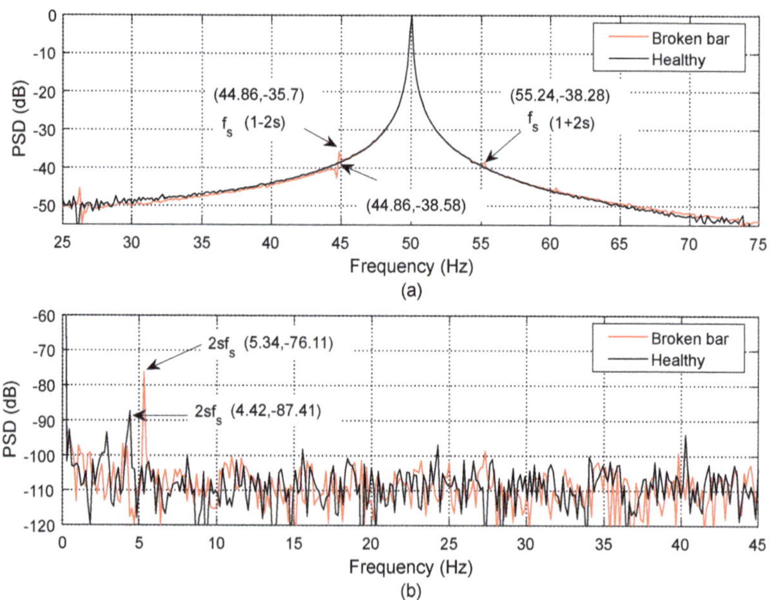

Figure 1. Spectra of the (**a**) stator current and (**b**) mechanical rotor speed of the motor supplied from the voltage source inverter and in stationary operation.

2.2. Time-Frequency Analysis of Startup Transient

At start-up transient or speed variations, neither the current nor the rotor speed can be considered as stationary or deterministic processes because their amplitude, frequency, and phase are not constant, besides the fact that behaviors and measurements are susceptible to many disturbances and unpredictable errors such as digital quantization, external vibrations, noise, and other environmental effects. It becomes necessary to consider the signals as random processes. In this work, the (t, f) analysis of the speed signal is computed by a high-resolution PSD estimation called multiple signal classification (MUSIC), which is a frequency estimation technique based on eigen analysis. MUSIC algorithm requires a short number of observed points to offer a high-resolution spectrum estimation, which makes it a suitable technique for VSI-fed induction machines analysis at startup transient [37].

The PSD of n-observation samples of the speed process is defined as the discrete-time Fourier Transform of its autocorrelation sequence:

$$P_{xx}(f) = T \sum_{k=-\infty}^{\infty} r_{nn}[k]e^{-j2\pi fkT},$$ (7)

where the variable $r_{nn}[k]$ is the autocorrelation function of the $n_r[k]$ and is defined as $n_r[k] = E[n_r[k]n_r[k+l]]$. The speed vector $\mathbf{n_r} = [n_r(0), ..., n_r(n-1)]$ can be written as $\mathbf{n_r} = \mathbf{x} + \eta = Sa + \eta$, where η is the additive noise in the measured signal and the $\mathbf{x}$ vector $[x(m), x(m+1), ..., x(m+n-1)] = \mathbf{Sa}$ is defined as:

$$\mathbf{x} = \begin{bmatrix} e^{j\omega_1 m} & e^{j\omega_2 m} & \cdots & e^{j\omega_q m} \\ e^{j\omega_1(m+1)} & e^{j\omega_2(m+1)} & \cdots & e^{j\omega_q(m+1)} \\ \vdots & \vdots & \ddots & \vdots \\ e^{j\omega_1(m+n-1)} & e^{j\omega_2(m+n-1)} & \cdots & e^{j\omega_q(m+n-1)} \end{bmatrix} \begin{bmatrix} a_1 e^{j2\pi\phi_1} \\ a_2 e^{j2\pi\phi_2} \\ \vdots \\ a_q e^{j2\pi\phi_q} \end{bmatrix},$$ (8)

The autocorrelation matrix of the measured mechanical speed can be written as the sum of the autocorrelation matrices of the signal $\mathbf{x}$ and the noise η as $\mathbf{R}_{nn} = \mathbf{R}_{xx} + \mathbf{R}_{\eta\eta}$.

An eigen decomposition of $\boldsymbol{R}_{xx}$ and $\boldsymbol{R}_{nn}$, $\boldsymbol{s}$ can be expressed as a linear combination of the principal eigenvectors $[v_{p+1}, ..., v_n]$ [38]. In the multiple signal classifications algorithm, the power spectral density is defined as:

$$P_{xx}(f) = \sum_{k=p+1}^{N} \left| \boldsymbol{s}^H(f)\boldsymbol{v}_k(f) \right|^2, \tag{9}$$

where $s(f)$ is the complex sinusoidal vector. Since $P_{xx}(f)$ has its zeros at the frequencies of the sinusoids, it follows that the reciprocal of $P_{xx}(f)$ has its poles at these frequencies. Therefore, the spectral estimation is computed as:

$$\widehat{P}_{xx}^q(f) = \frac{1}{\boldsymbol{s}^H(f)\mathbf{V}(f)\mathbf{V}^H(f)\boldsymbol{s}(f)}, \tag{10}$$

where $V = [v_p + 1, ..., v_n]$ is the matrix of eigenvectors of the noise subspace. The resulted spectrum displays sharp peaks at frequencies of the mechanical speed oscillations, hence the PSD estimation is used for a high-resolution time-frequency analysis.

3. Experimental Setup

A laboratory test bench that emulates a typical VSI-fed induction motor system was used for experimental investigation. The laboratory setup consists of a 0.75 kW three-phase induction motor (Model D-91056 by Siemens) fed by a voltage source PWM inverter (Model ACS355-03E-15A6-4 by ABB). Appendices A and B summarize the technical parameters and specifications. The mechanical load was provided by an electro-magnetic powder brake (Model SE2662-5R by Lucas-Nülle).

Five cases of a rotor bar condition were studied, including the healthy status and four different broken bar severity degrees (from an incipient fault condition to a full broken rotor bar). In the healthy case, all bars of the rotor are in healthy condition. For the first fault status, the damage degree was simulated by drilling a 2 mm-diameter hole in a rotor bar, the depth of the hole (dp) was $\frac{1}{4}$ of the bar height. In the second fault severity condition, the depth of the hole in the bar was increased to $\frac{1}{2}$ of the bar height. In the third fault severity condition, the depth of the hole was increased to $\frac{3}{4}$ of the bar height. Finally, the last fault severity condition was the fully broken rotor bar, where the depth of the hole was equal to the bar height, which is $h = 13$ mm. This results in five fault severity conditions, where s1 represents a healthy condition of the rotor, s2 a low level of degradation ($dp = 3.25$ mm), s3 the half-broken bar ($dp = 6.5$ mm), s4 a high level of degradation ($dp = 9.75$ mm), and s5 a fully broken rotor bar ($dp = 13$ mm). Figure 2a shows a cross-sectional view of the rotor cage with details of the different depth levels drilled into the bar and Figure 2b shows the drilled rotor bar.

The machine speed n_r was provided by a tachogenerator (Model SE2672-5U by Lucas-Nülle) with an output voltage range of 0–10 V proportional to its rotation speed. One phase current i_a was also measured for comparison purposes using a Hall effect transducer by LEM, model LV-25NP. A compact data acquisition system (NI cDAQ-9147) was used for data acquisition. This incorporated an analog input module, NI 9215, with 16 bits of resolution, and two independent input channels to capture the speed signal and stator current. The input scale of these two channels was 10 V. Therefore, the quality and reliability of the speed and stator current signals capture was the highest possible, as those signals were recorded using the high range of the input scale. Every isolated analog to digital converter was pre-programmed with a sampling frequency $f_n = 50$ kHz based on the Nyquist theorem (higher than twice of the highest PWM switching frequency) to prevent unrealistic alias frequency components in the measured signals at the acquisition stage. Figure 3 shows a photograph of the test ring.

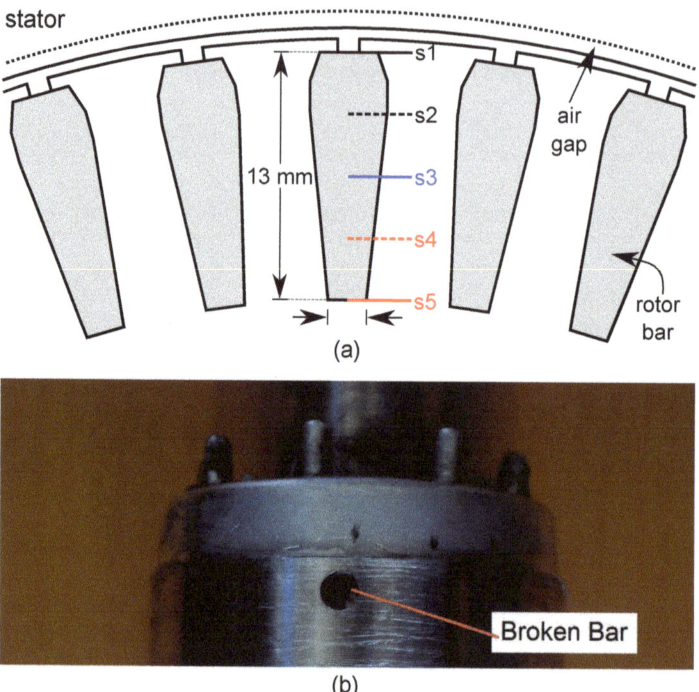

Figure 2. View of the rotor employed in the tests: (**a**) side view schematic for the transverse section of the squirrel cage rotor, and (**b**) squirrel cage rotor with a complete broken rotor bar condition.

Figure 3. Experimental test ring configuration: (1) asynchronous motor, (2) electro-magnetic powder brake, (3) powder brake control, (4) voltage source inverter, (5) signal condition stage, (6) data acquisition system, (7) PC, and (8) programmable function generator.

4. Experimental Results

The startup transient analysis examined the five case studies of different severity levels from healthy $s1$ to the faulty case $s5$, as explained in Section 3. The healthy case $s1$ and faulty case $s5$ were already analyzed in Section 2 (see Figure 1), but with the motor operating in a stationary regime. For every case study, low load and high load operation tests are studied. All the acquired signals were processed and plotted by the Matlab® software. The digital sequences of n_r and i_a are first decimated from the original f_n to a new sampling frequency f_n/M in a pre-processing stage for selecting the frequency band analysis $[0 - f_n/M)$ Hz and to decrease the computing requirements. The decimation is carried out by means of a low-pass filter and a sample rate compressor with the down-sample factor M. The Parks–MacClellan [39] algorithm was used to design the low-pass filter; its parameters are cutoff frequency $\omega_c = \frac{\pi}{M}$, pass-band attenuation $A_p = 0$ dB, and stop-band attenuation $A_s = 100$ dB. The filter design was chosen to provide an adequate response in the pass-band while limiting the high-frequency components. The decimation factor M is selected according to the spectral behavior of the fault signature. In the case of the stator current, the bandwidth of analysis (0, 62.5 Hz)contains the main spectral components and the fault-related side-band harmonics. In the case of speed signal, the bandwidth of analysis (0, 31.25 Hz)contains all the desired information for rotor fault diagnosis purposes. The decimation is processed with a multi-stage approach to avoid measurement errors and spectral aliasing at the signal processing stage. The decimation factors are M = 5 × 4 × 5 × 4 and M = 5 × 4 × 5 × 4 × 2 for the stator current and rotor speed, respectively. Figure 4 illustrates the proposed methodology.

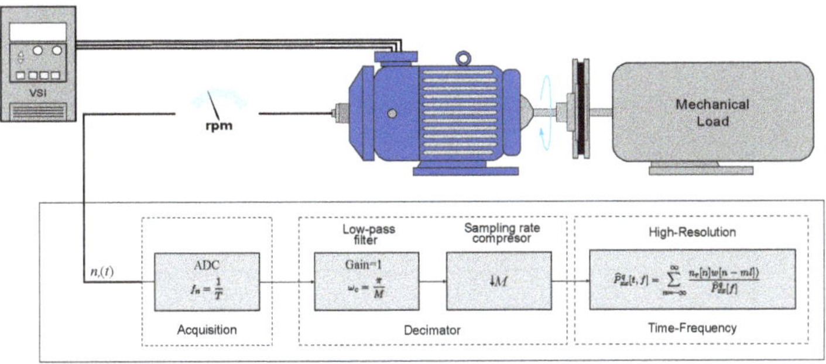

Figure 4. Simplified block diagram of the decimation process.

The measured signals consisted of a 10 s startup transient followed by 1 s of steady-state operation. During all the tests, $n_r(t)$ and $i_a(t)$ were recorded simultaneously to make a fair comparison between the stator current and the speed analysis. The precision and resolution of data acquisition are comparable to those used in industrial practice. Two series of tests were carried out at two different constant load torques, 2.6 and 3.8 Nm—low level (LL) and high level (HL), respectively. The VSI was programmed to provide a linear startup following a ramp from 0 Hz to the motor-rated frequency ($f_b = 50$ Hz) with a constant $\frac{df}{dt}$ during the first 10 s. Figure 5a presents the voltage and current waveforms of a startup transient (10 s) followed by steady-state operation (1 s); the signal amplitudes are normalized by their corresponding maximum value. As expected, $i_a(t)$ and $v_a(t)$ exhibit a larger harmonic content than a line-fed system.

In Figure 5b, the pulse-width-modulated waveform of $v_a(t)$ and the switching effect produced by the VSI can be seen, and the highest electrical magnitude of the stator current during the startup transient occurs around 2 s after energization. Figure 6 presents the spectral characteristics of a line-fed and VSI-fed motor; the comparison of these plots shows the harmonics and inter-harmonics injected by the VSI in the system.

The measured rotor speed waveforms for a healthy and faulty case (s5) under LL and HL are shown in Figure 7. The figure presents the time-varying synchronous speed $n_s = 120f_s(t)/p$ when the motor is supplied by a linear ramp of frequency voltage v/f (0–50 Hz/0–10 s).

The rotor speed is closer to the synchronous speed (dashed line) when the motor load is low due to the reduced motor slip. When the mechanical load is high, the slip also increases, and the rotor speed separates more from the synchronous speed. Between second 1 and 2.5, a dead zone is observed after starting due to the time the motor takes to overcome the load inertia at standstill.

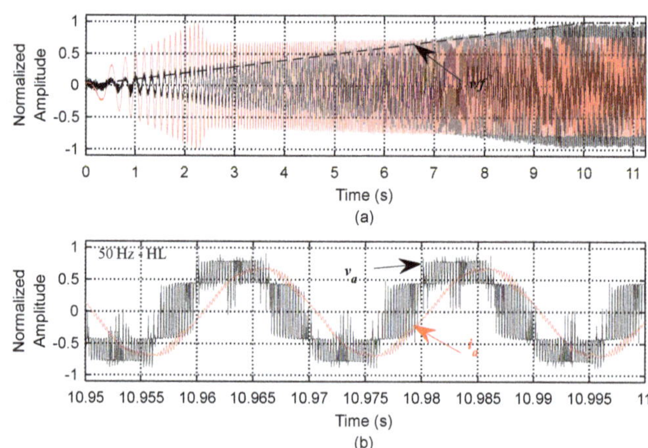

Figure 5. Voltage (v_a) and current (i_a) waveforms in the induction motor fed by the voltage source inverter during: (**a**) 10 s of startup transient and 1 s of steady state, and (**b**) close up of the steady state.

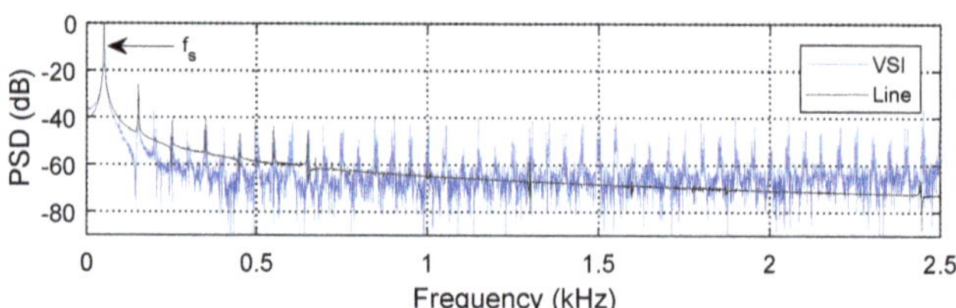

Figure 6. Spectra of the motor supply voltage when fed from the line or from the voltage source inverter and under stationary conditions.

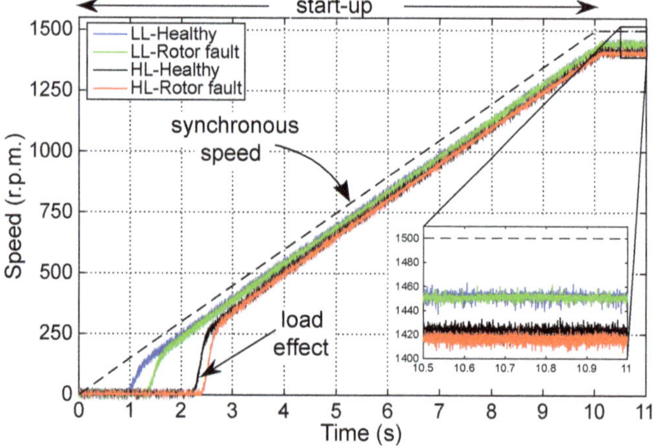

Figure 7. Startup transients performance of the speed in an induction motor fed by an inverter for healthy and faulty cases under low (LL) and high (HL) levels of mechanical load.

4.1. Time-Frequency Analysis of Startup

Time-frequency analyses for healthy and faulty tests are computed using the high-resolution estimation MUSIC; the PSD magnitude is presented in the z-axis of the (t, f) plane in a relative scale $20log_{10}|P_{xx}(e^{j\omega})|^2$. Rotor speed and stator current (t, f) decompositions are produced by applying a 81-point rectangular window with a 49 percent overlap.

4.1.1. Induction Motor Under High-Load Condition

The time-frequency decompositions of stator current and rotor speed for a high-loaded motor are displayed in Figures 8 and 9, respectively. Figure 8a illustrates the application of the stator current analysis for the healthy motor ($s1$), and Figure 8b presents the stator current analysis for the motor with a broken rotor bar ($s5$). The high-resolution technique is capable of realizing very sharp (t, f) spectral analyses, allowing the observation of individual spectral components. In both cases, Figure 8a,b, the maximum energy concentrations take place at the linear frequency variation $f_s(t) = 5t$ as a result of the dominant component of the voltage supply. The spectral harmonics of winding, eccentricity, and other components introduced by the VSI are also observed in the (t, f) decompositions. For non-sinusoidal voltage supplies, triplen harmonics are seldom present in three-phase induction motors; conversely, 2nd and 4th harmonics can be observed in the stator current analysis. In Figure 8b, the time-varying spectrum of the faulty motor is presented and shows the appearance of the fault sidebands $(1 \pm 2s)f_s$, which are close to the fundamental component and can be observed evolving with a positive slope. Although a visual comparison between Figure 8a,b shows the existence of the fault components in the vicinity of f_s in Figure 8b, a precise identification, isolation, and quantification of LSH or RSH is complicated, as its trajectories are partly overlapped with f_s.

The proposed motor speed signature analysis is presented in Figure 9 for the same experimental tests of Figure 8. As envisioned in Equation (4), Figure 9a shows that for a healthy motor the dominant spectral component in n_r is n_m and there is absence of a fault harmonic component. In contrast, Figure 9b clearly exposes the presence of a speed oscillation at the twice slip frequency during the startup transient, revealing a rotor defect in the motor. Notice that even though the electrical starting point is at $t = 0$ when the VSI energizes the induction motor, the fault oscillation $2sf_s$ rises about two seconds later when the shaft starts rotating. This is because the fault signature is directly related to the slip and consequently to the rotor movement. This speed oscillation trajectory can be easily tracked in the (t, f) domain; the average speed value does not produce a signature in the time-frequency domain and the trajectory of the speed oscillations are easily observed.

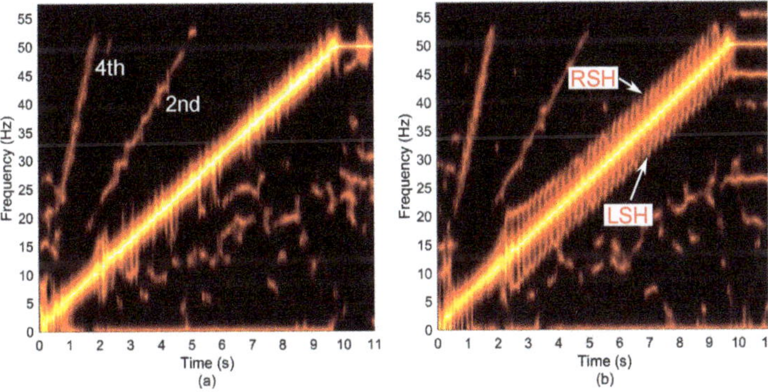

Figure 8. Time-frequency analyses of the IM stator current under startup transient and high-load level for: (**a**) healthy motor and (**b**) rotor with one full broken rotor bar.

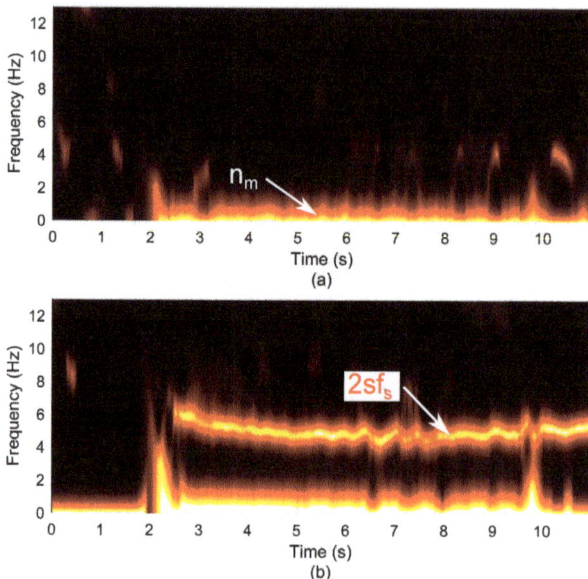

Figure 9. Time-frequency analyses of the motor speed under startup transient and high-load level for: (**a**) healthy motor and (**b**) rotor with one full broken rotor bar.

4.1.2. Induction Motor Under Low-Load Condition

Time-varying analyses resulting from the stator current for healthy $s1$ and a faulty $s5$ cases are compared in Figure 10, when the motor is operating under low-slip condition. Figure 10a illustrates the application of current signature analysis to the healthy motor; apart from noise, Figures 8a and 10a present a similar time-varying spectra, and both LL and HL condition (from a healthy machine) exhibit the presence of even harmonics, eccentricity harmonics, and the main energy component in the t-f decomposition concentrated at f_s. Conversely, Figure 10b shows a different energy distribution and depicts the appearance of the symmetrical fault-components (LSH and RSH), which are separated from the fundamental component and can be observed only after the second 10 at 45 and 55 Hz. However, its energy cannot be tracked in the startup transient (0 to 10 s), since they are very close to f_s and its trajectories are indistinguishable. The results presented in Figure 10 illustrate the main limitation of the motor current signature analysis to detect fault-harmonics when induction motors are operating under light load conditions. When the motor is operating under light load conditions, the slip approaches zero and the stator current analysis experience troubles in fault signature identification.

The proposed methodology, motor speed signature analysis, is presented in Figure 11 for the same experimental tests of Figure 10. In this case, the spectrum for the healthy motor with a constant mechanical load of 2.6 Nm is shown in Figure 11a. Although the result belongs to a startup transient under variable-frequency supply, it is found that n_r is composed only for a constant component n_m embedded in noise. On the other hand, the result shown in Figure 11b is different because the (t, f) analysis reveals a new harmonic component related to the broken rotor bar fault, now clearly the component appears at frequency $2sf_s$. One can also observe a frequency decrement in the separation between n_m and the fault-component trajectory due to the lower slip than when the motor load was high (in Figure 9b).

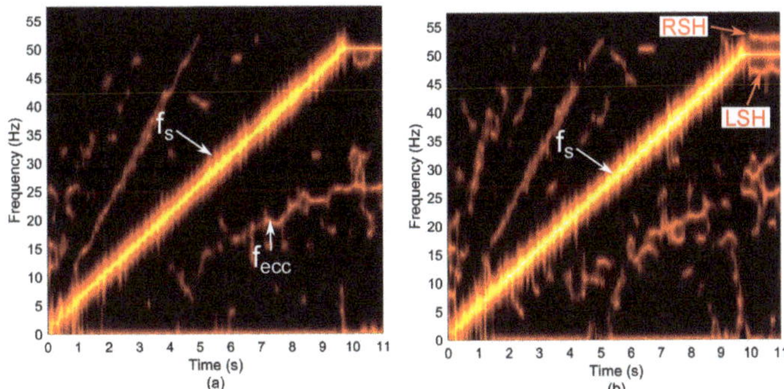

Figure 10. Time-frequency analyses of the IM stator current under startup transient and low load level for: (**a**) healthy motor and (**b**) a rotor with one full broken rotor bar.

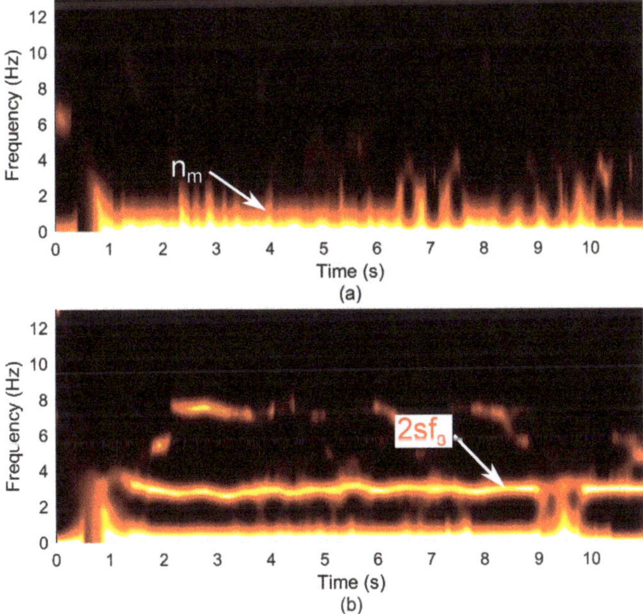

Figure 11. Time-frequency analyses of the motor speed under startup transient and low load level for: (**a**) healthy motor and (**b**) a rotor with one full broken rotor bar.

If Figures 9b and 11b are compared, in the latter in Figure 9b the fault-related component is far separated from the dominant component because the motor slip is higher due to a high load: The greater the load level, the greater the rotor slip and, in consequence, the higher the frequency speed fluctuation.

The results show that the fault components are better detected using the motor speed analysis, which gives a clear separation of the spectral components at the (t, f) plane and permits the extraction of the speed oscillation amplitude at the fault frequency. In the speed analyses given for a healthy and faulty cases, it can be observed that the dominant spectral components in the (t, f) decompositions are the constant trajectories n_m (Figures 9 and 11). Results of the stator current decompositions (Figures 8a and 10) are clearly different, where the dominant component is the frequency modulated component f_s (from 0 to 50 Hz). The low-frequency variation in n_m and $2sf_s$ in the motor speed spectrum improves the

PSD estimation by reducing bias and leakage in the (t, f) decompositions, thus giving an accurate localization and quantification of the fault signature.

4.2. Early Fault Detection

In addition to the fault detection analysis, the amplitude of the fault component is isolated and extracted from the (t, f) domain for diagnosis purposes. Figure 12 presents the results of the $2sf_s$ amplitude for the five severity levels from s_1 to s_5. The amplitudes evolution for HL and LL conditions are shown in Figure 12a,b, respectively. In the case of healthy condition s_1 and the s_2 severity level, the speed oscillation is small and it is difficult to determine any difference. However, the amplitude of the fault signatures for the cases s_3, s_4, and s_5 expose a clear increment in the speed fluctuation at $2sf_s$ as the severity of the fault increases.

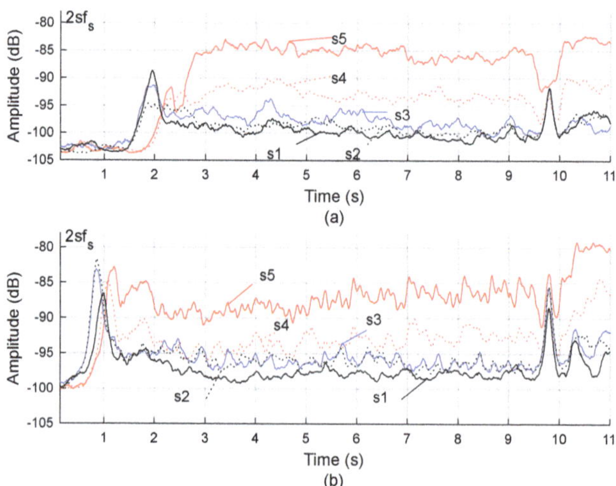

Figure 12. Time evolution of the $2sf_s$ amplitude during the startup transient for multiple severity cases under: (**a**) high load condition and (**b**) low load condition.

4.3. Induction Motor Under Load Oscillations

To study the performance of the proposed methodology, the (t, f) analysis is used to process the rotor speed of an IM under load oscillations. In particular, the low-frequency load oscillation phenomenon is interesting to consider because it is a common root cause of rotor fault false alarms when the classical steady-state analysis is performed. The oscillation in the load was produced by feeding the controller of a magnetic powder break with a low-frequency sinusoid, which is provided by a programmable function generator. This time the motor startup was programmed in the VSI with 5 s, to verify and validate the effectiveness of the analysis under short transient startup. Figure 13a shows the result of the healthy motor case under a normal condition without load oscillation, where there are not low-frequency trajectories present in the (t, f) decomposition. In Figure 13b, the result of the same healthy motor is shown but with a load oscillating by 4 Hz. A clear frequency component is present during the transient and the steady state at 4 Hz. The $2sf_s$ pattern induced by a rotor fault is shown in Figure 13c; with the measured speed signal, it is easy to corroborate that the slip at the stationary regime is s = 0.032, allowing the method to confirm that the trajectory which converge to 3.28 Hz corresponds to a rotor fault component. The (t, f) decomposition of the speed signal for the faulty motor operating under the oscillating load is shown in Figure 13d, where not only the oscillation component of the load at 4 Hz is observed in the (t, f) plane but also the presence of the rotor fault trajectory with different direction can be noticed. These experimental results show that the speed analysis of the startup transient is effective to identify rotor fault components and load oscillation components, even when in stationary regime they overlap at the same frequency.

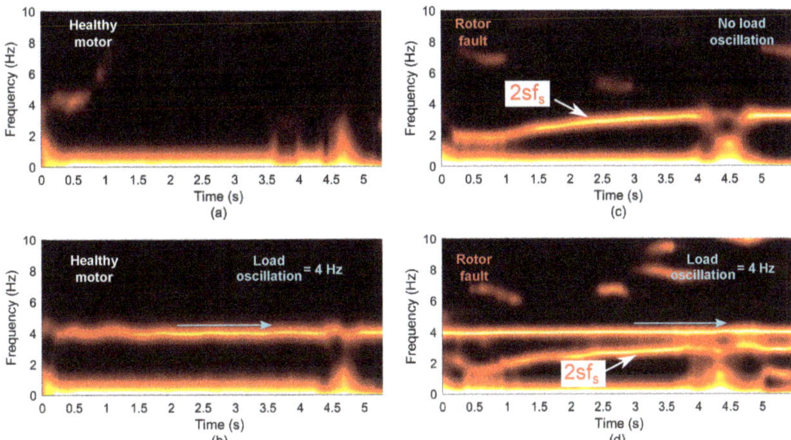

Figure 13. Time-frequency decomposition of the rotor speed under startup transient and high load level for: (**a**) healthy motor, (**b**) healthy motor and load oscillation, (**c**) rotor with a broken rotor bar without load oscillation, and (**d**) rotor with a broken rotor bar with load oscillation.

In this work, a tachogenerator was used to measure the speed, but in real applications other speed sensors can be used with a better reliability, robustness, and accuracy, such as optical encoders or speed resolvers. It should not be forgotten that a reliable and effective analysis of the stator current for fault detection is also based on a good estimate or accurate measurement of the motor speed, which is necessary for the location of fault-related components. The proposed method is a good complement to the analysis of the stator current and can help to avoid false alarms.

5. Conclusions

This paper presents a new methodology for the early detection of broken rotor bars in VSI-fed induction motors based on the analysis of motor speed in the time-frequency domain. Although speed monitoring has been considered in stationary conditions, it has not received attention under transient conditions. The proposed method has two advantages with respect to stator current analysis: the rotor speed shows fewer spectral components than the stator current and the average value of the rotor speed does not produce a frequency pattern. The study confirms the existence of the fault pattern $2sf_s$ in the speed during the startup transient and reports the diagnostic capabilities of the method to detect and distinguish between load oscillation components and rotor fault signatures during non-stationary conditions. The results prove that: (i) speed analysis under startup transients can be used as an effective and reliable technique for the diagnosis of broken rotor bars in inverter-fed induction motors during startups compared to the analysis of the stator current; (ii) broken rotor bars can be detected at early stages of the fault; (iii) the proposed analysis technique avoids false positives by low-frequency oscillations, as their trajectories in the time-frequency plane are different from those related to the fault. The analysis of the speed signal was carried using a high-resolution technique, MUSIC. However, much simpler techniques, such as short-time Fourier transform, can also be used if an appropriate window length is chosen to avoid spectral leakage around the average value of speed. As the development of a broken rotor bar is slow over time, this methodology can be incorporated into an on-line condition monitoring system. The computational burden of the proposed analysis is also low, so this is not an impediment either. One disadvantage of the proposed technique, compared to stator current monitoring, is the sensor used, which is not usually available in many adjustable speed drives. In future works, the authors will study if this methodology can be applied to the detection of other kinds of faults or machines.

Author Contributions: Conceptualization, T.A.G.-C., D.M.-S., and R.d.J.R.-T.; data curation, R.d.J.R.-T. and A.G.-P.; methodology, T.A.G.-C. and D.M.-S.; software, T.A.G.-C. and A.G.-P.; validation, A.G.-P and V.F.-C.; writing—original draft, T.A.G.-C. and V.F.-C.; writing—reviewing and editing, D.M.-S., V.F.-C., and R.d.J.R.-T. All authors have read and agreed to the published version of the manuscript.

Funding: This work was supported in part by the Mexican Council of Science and Technology (CONACYT) by the scholarship 487058, Mexico City, Mexico.

Institutional Review Board Statement: Not applicable.

Informed Consent Statement: Not applicable.

Data Availability Statement: Not applicable.

Conflicts of Interest: The authors declare no conflict of interest.The funders had no role in the design of the study; in the collection, analyses, or interpretation of data; in the writing of the manuscript, or in the decision to publish the results.

Abbreviations

The following abbreviations are used in this manuscript:

BRB	Broken rotor bar
f_s	Fundamental frequency
f_b	Motor rated frequency
f_n	Sampling frequency
FT	Fourier Transform
HL	High-load level
IM	Induction Motor
LL	Low-load level
LSH	Left Side Harmonic
MCSA	Motor current signature analysis
MUSIC	Multiple signal classification
PSD	Power spectral density
PWM	Pulse width modulation
RSH	Right side-band harmonic
sf_s	Slip frequency
t, f	time-frequency
VSI	Voltage source inverter

Appendix A

Three-phase induction motor rated characteristics: Rated power = 0.75 kW, Rated voltage = 230/460 V, Rated current = 3.2/1.86, Synchronous speed = 1500 r.p.m.

Appendix B

Voltage source inverter characteristics: Rated output voltage = 220/240 V, Rated power: 4 kW, Start-up mode = Linear, Control mode: scalar (v/f), Commutation method = pulse width modulation.

References

1. Tan, C.; Feng, Z.; Liu, X..; Fan, J.; Cui, W.; Sun, R.; Ma,Q. Review of variable speed drive technology in beam pumping units for energy-saving. *Energy Rep.* **2020**, *6*, 2676–2688. [CrossRef]
2. Park, Y.; Yang, C.; Kim, J.; Kim, H.; Lee, S.B.; Gyftakis, K.N.; Panagiotou, P.A.; Kia, S.H.; Capolino, G. Stray Flux Monitoring for Reliable Detection of Rotor Faults Under the Influence of Rotor Axial Air Ducts. *IEEE Trans. Ind. Electr.* **2019**, *66*, 7561–7570. [CrossRef]

3. Gyftakis,K.N.; Antonino-Daviu J.A.; Garcia-Hernandez, R.; McCulloch, M.D.; Howey D.A.; Marques-Cardoso A.J. Comparative Experimental Investigation of Broken Bar Fault Detectability in Induction Motors. *IEEE Trans. Ind. Appl.* **2016**, *52*, 1452–1459. [CrossRef]

4. Hassan, H.; Mahmood, F.; Akmal, M.; Nasir, M. Optimum operation of low voltage variable-frequency drives to improve the performance of heating, ventilation, and air conditioning chiller system. *Int. Trans. Electr. Energy Syst.* **2020**, *30*, 188–199. [CrossRef]

5. Martin-Diaz, I.; Morinigo-Sotelo, D.; Duque-Perez, O.; Arredondo-Delgado, P.A.; Camarena-Martinez, D.; Romero-Troncoso, R.J. Analysis of various inverters feeding induction motors with incipient rotor fault using high-resolution spectral analysis. *Electr. Power Syst. Res.* **2017**, *152*, 18–26. [CrossRef]

6. Lee, S.B.; Stone, G.C.; Antonino-Daviu, J.A.; Gyftakis, K.N.; Strangas, E.G.; Maussion, P.; Platero, C.A. Condition Monitoring of Industrial Electric Machines: State of the Art and Future Challenges. *IEEE Ind. Electron. Mag.* **2020**, *14*, 158–167. [CrossRef]

7. Gyftakis, K.N.; Cardoso, A.J.M. Reliable Detection of Stator Interturn Faults of Very Low Severity Level in Induction Motors. *IEEE Trans. Ind. Electr.* **2020**, *68*, 3475–3484. [CrossRef]

8. Antonino-Daviu, J.A.; Riera-Guasp, M.; Pons-Llinares J.; Park, J.; Lee, S.B.; Yoo, J.; Kral, C. Detection of Broken Outer-Cage Bars for Double-Cage Induction Motors Under the Startup Transient. *IEEE Trans. Ind. Appl.* **2012**, *48*, 1539–1548. [CrossRef]

9. Elbouchikhi, E.; Choqueuse, V.; Amirat, Y.; Benbouzid, M.E.H.; Turri, S. An Efficient Hilbert–Huang Transform-Based Bearing Faults Detection in Induction Machines. *IEEE Trans. Energy Convers.* **2017**, *32*, 401–413. [CrossRef]

10. Chen, X.; Feng, Z. Time-frequency space vector modulus analysis of motor current for planetary gearbox fault diagnosis under variable speed conditions. *Mech. Syst. Signal Process.* **2019**, *121*, 636–654. [CrossRef]

11. Riera-Guasp, M.; Antonino-Daviu, J.A.; Capolino, G.A. Advances in Electrical Machine, Power Electronic, and Drive Condition Monitoring and Fault Detection: State of the Art. *IEEE Trans. Ind. Electr.* **2015**, *62*, 1746–1759. [CrossRef]

12. Chen,S.; Yang, Y.; Peng, Z.; Dong, X.; Zhang, W.; Meng, G. Adaptive chirp mode pursuit: Algorithm and applications. *Mech. Syst. Signal Process.* **2019**, *116*, 566–584. [CrossRef]

13. Fernandez-Cavero, V.; Morinigo-Sotelo, D.; Duque-Perez, O.; Pons-Llinares, J. A Comparison of Techniques for Fault Detection in Inverter-Fed Induction Motors in Transient Regime. *IEEE Access* **2017**, *5*, 8048–8063. [CrossRef]

14. Hmida, M.A.; Braham, A. Fault Detection of VFD-Fed Induction Motor Under Transient Conditions Using Harmonic Wavelet Transform. *IEEE Trans. Instrum. Meas.* **2020**, *69*, 8207–8215. [CrossRef]

15. Pons-Llinares, J.; Riera-Guasp, M.; Antonino-Daniu, J.A.; Hableter, T.G. Pursuing optimal electric machines transient diagnosis: The adaptive slope transform. *Mech. Syst. Signal Process.* **2016**, *80*, 553–569. [CrossRef]

16. Yu, K.; Lin, T.R.; Tan, J.; Ma, H. An adaptive sensitive frequency band selection method for empirical wavelet transform and its application in bearing fault diagnosis. *Measurement* **2019**, *134*, 375–384. [CrossRef]

17. EMD-Based Analysis of Industrial Induction Motors With Broken Rotor Bars for Identification of Operating Point at Different Supply Modes. *IEEE Trans. Ind. Inform.* **2014**, *10*, 957–966. [CrossRef]

18. Rosero, J.A.; Romeral, L.; Ortega, J.A.; Rosero, E. Short-Circuit Detection by Means of Empirical Mode Decomposition and Wigner–Ville Distribution for PMSM Running Under Dynamic Condition. *IEEE Trans. Ind. Electr.* **2009**, *56*, 4534–4547. [CrossRef]

19. Garcia-Calva, T.A.; Morinigo-Sotelo, D.; Garcia-Perez, A., Camarena-Martinez, D.; Romero-Troncoso, R.J. Demodulation Technique for Broken Rotor Bar Detection in Inverter-Fed Induction Motor Under Non-Stationary Conditions. *IEEE Trans. Energy Convers.* **2019**, *34*, 1496–1503. [CrossRef]

20. Mendes, A.M.S.; Abadi, M.B.; Cruz, S.M.A. Fault diagnostic algorithm for three-level neutral point clamped AC motor drives, based on the average current Park's vector. *IET Power Electron.* **2014**, *7*, 1127–1137. [CrossRef]

21. Zhao, D.; Wang, T.; Gao, R.X.; Chu, F. Signal optimization based generalized demodulation transform for rolling bearing nonstationary fault characteristic extraction. *Mech. Syst. Signal Process.* **2019**, *134*. [CrossRef]

22. Duque-Perez, O.; Garcia-Escudero, L.A.; Morinigo-Sotelo, D.; Gardel, P.E.; Perez-Alonso, M. Condition monitoring of induction motors fed by Voltage Source Inverters. Statistical analysis of spectral data. In the Proceedings of the XXth International Conference on Electrical Machines, Marseille, France, 2–5 September 2012; IEEE: Danvers, MA, USA, 2012; pp. 2479–2484. [CrossRef]

23. Bazan, G.H.; Scalassara, P.R.; Endo, W.; Goedtel, A. Information Theoretical Measurements From Induction Motors Under Several Load and Voltage Conditions for Bearing Faults Classification *IEEE Trans. Ind. Inform.* **2020**, *16*, 3640–3650. [CrossRef]

24. Bazan, G.H.; Scalassara, P.R.; Endo, W.; Goedtel, A.; Palacios, R.H.C.; Godoy, W.F. Stator Short-Circuit Diagnosis in Induction Motors Using Mutual Information and Intelligent Systems *IEEE Trans. Ind. Electron.* **2019**, *66*, 3237–3246. [CrossRef]

25. Lei,Y.; Yang, B.; Jiang, X.; Jia, F.; Li, N.; Nandi, A.K. Applications of machine learning to machine fault diagnosis: A review and roadmap. *Mech. Syst. Signal Process.* **2020**, *138*. [CrossRef]

26. Liu, R.; Yang, B.; Zio, E.; Chen, X. Artificial intelligence for fault diagnosis of rotating machinery: A review. *Mech. Syst. Signal Process.* **2018**, *108*, 33–47. [CrossRef]

27. Burriel-Valencia, J.; Puche-Panadero, R.; Martinez-Roman, J.; Sapena-Bano, A.; Pineda-Sanchez, M.; Perez-Cruz, J.; Riera-Guasp, M. Automatic Fault Diagnostic System for Induction Motors under Transient Regime Optimized with Expert Systems. *Electronics* **2019**, *8*, 6. [CrossRef]

28. Valtierra-Rodriguez, M.; Rivera-Guillen, J.R.; Basurto-Hurtado, J.A.; De-Santiago-Perez, J.J.; Granados-Lieberman, D.; Amezquita-Sanchez, J.P. Convolutional Neural Network and Motor Current Signature Analysis during the Transient State for Detection of Broken Rotor Bars in Induction Motors. *Sensors* **2020**, *20*, 3721. [CrossRef]

29. Redon, P.; Rodenas, M.P.; Antonino-Daviu A. Development of a diagnosis tool, based on deep learning algorithms and infrared images. In Proceedings of the IECON 2020, The 46th Annual Conference of the IEEE Industrial Electronics Society, Singapore, 18–21 October 2020; IEEE: Danvers, MA, USA, 2020; pp. 2505–2510. [CrossRef]

30. Glowacz, A. Fault diagnosis of electric impact drills using thermal imaging. *Measurement* **2021**, *171*. 108815. [CrossRef]

31. Mohammed, A.; Melecio, J.I.; Djurović S. Stator Winding Fault Thermal Signature Monitoring and Analysis by In Situ FBG Sensors. *IEEE Trans. Ind. Electr.* **2019**, *66*, 8082–8092. [CrossRef]

32. Garcia-Calva, T.A.; Morinigo-Sotelo, D.; Garcia-Perez, A.; Romero-Troncoso, R.J. Rotor Fault Detection in Inverter-Fed Induction Motors Using Speed Analysis of Startup Transient. In Proceedings of the 12th International Symposium on Diagnostics for Electrical Machines, Power Electronics and Drives (SDEMPED), Toulouse, France, 27–30 August 2019; IEEE: Danvers, MA, USA, 2019; pp. 297–302. [CrossRef]

33. Elbouchikhi, E.; Choqueuse, V.; Auger, F.; Benbouzid, M.E. Motor Current Signal Analysis Based on a Matched Subspace Detector. *IEEE Trans. Instrum. Meas.* **2017**, *66*, 3260–3270. [CrossRef]

34. Nandi, S.; Toliyat, H.A.; Li, X. Condition Monitoring and Fault Diagnosis of Electrical Motors—A Review. *IEEE Trans. Energy Convers.* **2005**, *20*, 719–729. [CrossRef]

35. Filippetti, F.; Franceschini, G.; Tassoni, C.; Vas, P. AI techniques in induction machines diagnosis including the speed ripple effect. *IEEE Trans. Ind. Appl.* **1998**, *34*, 98–108. [CrossRef]

36. Romero-Troncoso, R.J. Multirate Signal Processing to Improve FFT-Based Analysis for Detecting Faults in Induction Motors. *IEEE Trans. Ind. Inform.* **2017**, *13*, 1291–1300. [CrossRef]

37. Morinigo-Sotelo, D.; Romero-Troncoso, R.J.; Panagiotou, P.A.; Antonino-Daviu, J.A.; Gyftakis, K.N. Reliable Detection of Rotor Bars Breakage in Induction Motors via MUSIC and ZSC. *IEEE Trans. Ind. Appl.* **2018**, *54*, 1224–1234. [CrossRef]

38. Garcia-Calva, T.A.; Morinigo-Sotelo, D.; Romero-Troncoso, R.J. Non-Uniform Time Resampling for Diagnosing Broken Rotor Bars in Inverter-Fed Induction Motors. *IEEE Trans. Ind. Electr.* **2017**, *64*, 2306–2315. [CrossRef]

39. McClellan, J.H.; Parks, T.W. A personal history of the Parks-McClellan algorithm. *IEEE Signal Process. Mag.* **2005**, *22*, 82–86. [CrossRef]

MDPI

St. Alban-Anlage 66

1052 Basel

Switzerland

www.mdpi.com

Energies Editorial Office

E-mail: energies@mdpi.com

www.mdpi.com/journal/energies